Contamination-Free Manufacturing for Semiconductors and Other Precision Products

Contamination-Free Manufacturing for Semiconductors and Other Precision Products

edited by

Robert P. Donovan

L & M Technologies, Inc., and
Sandia National Laboratories
Albuquerque, New Mexico

MARCEL DEKKER, INC.   NEW YORK · BASEL

ISBN: 0-8247-0380-4

This book is printed on acid-free paper.

Headquarters
Marcel Dekker, Inc.
270 Madison Avenue, New York, NY 10016
tel: 212-696-9000; fax: 212-685-4540

Eastern Hemisphere Distribution
Marcel Dekker AG
Hutgasse 4, Postfach 812, CH-4001 Basel, Switzerland
tel: 41-61-261-8482; fax: 41-61-261-8896

World Wide Web
http://www.dekker.com

The publisher offers discounts on this book when ordered in bulk quantities. For more information, write to Special Sales/Professional Marketing at the headquarters address above.

Current printing (last digit):
10 9 8 7 6 5 4 3 2 1

PRINTED IN THE UNITED STATES OF AMERICA

Preface

Semiconductor chip manufacturing, now a $200 billion per year industry, is carried out in the cleanest manufacturing areas on earth. As measured by the metric of aerosol particle concentration, chips are manufactured in rooms far cleaner than surgical operating rooms, because trace contamination can ruin products and thus can also ruin companies.

In the 1970s and early 1980s, product yield—the number of salable chips produced divided by the number of chips started through the production line—among U.S. manufacturers was on the order of 10–15%, while Japanese competitors achieved yields on the order of 60–90% and even higher. One factor thought to be significant in accounting for this striking difference was the greater awareness of the importance of contamination control characterizing Japanese manufacturers. The Japanese model, as detailed in the many Ohmi publications [1, 2], emphasized the complete elimination of all identifiable contaminants in the chip manufacturing environment regardless of cost or the absence of a clear, demonstrated link between these contaminants and yield. The axiom accepted without proof was that contamination was bad for—and potentially fatal to—products and had to be eliminated at all costs. The U.S. approach, on the other hand, often was "show me" before taking corrective action, and, since the link between contamination and yield is complex and dependent on many variables, making it difficult to establish clear proof of cause and effect, corrective action was slower in coming.

Nevertheless, chip yields for mature products in the United States are now also in the 70–90% range just like those in Japan, without any major change in contamination control philosophy by either group. Contamination control practices developed from both viewpoints have converged. However, these practices are dynamic and continue to change as chip technology and design rules change. The perennial contamination control question has now become the adequacy of today's control measures for manufacturing the next generation of chips, which undoubtedly will combine a larger number of components, all having smaller dimensions, on chips of larger total area. This trend implies chip sensitivity to smaller particle sizes and contaminant masses at the same time that larger chip areas must be free of these contaminants. The specter of chip processing technology and manufacturability being limited by contamination control technology thus continues to haunt the industry, although the high yields of today's production lines show that the state of the art in contamination control has adequately met the contamination challenges of the current mature product lines. Even so, contamination remains the leading cause of loss in chip yield for these mature products and always threatens to be even more costly in yield for future products in the absence of corresponding improvement in contamination control to match the increased contamination sensitivity of these coming products.

The contamination control technology described in this handbook has progressed far beyond that of the 1980s and will in turn be superseded by that of the coming decade. What is presented here is a contemporary snapshot of a continuously changing discipline/art. Present understanding and practices are both reviewed in the following sequence:

1. The contamination problem and target control levels (Chapters 1 and 2)
2. Detection/measurement of contamination (Chapters 3–5)
3. Transport and deposition of contamination (Chapters 6–8)
4. Removal of contamination (Chapters 9 and 10)
5. Sources of contamination; contamination prevention (Chapter 11)

Much of the information in this book grew out of a short course sponsored by the University of California Extension. This short course, offered under the same title as that of this book, was initially given in April 1995 in Albuquerque, New Mexico, and was repeated in April 1998 in Austin, Texas. Most of the chapter authors participated in one or both of these short courses and initiated their chapter preparation by updating their short course materials. Several of the chapters, however, were prepared independently of the short course and by authors who were not part of the short course faculty.

The industrial background of all the material presented is semiconductor chip manufacturing; indeed, that industry currently has the most demanding requirements for cleanliness in manufacturing. However, the understanding of contamination control developed by this leading-edge industry is directly applicable to contamination control in other industries, as is much of the technology. Presentation of the unprecedented high level of contamination control now routinely achieved in large semiconductor manufacturing facilities in a format that facilitates transfer to other industrial applications is one goal of the text. The primary audience, however, remains semiconductor manufacturers. This book seeks to provide them with a summary of the present understanding of contamination issues and the best contamination control practices in the industry and their justification.

ACKNOWLEDGMENTS

I wish to thank the authors for taking time from their busy schedules to prepare these materials, once again illustrating the maxim that the best way to get something done is to ask a busy person. I also wish to express my appreciation to Dr. Robert Blewer, who led the Contamination Free Manufacturing Center at Sandia National Laboratories under a cooperative research and development agreement between the Department of Energy and SEMATECH. It was he who encouraged both the offering of the short course and the subsequent decision to prepare a text based on the materials used in the short course.

Dennis Morrison and Julie (as in Jewel) Kestie of Sandia provided valuable support services during the assembly and proofing of the final text. Finally, I must acknowledge the many friendly exchanges with Moraima Suarez, Marcel Dekker's production editor for this volume. She turned what could have been a trying ordeal into an enjoyable venture. I will miss this camaraderie which has been part of my workweek over the past few years.

Robert P. Donovan

REFERENCES

1. R. W. Keeley and T. H. Cheney, eds., The Ohmi Papers. Canon Communications, Inc., Santa Monica, CA, 1990.
2. T. Ohmi, ed., Ultraclean Technology Handbook. Marcel Dekker, New York, 1993.

Contents

Contributors

Victor K. F. Chia, Ph.D.   Vice President of Marketing, ChemTrace, Fremont, California

Kurt Christenson, Ph.D.   FSI International, Chaska, Minnesota

Rodolfo E. Díaz, Ph.D.   Research Assistant Professor, Department of Mechanical and Aerospace Engineering, Arizona State University, Tempe, Arizona

Robert P. Donovan, M.S.   Process Engineer, L & M Technologies, Inc., and Contamination Free Manufacturing Research Center, Sandia National Laboratories, Albuquerque, New Mexico

Michael J. Edgell, Ph.D.   Director of Analytical Services, Charles Evans & Associates, Redwood City, California

Anthony S. Geller, Ph.D.   Principal Member of Technical Staff, Engineering Sciences Center, Sandia National Laboratories, Albuquerque, New Mexico

Allyson L. Hartzell, M.S.   Senior Staff Scientist, Micromachined Products Division, Analog Devices Incorporated, Cambridge, Massachusetts

E. Dan Hirleman, Ph.D.   Professor and Head, School of Mechanical Engineering, Purdue University, West Lafayette, Indiana

David Jensen, B.S.   Member of Technical Staff, Contamination-Free Manufacturing, Advanced Micro Devices, Inc., Austin, Texas

Brent M. Nebeker, Ph.D.   Visiting Assistant Professor, School of Mechanical Engineering, Purdue University, West Lafayette, Indiana

Daniel J. Rader, Ph.D.   Principal Member of Technical Staff, Engineering Sciences Center, Sandia National Laboratories, Albuquerque, New Mexico

Deborah J. Riley, Ph.D.   Member of Technical Staff, Advanced Micro Devices, Inc., Austin, Texas

Steven Verhaverbeke, Ph.D.   Technology Manager, Wet Processing, Transistor Gate and Substrate Division, Applied Materials, Santa Clara, California

1
Introduction

Robert P. Donovan
*L & M Technologies, Inc., and Sandia National Laboratories,
Albuquerque, New Mexico*

I. CONTAMINATION-FREE MANUFACTURING

Contamination-free manufacturing (CFM) means carrying out product manufacturing in environments and with processes that do not cause degradation in product performance or manufacturability attributable to contamination inadvertently introduced by these environments or processes. *Contaminants by this definition are any impurities or undesired materials that degrade product performance or reliability or adversely affect manufacturability.* "Contamination free" does not necessarily imply immeasurable quantities of any impurities or extraneous material on or in the product; it simply signifies that the concentrations of any of these impurities or extraneous materials are below the threshold of product impact. These thresholds will vary with product type; a manufacturing environment and sequence that is contamination free for one product may not be so for a second product, even one seemingly differing from the first by only minor changes in function, size, or application. Thus CFM is a moving target, the measure of adequacy being the performance and reliability of what is produced.

Other definitions of contaminants abound. For some, contamination in manufacturing includes the presence of unwanted energy sources such as electromagnetic fields, mechanical vibration, radioactivity, or static electrical charge, all of which can adversely affect product performance and manufacturability when present above certain thresholds. While these energy sources often represent serious problems that demand solutions, they are not part of the contamination issues to be discussed in this book. The definition of contamination used in this book, as given in italic type in the previous paragraph, is more restrictive and based strictly on some measure of mass—from molecules to films to particles.

The key to manufacturing in a contamination-free mode is to know what concentrations of what species are tolerable for what products and then to take those steps needed to ensure that these thresholds are not exceeded. For many industrial operations, including semiconductor manufacturing, knowledge of the tolerable thresholds of contamination is imprecise and successful manufacturing procedures are empirically determined. The chapters that follow will, in fact, present many empirical manufacturing procedures that have proven successful in semiconductor manufacturing. They will, however, also discuss present understanding of why these manufacturing procedures are successful, hopefully pointing the direction for more systematically developing the methods that will be required for the contamination-free manufacturing of the coming generations of products.

II. TYPES OF CONTAMINANTS

The contaminants to be considered in this book are of two general types: particles and molecular contaminants.

- A **particle** is a stable or quasistable agglomeration of molecules with overall dimensions in the range 2 nm to 1 mm.
- An **aerosol particle** is a particle suspended in air or gas with a settling velocity less than some arbitrary velocity, typically taken to be 0.01 cm/s.
- A **molecular contaminant** is any contaminant not classified as a particle such as a single molecule, a small collection of molecules, or a film.

All contaminants are thus either particles or molecular contaminants. The latter often condense on surfaces as films or trace deposits. They can be organic, volatile, and transient, or they can be metallic (elemental metals, metal oxides or silicides, other) and nonvolatile. In all forms molecular contaminants are, by definition, damaging to product performance, reliability, or manufacturability and, like particles, must be controlled to concentrations less than their thresholds of product impact.

Contaminants are a subclass of a larger class of manufacturing problems called defects. A **defect** is any material property, operational error, or environmental condition that degrades product performance or reliability or adversely affects product manufacturability. Typical properties or conditions making up the class of defects include structural imperfections, processing errors, or environmentally induced effects such as damage from electrostatic discharges. **Defect Reduction Technology** is the term now used in the 1997 NTRS to encompass all strategies, procedures, and practices designed to reduce the concentration of any of the large class of defects. Contaminants, however, are the primary type of defect to be discussed in detail in this book. Measurement methods are sometimes unable to distinguish between a contaminant and some other type of defect, and procedures for removing contaminants will sometimes eliminate other defects as well or should be designed to do so. Discussion of other defects or even other process requirements will be included whenever contamination control procedures must be modified or broadened to accommodate these other requirements.

III. MANUFACTURING ENVIRONMENTS

Manufacturing environments are highly variable among industries. Contamination-sensitive industries such as semiconductor manufacturing typically use cleanrooms for manufacturing, and indeed one definition of a **precision product** is a product that requires a cleanroom in order to be successfully manufactured. Cleanrooms are enclosed spaces in which aerosol particle concentration is controlled by high-efficiency particulate (HEPA) filters or ultra low particulate (ULPA) filters. Both FED-STD-209E and ISO14644-1 and -2 define classification systems based totally on aerosol particle concentration to describe the quality of air in a cleanroom and specify procedures for measuring only the concentration of aerosol particles in verifying that the air quality of a given cleanroom does indeed meet the classification claimed. Molecular contaminants in the cleanroom air do not affect this classification; their concentration can be any value and the cleanroom classification remains unchanged. The FED-STD-209E and ISO 14644-1 cleanroom classifications, restricted as they are to particle measurements only, are being recognized more and more as misnomers as appreciation of the deleterious consequences of molecular contamination continues to grow.

Cleanroom air is just one of the product environments to be considered. However, it applies to the manufacture of a wide spectrum of precision products. Other common manufacturing environments that products are exposed to include low-pressure chambers, high-temperature/high-pressure processes, and various liquid baths. Product contamination can occur during exposure to any of these environments and the generation, transport, and deposition of both particulate and molecular contaminants onto products in each of these environments make up a significant portion of the discussion to be presented.

Every bit as important, of course, are methods and instrumentation for measuring contaminants. Indeed, progress in contamination control has invariably been preceded by the introduction of improved measurement capability and instrumentation. This trend continues. This book reviews established measurement methods, as well as promising new approaches which emphasize on-line/in-line near-real-time performance. The different manufacturing environments generally require different instrumentation and hardware; and contamination measurement is addressed in each of the commonly encountered environments cited in the previous paragraph, in addition to measurements made directly on product surfaces which are ultimately the location of most importance. Surface contamination measurements are, of course, crucial in evaluating the performance of contamination removal technology.

The goal of the final chapter of the book is to link the understanding of contamination developed in the earlier chapters and its implications for the contemporary semiconductor manufacturing plant where the payoff must occur. Practices and procedures that are "contamination smart" follow from the basic framework presented earlier; they are formally summarized in the closing chapter along with the recognition of serious, outstanding contamination-related problem areas.

Table 1 Symbols and Units

Parameter	Symbol	SI unit
Area	A	meter^2 (m^2)
Boltzmann constant	k	1.38 E-23 joules/kelvin (J/K)
Capacitance	C	farad (F)
Diameter	d	meter (m)
Diffusion coefficient	D	meter^2/second (m^2/s)
Electric field	E	volts/meter (V/m)
Electrical current	i	ampere (A)
Electrical potential	V	volts (V)
Electrical resistance	R	ohms (Ω)
Electrostatic charge	q	coulomb (C)
Energy	E	joules (J)
Force	F	newton (N)
Frequency	f	hertz (Hz)
Kinematic viscosity	v	meter^2/second (m^2/s)
Length	l	meter (m)
Mass	m	kilogram (kg)
Mass density	ρ	kilograms/meter^3 (kg/m^3)
Number density	c	number/meter^3 (m^{-3})
Power	P	watts (W)
Pressure	P	pascal (Pa)[N/m^2]
Resistivity	ρ	ohm-meter (Ω-m)
Temperature	T	kelvin (K)
Time	t	seconds (s)
Velocity	v	meters/second (m/s)
Viscosity	μ	pascal-seconds (Pa-s)
Volume	V	meter^3 (m^3)*

*This text also uses the non-SI units: liter (L) and milliliter (mL).

IV. SYMBOLS/UNITS

The units and symbols of the Système International d'Unites (SI) are used throughout the book with English equivalents, or other familiar units, inserted parenthetically in those discussions in which common usage still favors traditional units. Table 1 lists the symbols used and the SI units for a number of the more frequently used parameters appearing in the book. The same symbol is sometimes used to represent different parameters in different chapters or contexts. In general, the intended parameter will be clear from the text with which it appears. In addition, most of the discussions include explicit definitions of the parameters as they are introduced so that misunderstanding or confusion from similar symbology should not be a serious problem.

The SI allows easy scaling from the units given in Table 1 by the insertion of prefixes, which multiply the basic SI unit. A listing of these prefixes is reproduced in Table 2.

Table 2 SI Prefixes

Prefix	Multiplier	Symbol
exa	10^{18}	E
peta	10^{15}	P
tera	10^{12}	T
giga	10^{9}	G
mega	10^{6}	M
kilo	10^{3}	k
hecto	10^{2}	h
deka	10	da
deci	10^{-1}	d
centi	10^{-2}	c
milli	10^{-3}	m
micro	10^{-6}	μ
nano	10^{-9}	n
pico	10^{-12}	p
femto	10^{-15}	f
atto	10^{-18}	a

2

National Technology Roadmap for Semiconductors
Basis and Alignment*

David Jensen
Advanced Micro Devices, Inc., Austin, Texas

The need for promoting and extending the science of contamination-free manufacturing (CFM) is directly linked to the defect reduction (DR) requirements outlined in the National Technology Roadmap for Semiconductors (NTRS) [1]. This chapter is a brief overview of the DR section of the NTRS with particular emphasis on the technology requirements of CFM and contamination control, the subjects of this book.

I. NTRS BACKGROUND

The first semiconductor technology roadmap was published in 1993 under the sponsorship of the Semiconductor Industry Association (SIA) to provide a consensus view of the major problems to be solved and to increase cooperation in precompetitive semiconductor research and development. The NTRS was motivated by the rapid pace in increasing complexity in semiconductor technology [1], and the need to concentrate limited research on those developments deemed most critical to continued growth. Supporting organizations included:

---

*Portions of this chapter have been adapted from the "Mapping the Roadmap" series appearing in the January, March, June, July, and October issues of MICRO Magazine, copyright 1998, used with permission.

- National Institute of Standards and Technology
- National Science Foundation
- SEMATECH
- Semiconductor Industry Association
- Semiconductor Research Corporation
- U.S. Department of Commerce
- U.S. Department of Defense
- U.S. Department of Energy

The roadmap was revised in 1994 [2], 1997, and 1999. Each revision takes a 15-year view of semiconductor technology requirements for research, development, and manufacturing.

II. DEFECT REDUCTION CROSS-CUT TECHNOLOGY WORKING GROUP

The most critical factors affecting profitability in leading edge, high-volume, integrated circuit manufacturing are 1) customer satisfaction, or generally speaking, designs which meet performance, functionality, price, quality, and reliability requirements, 2) cost control/reduction, and 3) time to market. In today's integrated circuit marketplace, products that are priced moderately higher because of higher manufacturing costs associated with low-yielding processes and/or those that enter the market just months late can dramatically reduce revenue and become the difference between profit and loss. Significant factors affecting both time to market and cost are the level of defects generated by manufacturing processes and the rate at which these defects are reduced. Defects can be defined in the broadest sense as any actual outcome that deviates from the expected outcome. Thus defects include contaminants, ineffective control of physical and electrical parameters, device structure-related issues, process-to-process interactions including geometric effects, and design-process interactions. The chapters that follow consider just contamination. The NTRS considers all types of defects. Contamination is believed to be the major source of random defect yield loss in semiconductor manufacturing today. Profitable IC manufacturers understand how big a lever yield (yield $\cong$ functional chips/total number of chips completing production) is with respect to ongoing profitability. These companies are well known for their comprehensive programs in defect reduction (or yield engineering) and a general philosophy that all operators, technicians, and engineers must continually work to reduce defect levels. Some manufacturing facilities even operate under the philosophy that, in part, IC manufacturing is the business of defect control and reduction. Based on the critical nature of defect reduction activities within all aspects of IC design and manufacturing, the Roadmap Coordinating Group (RCG) of the Semiconductor Industry Association (SIA) elevated Defect Reduction (DR) technology to its own Cross-Cut Technology Working Group (CCTWG) for the 1997 revision of the SIA National Technology Roadmap for Semiconductors (NTRS).

III. IMPORTANCE OF DEFECT REDUCTION

In 1996, while analyzing process diagnostics, the VLSI Research, Inc. market analysis group reported that yield improvement rates were accelerating as evidenced by the decreasing time required by the industry to achieve ~98% DRAM (Dynamic Random Access Memory) yields [3]. Figure 1 shows this trend.

For these accelerated learning rates to continue, significant improvements in yield will be necessary early in the life cycle of products, while products are in research and pilot production. It's hard to envision near 100% yield on a 1 Gbit DRAM in its first year of production, without having transferred the product and associated processes from development to manufacturing at >50% yield, and yet this is what continued acceleration in yield learning rate implies. Actually, VLSI Research predicted that yields of ~90% will be required for the transfer of a 1 G DRAM [3] from development to pilot production. This farreaching goal will not be realized without significant improvements in yield modeling, defect detection, defect sources and mechanisms identification, and defect prevention and elimination—areas addressed in the Defect Reduction section of the 1997 NTRS.

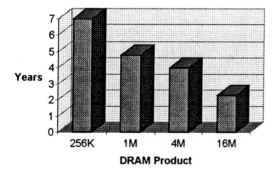

Figure 1   Time duration to achieve 98% yield.

IV. DEFECT REDUCTION CYCLE

The DR CCTWG roadmap was structured to align with industry typical yield learning and defect reduction methodology, focusing on four key topics: Yield Model and Defect Budget (YMDB), Defect Detection (DD), Defect Sources and Mechanisms (DSM), and Defect Prevention and Elimination (DPE). This cycle of learning and yield improvement is pictured in Figure 2.

Given the start-up of a new fab or the introduction of a new process or product, yield targets are typically established to account for time to market, profitability, and other business factors. From these targets, defect budgets are calculated through the use of a yield model algorithm. Ultimately, defect budgets help set control limits for in-line defectivity at each of the critical process steps. Next, the yield program will focus on detecting defects, or events that lead to defects, through the use of in-line and in situ monitoring equipment.

Before defects can be eliminated, they need to be sourced and their creation mechanisms fully understood. In this phase, the yield program will employ a great deal of modeling and data reduction technologies in conjunction with all in-line data to not only characterize the defect mechanism, but also attempt to correlate the defect to some in-line parameter that is typically measured. Finally then, once a defect source is identified, measures can be taken to prevent and or eliminate it. The topic of defect prevention and elimination includes the determination of allowable levels of microimpurities in the manufacturing materials or the wafer environment. It is in this last section of defect prevention and elimination that many of the technology requirements that drive CFM science are found. For example, 1) Chapters 3 and 4 discuss off-wafer and in situ

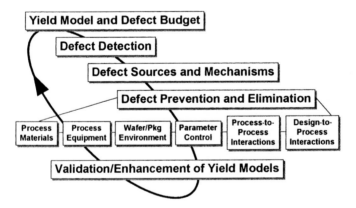

Figure 2   Defect reduction cycle.

measurement technologies which assist in characterizing and reducing levels of contaminants in critical processing environments (i.e., gases and chemicals); 2) Chapters 6 and 7 review deposition and transport of particulate matter which must be understood in order to control these contaminants to roadmap levels necessary for chemicals, gases, water, and air; and 3) Chapters 8–10 deal with deposition and transport of molecular contaminants which also have levels of increasing purity as outlined in the roadmap. The yield learning cycle (Fig. 2) is repeated after yield models are adjusted to account for new yield learnings.

V. KEY MESSAGES RELATED TO CFM

The DR section of the NTRS and the supplemental material described by Jensen et al. [4] in the MICRO "Mapping the Roadmap" series highlights two key messages that have significant CFM relevance:

- Contamination control within equipment is paramount to achieve tool defect densities targeted for 60% first year yield and 85–95% yield in mature products. The ability to detect these contaminants (particulate and nonparticulate) within processing environments will be enabled by developments in the areas of in situ monitoring during process (and process tool) development and in situ chamber cleaning and defect "transparent" materials—both believed to require fundamental study for future technologies. The role of modeling associated with contamination formation, transport, and deposition must be emphasized to help focus tool defect reduction learnings.
- Order-of-magnitude improvements in the purity of process fluids (gases, liquid chemicals, and ultrapure water) do not appear to be necessary over the next two to three technology generations. Fundamental understanding of the impact of contaminant(s) on product yield and device performance is needed to justify the additional improvements in chemical/gas manufacturing, distribution, and control technologies. Today's process fluid technologies deliver chemicals/gases of purity that exceed that measured within processing environments by 2 to 3 orders of magnitude.

Through the use of advanced test structures and modeling techniques, the fundamental challenge in the area of process critical materials is to understand the correlation between impurity concentration and device yield, reliability, and performance in order to assess whether increasingly stringent contamination specifications are truly required and to provide early warning of needs for tighter specifications in those cases where they are warranted.

VI. CROSS-CUT REQUIREMENTS

DR in the NTRS is considered a cross-cut technology. This means that DR has relevance in virtually all the other thrusts (e.g., Litho, Interconnect, etc.). The following is a brief sampling of the some of these cross-cut technology needs as outlined in the NTRS.

A. Process Integration, Devices, and Structures

Predictive and diagnostic test structures will be necessary to understand the impact of trace impurity levels (e.g., metallics, ionics, etc.) within input materials (i.e., water, chemicals, gases, wafers, etc.) and to detect, in-line, very small levels of contaminants.

B. Front End Processes

The front end of a chip process includes steps in which the active elements (gate, source, and drain) of a transistor are created and electrical contacts to these elements are established.

Impurity specifications for critical materials in front end processes need to be linked to known fault mechanisms so as to achieve economic viability of raw materials.

Research into particle size distributions below ~50 nm and into the impact of particle size upon yield will aid in developing appropriate line monitoring and yield engineering defect detection tools.

Surface termination control will continue to drive requirements in isolation technology for front-end-of-line (FEOL) processes. Outgassing from cleanroom materials, cassettes, pods, filters, etc. and the presence of airborne organic and ionic species have been shown to degrade device performance and yield. Relating these sources of contaminants to yield and performance will be necessary to optimize the insertion of appropriate control technology.

C. Lithography

Control of ambient base gas (ammonia, amines, etc.) concentrations will be necessary to control neutralization of photogenerated acids. To support this need, development in chemistry models, nonchemically activated deep ultraviolet (DUV) resists, and portable, easy-to-use, high-sensitivity detection tools will be necessary. Continued advancements in molecular filtration, removal and trapping techniques, as well as wafer isolation technologies, will also aid in minimizing the impact of airborne base gases as noted above and in minimizing lens clouding. Backside particle reduction is necessary for improving focus requirements. Improved wafer handling (chucks, edge handling, etc.) and backside particle detection capabilities will provide the greatest benefit to minimizing backside particle impact in lithography.

D. Interconnect

Issues with new interconnect materials and structures present problems such as novel spin-on dielectrics, including bubbles, striations, shrinking, cracking, and delamination. Implementation of copper interconnect technology will also require trace-level real-time metrology to establish acceptable copper levels and copper contamination vectors in the fab and backside to front-side metal contamination transport.

Inherently cleaner tools and processes will be required through effective particle repulsion techniques such as thermophoresis and avoiding gas phase particle nucleation in reactors as examples. Better understanding of process chemistry regimes will be needed to help reduce the rate of residue build-up on chamber walls and subsequent flaking, thus increasing mean-time-before-cleaning.

E. Factory Integration

The most critical requirements in factory integration are determining the variation tolerance of critical process parameters and the interaction between processes that could lead to better process control and reduced reliance on end-of-line inspection. Data on the threshold at which impurities in critical process fluids impact process (e.g., etch rate, film uniformity) and device performance (e.g., threshold voltage, gate oxide integrity) can be used to guide in situ sensor development and help reduce time to high yield by eliminating the source of defects in advance.

VII. YIELD MODEL AND DEFECT BUDGET

The basis for any advancement or experimentation in CFM is theoretically based on the assumption that reducing particle or metallic levels will increase yield. These assumptions are often simulated with the assistance of a yield model capable of taking into account process flow, defect sensitivities, defect distributions, product design, etc., in predicting allowable defect budgets for a given process sequence or a given process tool in the manufacturing process. The yield model

Table 1 1997 NTRS Defect Budgets at 60% Yield (First Year of Production)

Year of first product shipment technology generation		1997 250 nm	1999 180 nm	2001 150 nm	2003 130 nm	2006 100 nm	2009 70 nm	2012 50 nm
FEOL	Doping	860	376	231	149	70	27	11
	Interconnect	1076	471	289	186	87	33	14
	Surface prep	1642	718	441	284	133	51	21
	Thermal/Thin film	850	372	228	147	69	26	11
BEOL	Interconnect	605	265	162	105	49	19	8
	Planarization	1418	620	380	245	115	44	18
	Surface prep	1718	751	461	297	140	53	22
FEOL/BEOL	Lithography	648	284	174	112	53	20	8
	Metrology inspection	1195	523	321	207	97	37	15
	Wafer handling	30	13	8	5	2	1	0.4

BEOL, back-end-of-line; FEOL, front-end-of-line.

and defect budget (YMDB) section of the roadmap describes this methodology and its associated difficult challenges, technology requirements, and potential solutions.

Here the purpose is simply to reference the defect budget portion of the YMDB section of the NTRS. Requirements based on the SEMATECH 0.25 μm yield model and follow-up validation activity with SEMATECH member companies [5] are shown in Table 1. In this table the light screen indicates that solutions are being developed and the dark screen indicates there is no known solution.

These targets have been set under the assumption that the critical ("killer") particle size is 50% of the minimum dimension of elements on the chip (noted as the technology generation row in Table 1). For CFM-related activity, then, efforts to reduce particles which contribute to these budgets can be directly linked to yield impact through the use of a model. Chapter 11 presents a brief review of the model and lists the variables which impact particle reduction efforts.

Dance et al. [6] provide a more comprehensive review in the basis for the yield model and supplementary material for further study in the MICRO "Mapping the Roadmap" series.

VIII. DEFECT DETECTION

Weber et al. [7] in the MICRO "Mapping the Roadmap" series provide a good summary and detailed background of the factors which impact defect detection (DD): The shrinking feature sizes of integrated circuits mandate sensitivity to smaller defects; throughput requirements increase as semiconductor processes mature; and more rapid yield learning requires shorter cycles of learning.

The most obvious synergy with CFM is the factor of shrinking feature size. Integrated circuit feature sizes have been shrinking for more than 30 years, and the 1997 NTRS expects this trend to continue [1,8]. Circuits with smaller feature sizes are susceptible to electrical faults induced by smaller defects, requiring defect detection technology to become increasingly more sensitive. Suppliers of defect detection equipment introduce new technology to keep up with these requirements. For example, suppliers of optically based defect inspection tools introduce new light sources with shorter wavelengths and optimize the lens optics for the new physical conditions. This trend is likely to continue until inspection equipment may need to capture defects so small that optically based detection will no longer be suitable for the job. Table 2 indicates select defect detection sensitivity requirements for patterned, unpatterned, and in situ applications [1]. Here again the light screen indicates solutions being pursued and the dark screen is an indication of no known solutions for that particular requirement.

Table 2 Defect Detection Particle Sensitivity Requirements (1997 NTRS)

Year of first product shipment technology generation	1997 250 nm	1999 180 nm	2001 150 nm	2003 130 nm	2006 100 nm	2009 70 nm	2012 50 nm
Patterned wafer inspection, PSL spheres at 90% capture, equivalent sensitivity (nm)							
Process research and development	83	59	50	43	33	23	17
Unpatterned, PSL spheres at 90% capture, equivalent sensitivity (nm)							
Wafer backside	200	200	200	200	100	100	100
Metal film	125	90	75	65	50	35	25
Nonmetal films	83	60	50	43	33	23	17
Bare si	83	60	50	43	33	23	17
In situ particle monitoring							
Sensitivity (nm) (@2 : 1 signal/noise ratio)	250	180	150	130	100	70	50

PSL, polystyrene latex.

Detection of particles on patterned wafers is complicated by the presence of patterns which can alter the signature of the particles dependent on the local topography and material characteristics of the pattern. Detection is further exacerbated by the presence of normal processing conditions (grains in polysilicon or metal surfaces) and process variations (film thickness variations), which interact to increase the background signal and thus reduce the overall signal-to-noise ratio for the particle. Particles and defects that reside below the pattern surface are even more difficult to detect as detection energy (photons, electrons, ions, etc.) must penetrate into these structures and the exiting energy must be of sufficient enough magnitude to exit the structure and then be detected. This is the problem described in the roadmap as high aspect ratio inspection, a formidable challenge for particle/defect detection scientists and technologists.

In situ sensors have the potential of measuring defects and contaminants in real time. Sensors can detect deviations from the normal process conditions in a chamber and either emit a fault signal or induce automatic corrective action. Applications for in situ sensors exist in many process tools, and are needed for detecting more than just particles. Sensors detect many potentially damaging chemicals, including metal ions that can alter the performance of MOS devices.

IX. DEFECT SOURCES AND MECHANISMS (DSM)

Determining the source of defects and their formation is extremely complex. Process-borne particles, organics, metallics, etc. are generated off-wafer and the sourcing of the defects must also encompass knowledge of their transportation and deposition onto wafer surfaces. Other chapters address detailed models and fundamentals in particulate and nonparticulate transport and deposition in various media onto wafer surfaces. However, not all defects can be modeled as physical remnants on wafer surfaces. Oftentimes upon failure analysis of a nonfunctional circuit, no physical remnant is visible. Fault sourcing, already a difficult challenge, will become orders of magnitude more difficult in the future. The NTRS [1] predicts that for microprocessors the number of process steps will increase nearly linearly and the number of transistors in a circuit will increase exponentially from one process generation to the next over the next 15 years. Accordingly, as minimum feature size decreases from one generation to the next, killer defect size (or critical size, defined as $\frac{1}{2}$ the minimum feature size) decreases proportionately. To achieve a 60% yield target on a complex microprocessor in its first year of manufacturing, defect density must also decrease from one generation to the next.

The challenge is to isolate a shrinking needle in a growing haystack. The needle is a killer defect, whose size decreases over time. The haystack

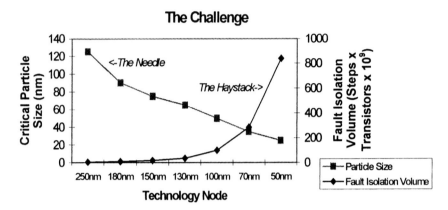

Figure 3   Fault isolation volume and critical particle size versus technology.

is a chip whose volume is constantly expanding, within which the number of circuit elements grows exponentially. The Fault Isolation Volume is defined as the product of the number of process steps and the number of transistors. This volume represents the space within which an electrical fault must be isolated in order to assign root cause.

Figure 3 illustrates the challenge, by graphically depicting how quickly the needle shrinks and the haystack expands. When we compute the ratio of the haystack (Fault Isolation Volume) to the needle (critical size in nm) in the 250 nm generation, we see it is on the order of 10^8 : 1. By the 150 nm generation, that ratio will grow to 10^9 : 1. And by the 50 nm generation, the challenge will grow by another thousand-fold to 10^{12} : 1.

Gross et al. [9] provide details of what developments are necessary in order to meet these sourcing challenges in the MICRO "Mapping the Roadmap" series.

X. DEFECT PREVENTION AND ELIMINATION (DPE)

As outlined in the NTRS [1] and indicated in Figure 2, means of preventing and/or eliminating defects can encompass improvements/control of contaminants in process materials (chemicals, gases water, starting material, etc.), process equipment, wafer environment control (including airborne molecular contaminants), process (parameter) control, and interactions between processes, as well as interactions between the actual circuit design and individual processes. The areas of process liquid chemicals, airborne molecular contaminants, and wafer environment control are most relevant for CFM and topics in this text. Table 3 indicates a few of these requirements from the DPE section of the NTRS [1]. It should be noted in contrast to previous technology requirements, this table has no dark-screened squares, indicating that most of these requirements are either current state of the art (no screen) or have solutions being pursued (light screen).

The percentage of process steps affected by nonparticulate or molecular contamination is expected to increase. The roadmap indicates target levels of ambient acids, bases, condensables, and metals for specific process steps assuming the noted exposure times and sticking coefficients.

As noted in sections above, little understanding exists today regarding impurity specifications in novel materials such as sputter targets, plating solutions, CMP slurries, and chemical vapor deposition (CVD) precursors, and future focus on this is required.

In process liquid chemicals, particle concentrations have been held constant at critical particle size. For these process chemicals, prediffusion cleaning requirements drive the most aggressive impurity levels. These levels have been relaxed compared with the 1994 SIA Roadmap [2] to correspond with the specification of surface levels of metallics. Some of the NTRS surface preparation technology requirements are indicated in Table 4. This evolution (FEOL metallics) shows only a 10× improvement required over the next 15 years.

To detect, to understand, and to eliminate unwanted process interactions, process monitoring will play a key role in the future. The appropriate sensors and data must be available, along with an appropriate information management system to correlate process parameters to upstream/downstream parameters and yield and provide smart intertool and intratool statistical process control (SPC).

The NTRS also outlines numerous opportunities for potential solutions in many of these challenging CFM areas. They include a number of topics described and reviewed in this text:

- In-line trace impurity analytical technology for process critical materials
- Analytical technology to characterize ultratrace levels of bacteria
- Rapid online analytical technology for ultrapure water systems
- In situ chamber monitoring
- Particle avoidance techniques
- Fundamental understanding of reactor contamination formation, transport, and deposition
- In situ process control
- Affordable, accurate, repeatable, real-time sensors for airborne molecular contaminants (AMC)
- Design and material selection of wafer carriers and enclosures

Fosnight and Jensen [10] provide review and supplementary material in these topics in the MICRO "Mapping the Roadmap" series.

Table 3 CFM-Related Technology Requirements from the 1997 NTRS

Year of first product shipment technology generation	1997 250 nm	1999 180 nm	2001 150 nm	2003 130 nm	2006 100 nm	2009 70 nm	2012 50 nm
Wafer environment control							
Critical particle size (nm)[a]	125	90	75	65	50	35	25
Particle $\geq$ crit size (/m^3)[b]	27	12	8	5	2	1	1
Airborne molecular contaminants (pptM)[c]							
Litho—Bases (as amine)	1000	1000	1000	1000	1000	1000	1000
Gate—Metals (as Cu, $E = 2 \times 10^{-5}$)	0.7	0.3	0.3	0.2	0.1	0.07	<0.07
Gate—Organics (as MW = 250, $E = 1 \times 10^{-3}$)	300	200	200	100	100	70	50
Process critical materials							
Critical particle size (nm)[b]	125	90	75	65	50	35	25
Liquid chemicals[d]							
Particles $\geq$ critical size (/mL)	<0.5	<0.5	<0.5	<0.5	<0.5	<0.5	<0.5
HF$^-$, H$_2$O$_2$, NH$_4$OH: Fe, Cu (ppt, each)	<500	<250	<200	<150	<100	<50	<50
Other metals (ppt, each)	<1000	<500	<400	<300	<200	<100	<100
Bulk ambient gases							
N$_2$O$_2$, Ar, H$_2$: H$_2$O, O$_2$, CO$_2$, CH$_4$ (ppt, each)	100–1000	100–1000	<100	<100	<100	<100	<100
Particle $\geq$ critical size (L)	<0.1	<0.1	<0.1	<0.1	<0.1	<0.1	<0.1

[a] Critical particle size is based on 1/2 design rule. All defect densities are "normalized" to critical particle size. Critical particle size does not necessarily mean "killer."

[b] Airborne particle requirements are based on deposition velocity of 0.01 cm/s resulting in 1 particle/m^2/h for an ambient concentration of 3 particles/m^3. Values are back-calculated assuming: the "wafer-handling" defect target, 300 process steps (increasing by 10 per generation), and a wafer exposure time of 1000 h. As an example, the 250 nm requirement is calculated as (30 particles/m^2/step) × (300 steps)/(1000 h) × [(3 particles/m^3)/(1 particle/m^2/h)] = 27 particles/m^3.

[c] Ion indicated is basis for calculation. Exposure time is 60 min with starting surface concentration of zero. Basis for lithography is defined by lithography roadmap. Gate metals and organics scale as surface preparation roadmap metallics and organics, respectively. Salicidation and contact acids and bases scale as surface preparation BEOL anions and metals, respectively. All airborne molecular contaminants are calculated as $S = E * (N * V/4)$, where S is the arrival rate (molecules/cm^2/s), E is the sticking coefficient (between 0 and 1), N is the concentration in the air (molecules/cm^3), and V is the average thermal velocity (cm/s).

[d] Particle targets apply at POU, not incoming chemical. Point-of-tool connection chemical metallic targets are based on Epi starting material, sub-ppb contribution from bulk distribution system 1 : 1 : 5 standard clean 1 (SC-1) and elevated temperature 1 : 1 : 5 standard clean 2 (SC-2) final clean step. "HF last" or "APM last" cleans would require ~10× and ~100× improved purity HF (mostly Cu) and APM chemicals, respectively.

Table 4 Surface Preparation 1997 NTRS Technology Requirements

Year of first product shipment technology generation	1997 250 nm	1999 180 nm	2001 150 nm	2003 130 nm	2006 100 nm	2009 70 nm	2012 50 nm
Front end of line[a]							
Light scatters[b]							
DRAM	0.3	0.15	0.1	0.075	0.03	0.015	0.01
Logic	0.75	0.5	0.45	0.4	0.25	0.2	0.15
Particle size (nm)	125	90	75	65	50	35	25
Critical metals (atoms/cm^2)[c]	5×10^9	4×10^9	3×10^9	2×10^9	1×10^9	$<10^9$	$<10^9$
Other metals (atoms/cm^2)[d]	5×10^{10}	2.5×10^{10}	2×10^{10}	1.5×10^{10}	1×10^{10}	5×10^9	$<5 \times 10^9$
Organics/polymers (C atoms/cm^2)[e]	1×10^{14}	7×10^{13}	6×10^{13}	5×10^{13}	3.5×10^{13}	2.5×10^{13}	1.8×10^{13}
Oxide residue (O atoms/cm^2)[f]	1×10^{14}	7×10^{13}	6×10^{13}	5×10^{13}	3.5×10^{13}	2.5×10^{13}	1.8×10^{13}
Back end of line[g]							
Particles (cm^{-2})	0.3	0.15	0.13	0.1	0.06	0.045	0.03
Particle size (nm)	125	90	75	65	50	35	25
Metals (atoms/cm^2)[h]	1×10^{11}	5×10^{11}	4×10^{11}	2×10^{11}	1×10^{11}	$<10^9$	$<10^9$
Anions (atoms/cm^2)[i]	1×10^{11}	1×10^{11}	1×10^{11}	1×10^{11}	1×10^{11}	1×10^{11}	1×10^{11}
Organics/polymers (C atoms/cm^2)	1×10^{14}	7×10^{13}	6×10^{13}	5×10^{13}	3.5×10^{13}	2.5×10^{13}	1.8×10^{13}
Oxide residue (O atoms/cm^2)[j]	1×10^{14}	7×10^{13}	6×10^{13}	5×10^{13}	3.5×10^{13}	2.5×10^{13}	1.8×10^{13}

[a] Starting wafer up to deposition of the premetal dielectric.
[b] Acceptable gate oxide integrity (GOI) defect densities and a kill ratio of 20% measured postcritical clean; tighter levels may be required if critical, nongate area is considered.
[c] DRAM requirement for Ca, Co, Cu, Cr, Fe, K, Mo, Mn, Na, Ni, W measured postcritical clean for a gettered wafer.
[d] DRAM requirement for Al, Ti, V, Zn (Ba, Sr, and Ta if present in the factory measured postcritical clean for a gettered wafer).
[e] Measured postcritical clean including pregate, prepoly, premetal, presilicide, precontact, and pretrench fill.
[f] Measured premetal, presilicide, and precontact.
[g] Polysilicide metal dielectric deposition through passivation.
[h] K, Li, Na, measured postcritical clean.
[i] Cl, N, P, S, F measured postcritical clean. Assumes no fluorinated oxide.
[j] Measured postcritical clean of a metallic surface region.

XI. SUMMARY

This short introduction to the DR technology requirements from the NTRS by no means encompasses the breadth or depth of opportunity and challenge for CFM scientists and technologists. After significant advancements have been made in integrated circuit design and integration, the natural occurrence of both particulates and nonparticulates is still expected to be a dominant concern for maintaining high levels of profitable yield within the semiconductor manufacturing industry. This chapter should form the basis for the importance of the fundamental and applied topics that will be covered throughout this text, as well as motivate the reader to deeper study of the impacts of contaminants, and the science and technology which describe their formation, transport, and deposition.

ACKNOWLEDGMENTS

Significant numbers of people contributed to the roadmap material and in particular to the articles referenced herein. The author would like to acknowledge these:

TWG Participants

Bob Blewer, Ken Tobin, Susan Cohen, Bill Fil, Venu Menon, Sanjiv Mittal, Milt Godwin, Ron Harris, Dinesh Mehta, Brian Duffy, Terry Francis, Farhang Shadman, Dan Hirleman, Daren Dance, Bill Fosnight, Keith Dillenbeck, Ralph Richardson, Zach Hatcher, Brian Trafas, Paul Proctor, Fred Lakhani, Randy Collica, Jim McAndrew, Mike Grobelny, Charlie Peterson, Jieh Hwa Shyu, Robert Alexander, Val Rio, Lindsey Hall, Pat Lamey, Matt Ivanis, Jennifer Sees, Bobby Bell, Chris Gondran, Dan Clark, Steve Lakeman, Devon Kinkead, Pat Gabella, Ram Akella, Hank Walker, Wojciech Maly, Tom Larson, Charlie Gross, David Jensen

TWG Advisory Groups

SEMATECH Contamination Free Manufacturing FTAB
SEMATECH Defect Detection and Analysis PTAB
SEMATECH Defect Reduction in Equipment PTAB

REFERENCES

1. The National Technology Roadmap for Semiconductors, San Jose, CA, Semiconductor Industry Association, 1997.
2. The National Technology Roadmap for Semiconductors, San Jose, CA, Semiconductor Industry Association, 1994.
3. "Change in Chip Making and How it is Driving Process Diagnostics," VLSI Research, Inc., June, 1996.
4. D. Jensen, C. Gross, and D. Mehta, "Mapping the Roadmap: New Industry Document Explores Defect Reduction Technology Challenges," MICRO, January 1998.
5. F. Lakhani, D. Dance, and R. Williams, "0.25 Micron Integrated Circuit Yield Model Design and Validation," presented at the International IEEE Symposium on Semiconductor Manufacturing, San Francisco, October 1997.
6. D. Dance, D. Jensen, and R. Collica, "Mapping the Roadmap: Developing Yield Modeling and Defect Budgeting for 0.25 μm and Beyond," MICRO, March 1998.
7. C. Weber, D. Jensen, and D. Hirleman, "Mapping the Roadmap: What Drives Defect Detection," MICRO, June 1998.
8. G. Moore, "Progress in Digital Integrated Circuits," IEDM Tech Dig, 1975, p. 11.
9. C. Gross, K. Tobin, D. Jensen, and D. Mehta, "Mapping the Roadmap: Future Technology Requirements for Rapid Isolation and Sourcing of Faults," MICRO, July 1998.
10. B. Fosnight and D. Jensen, "Mapping the Roadmap: Defect Prevention and Elimination," MICRO, October 1998.

3

Off-Wafer Measurement of Contaminants

Robert P. Donovan
L & M Technologies, Inc., and Sandia National Laboratories, Albuquerque, New Mexico

All wafer-processing environments invariably include contaminants, some of which end up on the wafer surface and potentially contribute to product failure. The areal density of surface contaminants deposited on a wafer following exposure to a contaminated environment depends on, among other variables, the concentration of contaminants in that environment. A simple model to predict the areal density of the surface contamination following exposure to a contaminated environment is

$$\frac{N}{A} = c v_d t \tag{1}$$

where

N/A = areal density of contaminants deposited on the wafer surface (number or mass/area)

c = concentration of the contaminant in the environment surrounding the wafer (number or mass/volume)

v_d = a contaminant deposition velocity (length/time)

t = time of wafer exposure to the environment.

This chapter considers only methods of measuring the variable c. Chapters 4 and 5 discuss measurement of N/A. Chapters 6 to 8 discuss the deposition velocities associated with the various mechanisms whereby particles and molecular contaminants in wafer-processing environments are transported and deposited on wafer surfaces (as well as other factors affecting the deposition of contaminants).

I. ANALYZER/SENSOR CATEGORIES

While *sensors* are typically considered a component of an *analyzer* (a device that qualitatively and/or quantitatively separates an unknown into its constituent parts) or a *monitor* (a device that quantitates a specific, predetermined component), the distinction among these three terms is fuzzy. They are often used interchangeably in the literature, a practice that will followed be in this chapter. The chapter emphasis is on measurement methods irrespective of the terms used to describe the hardware.

Measurement methods can be classified by the relationship between the sample actually measured and the system being characterized by that measurement. Five general categories are typical [1,2]:

1. *Off-line.* Discrete samples are withdrawn from the system and shipped/taken to a laboratory for analysis.
2. *At-line.* Discrete samples are withdrawn from the system and are immediately analyzed on site.
3. *On-line.* Analyses are conducted on a system slipstream, which is discarded.
4. *In-line.* Analyses are conducted continuously by sensors inserted directly into the system or into a slipstream that is returned to the system.
5. *Noninvasive.* Continuous, in situ analyses of the system are carried out by sensors external to the system (for example, optically through a window).

Sample disposition is explicitly incorporated into several of these definitions and is implicit in the others. In categories 1 and 2 the sample evaluated is invariably lost to the system. The major distinction between category 3 (on-line sampling from a slipstream) and category 4 (in-line sampling which may or may not use a slip stream) is sample disposition. In category 4, the sample remains a part of the system; in category 3, it does not. Sample coupling to the analyzer, rather than sample disposition, is the property distinguishing category 5 sampling from category 4 sampling.

Most of the sensors/measurement methods discussed in this chapter fall into categories 1 to 3, although a few belong in category 4 or 5. The measurement methods discussed in Chapter 4 fall primarily into category 1 or 2. A major advantage of sensors in categories 3 and higher is speed of response to system changes and the attendant capability to rapidly provide data for either feedback or feedforward process control. The scarcity of category 4 or 5 sensors represents an opportunity for developers of innovative sensors.

The two general contaminant types defined in Chapters 1 and 2 are particulate contaminants and non-particulate contaminants (molecular contaminants).

As the names imply, these categories are mutually exclusive. Not surprisingly then, the measurement methods for these two contaminant types differ significantly and are discussed in separate sections of this chapter.

II. MEASUREMENT OF PARTICLE CONCENTRATION

By far the most common and sensitive instruments for measuring particle concentration are those based on light scattering. The strong dependence of the intensity of light scattering upon particle size, especially submicrometer-sized particles, makes light scattering a high-resolution method for both sizing and counting particles. Today's instrumentation can count single particles with effective light-scattering diameters as small as 0.05 μm. Light-scattering principles as reviewed here apply to both off-wafer measurements in a wafer environment and to the on-wafer measurements discussed in Chapter 4.

A. Basics of Light Scattering

The discussion in this section closely follows that presented by Hinds [3]. Light scattering is the term used to describe the interaction of a light beam with an intercepted particle, as sketched in Figure 1. The incident beam approaches the particle from the left and undergoes the following interactions:

- *Absorption.* Electromagnetic radiation incident on the particle is converted to other forms of energy, mostly heat.

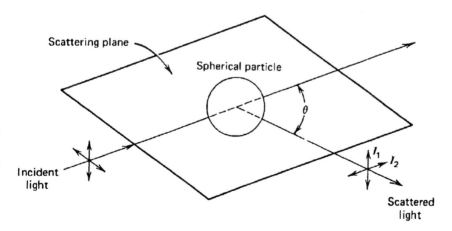

Figure 1 Light scattering by a particle. (From W. C. Hinds, Aerosol Technology. Copyright 1982 by John Wiley & Sons, Inc. Reprinted by permission of John Wiley & Sons, Inc.)

- *Inelastic scattering.* Electromagnetic radiation is absorbed by the particle and reradiated diffusely at a wavelength different from that of the incident beam but characteristic of the particle composition.
- *Elastic scattering.* The incident electromagnetic radiation is reflected, refracted, or diffracted by the particle without change of wavelength—the kinetic energy and momentum of the scattered light is equal to that of the incident light.

Elastic scattering is the interaction mode in which today's optical particle counters operate and is the primary mode to be discussed in this chapter.

The scattering plane sketched in Figure 1 contains the direction of the incident beam and the direction of the detector from the particle. The scattering angle θ is measured in the scattering plane and is the angle between the direction of the incident beam and the direction of the detected beam. The intensity of both the incident beam and the scattered beam can be separated into two components: 1) a component perpendicular to the scattering plane and 2) a component parallel to the scattering plane and perpendicular to the direction of beam propagation. The scattering problem becomes one of expressing the intensity components of the scattered beam in terms of the intensity of the incident beam and those properties of the scattering center that affect the scattering interaction. Since particle size is one of the primary properties affecting light scattering, the intensity of the light scattered by a particle can then be used to deduce the particle size.

The classical solutions to the scattering problem are obtained by solving Maxwell's electromagnetic equations with boundary conditions appropriate to the particle scattering geometry illustrated in Figure 1. Two general solutions result, each applying to a specific range of particle sizes. *Rayleigh scattering* describes the light scattering by a spherical particle of diameter small compared to the wavelength of the incident light:

$$I(\theta) = I_1 + I_2 = \frac{I_0 \pi^4 d_p^6}{8R^2 \lambda^4} \left(\frac{m^2 - 1}{m^2 + 2} \right)^2 + \frac{I_0 \pi^4 d_p^6}{8R^2 \lambda^4} \left(\frac{m^2 - 1}{m^2 + 2} \right)^2 \cos^2 \theta$$

$$= \frac{I_0 \pi^4 d_p^6}{8R^2 \lambda^4} \left(\frac{m^2 - 1}{m^2 + 2} \right)^2 (1 + \cos^2 \theta) \tag{2}$$

where

$I(\theta)$ = total intensity of the light scattered in the direction θ
θ = scattering angle, measured in the scattering plane
I_1 = intensity component of scattered light perpendicular to the scattering plane

I_2 = intensity component of scattered light parallel to the scattering plane and perpendicular to the propagation direction of the scattered beam

I_0 = intensity of the incident light beam

d_p = diameter of the spherical particle

R = distance from the particle

λ = wavelength of the incident and scattered light

m = index of refraction of the particle

Figure 2 is a polar plot of Eq. (2). Because of the $\cos^2 \theta$ term, the I_2 component disappears at a scattering angle of 90°. The I_1 term has no angle dependence and plots as a circle. Light scattered at an angle of 180° (backscattering) has the same intensity as light scattered in the forward direction at 0°, a distinctive feature of Rayleigh scattering.

Another notable feature of the Rayleigh scattering equation is the d_p^6 / λ^4 dependence of the intensity of the scattered light. This strong dependence on particle diameter predicts high size-resolution for optical scattering. However, it also implies that signal strength falls off rapidly with decreasing particle size. Using shorter wavelengths of electromagnetic radiation to illuminate the particle, on the other hand, greatly increases scattering intensity. The Rayleigh scattering analyses generally apply to spherical particles with diameters less than 0.1 μm.

Rayleigh scattering also describes light scattering from individual molecules in the air or other wafer environments. While the scattering intensity from a single air molecule, for example, is too small to be detected, the background signal from the entire collection of atmospheric air molecules that are illumi-

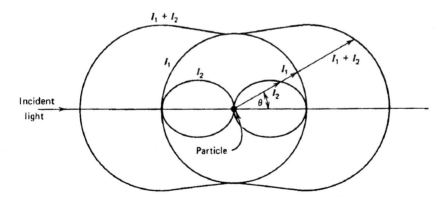

Figure 2 Dependence of Rayleigh scattering intensities, I_1 and I_2, upon scattering angle. (From W. C. Hinds, Aerosol Technology. Copyright 1982 by John Wiley & Sons, Inc. Reprinted by permission of John Wiley & Sons, Inc.)

nated in the sampling chamber of an optical particle counter is a primary factor limiting the size sensitivity of optical particle counters to particles with diameters of 0.05 μm or greater. Molecular scattering dictates small detection volumes in order to minimize this background scattering signal. Unfortunately, small detection volumes also mean low sample flow rates and poor counting statistics when particle concentration is low. As it is, even optically detecting particles with 0.05 μm diameters in air at practical sample flow rates requires nonstandard, clever instrument design. One technique is to divide the total sample flow into parallel channels so that the volume of each individual detection volume is much smaller than if all the sample flow passed through a single counting volume. Assuming a particle appears in only one of the parallel channels allows detection with a background signal characteristic of the smaller volume, but at the same time the total sample flow remains the sum of the flows through all the parallel channels.

Mie scattering, the second general solution to the optical scattering problem, characterizes light scattering from spherical particles with diameters of the same order of magnitude as that of the wavelength of the incident electromagnetic radiation:

$$I_1(\theta) = \frac{I_0 \lambda^2 i_1}{4\pi^2 R^2} \tag{3}$$

$$I_2(\theta) = \frac{I_0 \lambda^2 i_2}{4\pi^2 R^2}$$

where all symbols are the same as in Eq. (2) except that i_1 and i_2 are the Mie intensity parameters for scattered light with perpendicular and parallel polarization, respectively.

The Mie intensity parameters are complex functions of particle size, index of refraction, and scattering angle so that the compact expressions of Eq. (3) actually represent much more complicated behavior than the lengthier expressions of Eq. (2). When particle diameter is of the same order of magnitude as the wavelength of the illuminating radiation, multiple interferences occur among the various components of the beam within the particle, and the net result is that the scattering intensity exhibits an oscillatory behavior with both scattering angle and particle diameter for particles in this size range. Figure 3 illustrates typical dependence of scattering intensity as a function of scattering angle. It also shows that for particles with diameters close to the Rayleigh size range, the Mie expressions of Eq. (3) approach the Rayleigh expressions of Eq. (2); that is, at the smallest size parameter shown in Figure 3 ($\alpha \equiv \pi d_p/\lambda = 0.8$), I_1 has only a small θ dependence, similar to that of I_1 in Figure 2 and, for the same size parameter, I_2 in Figure 3 vanishes in the region where $\theta = 90°$, similar to the behavior of I_2 in Figure 2.

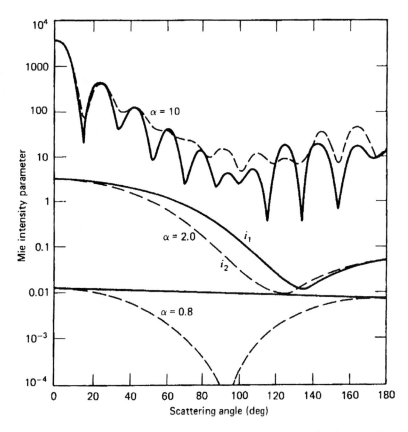

Figure 3 Dependence of Mie intensity parameters, i_1 (solid lines) and i_2 (dashed lines), upon scattering angle. (From W. C. Hinds, Aerosol Technology. Copyright 1982 by John Wiley & Sons, Inc. Reprinted by permission of John Wiley & Sons, Inc.)

Figure 4 shows that Mie scattering intensity also oscillates with size parameter. For values of $\alpha \geq 6$, scattering intensity at any given scattering angle is not a unique function of particle diameter. An initial reaction to this dependence is that optical scattering, in spite of its high size-sensitivity and resolution, may not be a suitable basis for uniquely sizing particles. However, optical particle counters are designed to collect light scattered over a range of scattering angles, and this averaging effect results in essentially a monotonic relationship between particle size and the intensity of the scattered light (Fig. 5). So optical particle counters, based on light scattering, are indeed the primary analyzers used to size and count particles in wafer environments and on wafer surfaces.

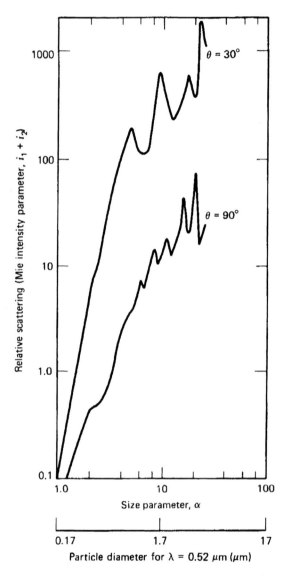

Figure 4 Typical dependence of $i_1 + i_2$ upon particle diameter. (From W. C. Hinds, Aerosol Technology. Copyright 1982 by John Wiley & Sons, Inc. Reprinted by permission of John Wiley & Sons, Inc.)

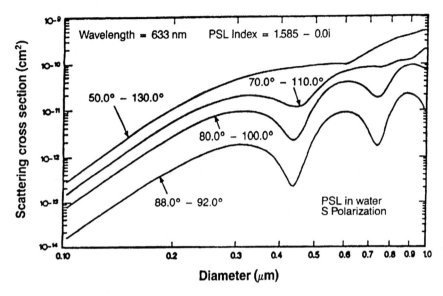

Figure 5 Increase of monotonic behavior of scattering cross section at large collection angles. (From Ref. 4.)

B. Other On-Line Methods of Counting Particles

There are a number of methods for measuring particles that use light scattering indirectly or differently than in the standard particle counters. One of these is the *aerodynamic particle sizer*, which uses a split laser light beam to measure the time of flight of a particle entrained in a known airflow (Fig. 6). Particles at rest entering a gas flow require time to accelerate to the velocity of the gas flow. The time required for the particle to reach 63% of its final velocity is the *particle relaxation time* which depends on *particle aerodynamic diameter* (the aerodynamic diameter is the diameter of a unit density sphere having the same settling velocity as the particle) among other variables [Eq. (4)]:

$$\tau = \frac{\rho_p d_a^2 C_c}{18\eta} \tag{4}$$

where

τ = particle relaxation time
ρ_p = particle density, assumed to be 10^3 kg/m^3 (1 g/cm^3)
d_a = particle aerodynamic diameter
C_c = Cunningham slip correction factor [an empirical multiplier > 1, extending the use of Eq. (4) to particles with diameters < 1 μm in which size range the Stokes' law assumptions of continuum flow and

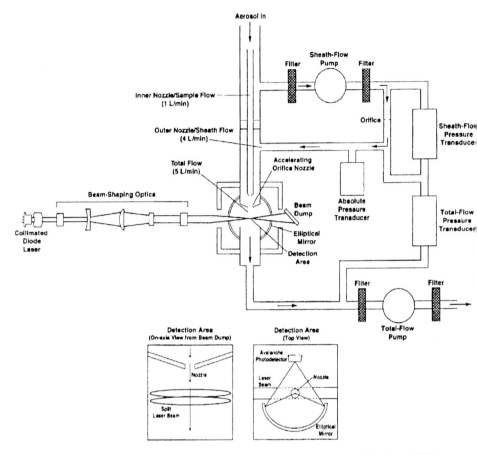

Figure 6   Schematic of the aerodynamic particle sizer. (Courtesy of G. Sem, TSI Inc., St. Paul, MN.)

zero fluid velocity at the particle surface, used in deriving Eq. (4), begin to break down]. (See Chap. 6, Sec. II.A, for a more detailed discussion of the Cunningham slip correction factor.)

$$= 1 + \left[\frac{3}{2Pd_a} \right] [6.32 + 2.01 \exp(-146Pd_a)]$$

P = absolute pressure, N/m^2

d_a = particle aerodynamic diameter, μm

η = fluid viscosity, taken to be that of air at 20°C and 1 atmosphere

= 0.181 × 10^{-5} N-s/m^2

At $t = 3\tau$ the inserted particle velocity has reached 99% of the fluid velocity.

Particle sizing in an aerodynamic particle sizer derives not from the size dependence of light scattering by the particle but rather from the aerodynamic properties of the particle. The two branches of the split laser beam are simply triggers for timing circuits that measure the particle time of flight over a fixed distance in which the particle is accelerating in a fluid flow. This measured time is used to calculate a τ value which is related to particle aerodynamic diameter by Eq. (4). The instrument-measured times are in the micro- to millisecond range, corresponding to particle aerodynamic diameters in the size range of 0.5–20 μm.

The *condensation particle counter (CPC)* is an example of an instrument that uses light scattering as an event counter only rather than a particle sizer and counter. This instrument incorporates upstream conditioning to grow particles that are too small to be detected by conventional light-scattering instruments into larger particles that are easily detected by a simple light-scattering particle counter. The upstream-conditioning step is to saturate the gas stream with a vapor at a temperature somewhat above ambient and then cause vapor condensation on the small particles by passing the stream into a region of temperature below ambient (Fig. 7). By this modification some CPCs can detect particles as small as 3 nm, more than an order of magnitude smaller than the 50 nm detection limit of

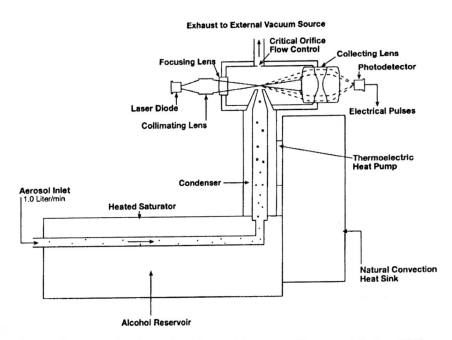

Figure 7 Schematic of a condensation particle counter. (Courtesy of G. Sem, TSI Inc., St. Paul, MN.)

the most sensitive optical particle counter (OPC). However, CPCs have no sizing capability. All particles that are detected have grown to a size either unrelated or complexly related to that of the initiating condensation center and so are simply counted without any sizing information. Additional upstream conditioning is required to obtain sizing information. Two such conditioning stages are the *differential mobility particle sizer (DMPS)* and the *particle diffusion battery*. The first uses electrical forces to separate particles according to electrical charge/mass and the second, particle diffusion. While these devices are widely used by aerosol technologists, they have had little impact on semiconductor practices and in cleanrooms and will not be discussed further. Hinds [3] describes these devices more fully and provides references for additional details.

C. Other Particle-Measuring Methods (Off-Line)

While light scattering underlies the operating principles of most particle-measuring instrumentation used in semiconductor manufacturing, other methods, based on other principles, do exist. *Counting of particles on a filter* through which a known volume of fluid has been passed yields a direct measurement of aerosol particle number concentration. Similarly, measuring the weight gain of that filter provides a measurement of aerosol particle mass concentration. Particle number concentration is the more common unit of concentration used in measuring and specifying particulate contamination in semiconductor cleanrooms and manufacturing environments. Mass concentration is the typical unit used by EPA and other groups in specifying exposure and emission standards.

Procedures for conducting filter measurements are spelled out in the following standards:

ASTM F25–68 (Reapproved 1995), "Standard Method for Sizing and Counting Airborne Particulate Contamination in Clean Rooms and Other Dust-Controlled Areas Designed for Electronic and Similar Applications" (ASTM, 100 Barr Harbor Dr., West Conshohocken, PA 19428)
FED-STD-209E, "Airborne Particulate Cleanliness Classes in Cleanrooms and Clean Zones," Sept. 11, 1992 (General Services Administration, General Products Commodity Center, Federal Supply Service, 819 Taylor Street, Fort Worth, TX 76102)

In this method the particle concentration c is calculated from Eq. (1) by measuring the N/A on the filter, knowing the time of exposure t, and setting the deposition velocity v_d equal to the linear flow velocity of the environmental medium through the filter. This last assignment assumes that all particles in the flow are captured and retained by the filter, which, for particles of sufficient size to be counted under a microscope, is a good assumption. While in principle this method works for particles over a broad size range, both the ASTM Standard

F25 and the FED-STD 209E explicitly specify that it should be used only for counting particles greater that 5 μm in diameter, particle counters based on light scattering being preferred for counting smaller particles.

A related method is that employing an *impactor* to collect particles from an environment by depositing them on a stub for insertion into a scanning electron microscope (SEM) or on another surface suitable for counting by any of various other techniques. The SEM method allows the counting of particles smaller than what is possible with an optical microscope and relies on particle capture by the mechanism of inertial impaction (Chap. 6, Sec. III). Flow from the environmental medium being sampled is constricted through a nozzle or orifice with the flat collection surface of the SEM stub placed perpendicular to the flow immediately downstream of the nozzle exit (Fig. 8). This positioning allows the fluid to flow around the stub, but the inertia of many of the particles entrained in the fluid causes them to strike and adhere to the stub where they can be counted and examined under the SEM. Particles with aerodynamic diameters as small as 0.2–0.3 μm can be captured and counted by the impactor method. When operated at subatmospheric pressures, the cutoff diameter for particle collection is even lower.

SEM samples can also be collected electrically in a *point-to-plane electrostatic precipitator*. This device captures particles by the mechanism of electrophoresis—the drift of electrically charged particles in an electric field [see Chap. 6, Eq. (16)]. A sharpened electrode, protruding into the aerosol flow

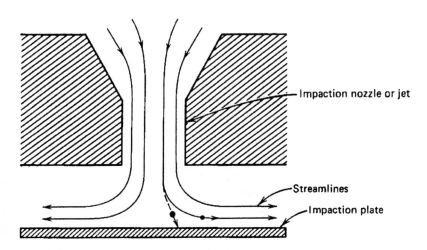

Figure 8   Cross section of an inertial impactor. (From W. C. Hinds, Aerosol Technology. Copyright 1982 by John Wiley & Sons, Inc. Reprinted by permission of John Wiley & Sons, Inc.)

opposite the SEM collection stub, both charges the aerosol particles to be collected and creates an electric field to drift those charged particles toward the SEM stub where they are captured (Fig. 9). Even for particles which have just been "neutralized" by passing through a radioactive cell, electrophoretic forces become dominant in the submicrometer particle size range. At the high levels of particle charge achievable in a corona discharge, electrical forces also dominate the transport of much larger particles.

Finally, particle samples can also be collected by *settling plates*—plates or SEM stubs placed with collection surfaces perpendicular to the gravitational force. In the absence of significant convective flow (still air), the deposition velocity of unit density particles larger than 0.2–0.3 μm is dominated by gravitational settling and is simply the terminal settling velocity of the particle. By equating the Stokes drag force on a spherical particle to the gravitational force on the particle, the terminal velocity becomes:

$$V_T = \tau g \tag{5}$$

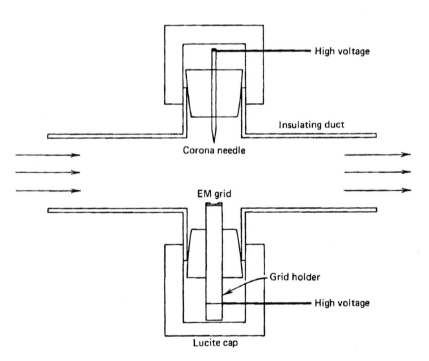

Figure 9 The point-to-plane particle precipitator. (From W. C. Hinds, Aerosol Technology. Copyright 1982 by John Wiley & Sons, Inc. Reprinted by permission of John Wiley & Sons, Inc.)

where

V_T = terminal velocity of a particle attributable to gravitational settling
τ = particle relaxation time
g = gravitational acceleration constant

A major advantage of all off-line particle measurement methods that consist of collecting particle samples is that the particle sample is retained and available for detailed analyses, including composition, unlike those on-line methods that simply capture a light-scattering signal from a passing aerosol stream. This latter approach is the common method used in monitoring. For troubleshooting, however, the off-line methods that retain the sample are usually more valuable even though they are not the preferred routine monitoring method. Both scanning and transmission electron microscopy are powerful tools in the postcollection analyses of particles on filters [5].

Sample collection technique is especially important in the measurement of airborne bacteria. Gravitational settling, filtration, impaction, or impingement don't always yield the same results [6]. Selecting the "best" sampling method depends on matching the collection mechanism to the environment from which the sample is being drawn. Fragmentation of bacteria by an energetic collection mechanism can distort the size distribution. Damage to bacteria during collection can adversely affect the accuracy of subsequent counting methods. Keeping the bacteria alive between collection and counting is often an important consideration. A gelatin membrane filter avoids desiccation losses and minimizes many sampling errors [6].

D. Novel Methods for Measuring Particles

Aerosol mass spectrometers cannot only count particles but can also analyze particle composition. These instruments operate by ablating particles in a pulsed laser beam, forming ion fragments which are then mass-analyzed. Such an instrument can both count and analyze aerosol particles but the mass analyzer requires subatmospheric pressure. Aerosol particles with diameters as small as 10 nm have been analyzed by laboratory versions of this type of instrument [7].

A related technique, capable of operating at atmospheric pressure, that has also shown submicrometer particle detection capability in the laboratory, is *particle pyrolysis* followed by surface ionization [8,9]. The method consists of impacting a particle on a hot filament which dissociates it. The composition of many particles is such that at least some of the dissociated molecules will be ionized by contact with the filament surface. These ionized species are then detected and counted in an appropriate downstream stage such as an electrometer or a mass spectrometer (Sec. III).

While no commercial instrument based on particle pyrolysis is used in the semiconductor industry at present and indeed the impact of the lone commercial version of the laboratory prototype on industrial practice in general has been minor, it is mentioned here because it is a particle-measuring method based on the conversion of particles into molecular contaminants, the other major contaminant type of manufacturing concern. Particle pyrolysis followed by detection of molecular fragments implies that the metrology for molecular contaminants is more sensitive than that of particles. In principle this type of conversion extends the detection range for contaminants from that of single particles to that of single ions!

E. Environmental Constraints on Particle Measurements

Detection of particles by light scattering can be adapted to all common processing environments, including atmospheric pressures, high and low pressures, and liquids; and indeed, instrumentation that uses light scattering for particle measurement in all these environments is commercially available.

1. Particle Measurement at Atmospheric Pressure

The initial application of light scattering to count and size particles was, not surprisingly, an optical particle counter designed to operate in benign atmospheric pressure environments—the ambient air either inside or outside a cleanroom—and near atmospheric pressure environments made up of inert or other process gases such as nitrogen or oxygen. This application remains a major use for optical particle counters, verification of particulate contamination levels in a cleanroom per FED-STD-209E [10] or ISO 14644-1 [11] being prime examples of this application.

Two general types of optical particle counters are available: the basic *optical particle counter (OPC)* and a higher resolution instrument often referred to as a *particle spectrometer* [4]. Various OPC configurations are available commercially, Figure 10 illustrating two extremes. The forward scattering configuration takes advantage of the increased scattering intensity associated with Mie scattering; the 90° scattering angle configuration sacrifices scattering intensity for reduced background noise.

The basic OPC typically counts and classifies particles into two, four, or eight size bins between 0.1 and 5 μm. Particle size is determined by matching the intensity of light scattered by the particle to that scattered by a polystyrene latex sphere (PSL) of known diameter. Particle size is thus expressed in terms of the diameter of a PSL sphere that scatters the same light intensity as the particle—the so-called equivalent optical scattering diameter of the particle. Initially the most common light source was the 633 nm line of a helium-neon laser. Semiconductor

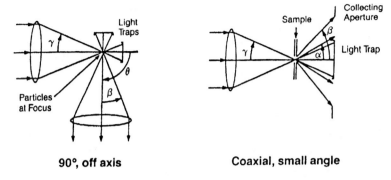

90°, off axis   Coaxial, small angle

Figure 10   OPC scattering configurations. (From Ref. 12.)

laser diodes (780 nm) have now become popular because of their reliability, ruggedness, and low cost. Shorter wavelength argon ion lasers (458 nm) are also being used in order to capitalize on the enhanced scattering intensity at the shorter wavelengths [Eq. (2)].

Flow rates between 0.28 and 28 liters/min (0.01 and 1 cfm) are available in off-the-shelf optical particle counters. High flow rates reduce the sampling time required to obtain statistically significant counts within state-of-the-art clean-rooms and other low-concentration volumes. Low flow rates are needed when sampling high-concentration sources. Otherwise the counter saturates, miscount-ing and distorting the size distribution. Typical maximum recommended particle concentrations to preserve less than 10% coincidence loss in an OPC are in the 10^4–10^6 particles/cm^3 (10^5–10^7 particles/ft^3) range, depending on sampling volume and flow rates.

A *particle spectrometer* classifies a particle distribution into eight, sixteen, or even more size bins, taking advantage of the huge dependence of light scatter-ing from small particles upon particle diameter [a sixth-power dependence in the Rayleigh scattering regime, see Eq. (2)] to achieve fine particle size resolution. Spectrometers achieve the greatest size sensitivity among the family of optical scattering instruments but also operate at the lowest flow rates—as low as 0.06 liters/min (0.002 cfm). They are well suited for measuring the size distributions of particles emitted by highly concentrated sources. They are not well suited for characterizing cleanrooms or other low concentration sources and hence are not generally used to monitor manufacturing environments.

The *condensation particle counter (CPC)* extends the size range over which aerosol particles can be counted to as low as 3 nm but at the cost of losing sizing information—all particles are counted but not classified into size bins, since the CPC achieves its lower size detection limit by increasing the size of each particle to the range easily detected by an OPC. Butanol is the typical working fluid for CPCs designed to operate in inert environments, including air (Fig. 7). For reactive environments, inert working fluid CPCs are also on the market [13,14]. Alternatively, McDermott [15] has developed a diluter tube based on nonturbulent mixing of an inert gas with the reactive gas. This added inert gas dilutes the reactive gas species without diluting the aerosol particle concentration in the laminar streamlines of the tube. The dilution of the reactive gas is by diffusion of the inert gas molecules into the particle streamlines and the simultaneous outdiffusion of the reactive gas molecules. No turbulent mixing occurs. The more massive aerosol particles do not appreciably diffuse out of their streamlines, and thus an accurate measurement of the particle concentration in the reactive gas can be made by feeding the undiluted particle streamline into an inert gas CPC.

The maximum gas flow rates through both the inert and the reactive gas CPC instruments now commercially available are typically 2.8 liters/min (0.1 cfm), an order of magnitude lower than what is available with OPCs. This reduced flow rate is a distinct disadvantage in monitoring the low aerosol particle concentrations typical of contemporary cleanrooms—collecting a statistically significant sample requires an order of magnitude more time. However, as product sensitivity to particles in the 5–50 nm size range increases, the importance of the CPC increases, since it remains the only commercially available particle counter capable of on-line detection of particles in this size range. In addition, particle concentrations often exhibit an inverse power law dependence on particle size so that the particle concentrations to be measured in the 5–50 nm range are usually higher than those in the >50 nm range, partially (or totally) compensating for the lower flow rate penalty.

2. Particle Measurement at Subatmospheric Pressures

In situ particle monitors (ISPMs) is the name now generally assigned to those light-scattering instruments that are located within processing environments and are used to measure particle concentrations in vacuum-processing equipments and other low-pressure systems. Light scattering principles, as described in Section II.A, apply at subatmospheric pressures as well as at atmospheric pressures. What does change at these lower pressures is control of the sample entering the scattering region of the particle counter. At lower pressures, little or no aerodynamic flow control is available to convect a representative aerosol sample into a scattering chamber for counting and sizing. The usual mode of ISPM operation is one of letting whatever particle transport mechanisms are present in the environment being monitored bring the particles from their point of entry into the environment or their point of generation within the environment to the detection region of the ISPM. Since these particle transport mechanisms are generally complex and not fully analyzed for most subatmospheric processing environments, obtaining a realistic, representative sample most often depends upon imperfect "engineering judgment" for optimal ISPM location. In addition, the placement of the ISPM cannot interfere with the process operation.

A common site chosen for locating an ISPM is in the vacuum pump line. Even here, however, the area in which particles are counted and sized is typically only a small fraction of the pump line cross section. Unfortunately, particle flows in the pump line are not well mixed so that even in this location delivering a representative sample to the ISPM is not assured. In addition, the pump line is removed from the wafer location where the measurement of particle concentration is of highest value. Nonetheless relative changes in the particle concentration at an arbitrary fixed location in the pump line or other in situ location can provide useful process information. Significantly improved equipment design and process operation have already resulted from the real-time in situ measurements provided by ISPMs.

While accurate measurements of particle concentration within subatmospheric processing equipment remain elusive because of the problems discussed in the previous two paragraphs, progress is being reported in understanding and controlling the uncertainties in particle transport that confound the measurements. For example, researchers at the University of Minnesota have developed aerodynamic lenses that operate at 1 torr (130 Pa) and can focus particles from any point in an exhaust flow of that pressure into well-defined narrow beams that can be directed into the detection region of an ISPM [16]. Incorporation of such lenses into ISPMs would appear to significantly upgrade the quality of the data generated by these already useful monitors. Bear in mind, however, that the performance of such aerodynamic lenses deteriorates as operating pressures decrease below 1 torr.

Using optical fibers to introduce the primary light beam into the subatmospheric processing chamber and to collect the scattered light signal allows the sensitive optical components to be located external to the process and thus not subject to incompatible processing environments [17]. This configuration also increases monitoring options by making it possible to sample a large number of internal equipment sites by inserting multiple fibers. Fiber flexibility enables sampling of locations that would be inaccessible or just marginally accessible using the standard ISPM hardware.

3. Particle Measurement at Above Atmospheric Pressures

The measurement of particles at above atmospheric pressures, in high-pressure gas flows, for example, has been addressed by noninvasive optical scattering instruments that use one window for admitting an interrogating laser beam into the high-pressure region of interest and a second window for detecting the signal optically scattered by any intercepted particles. Since focusing of the particle flow stream is usually not part of the arrangement, the question of collecting a representative sample arises here, even as it does for sampling at subatmospheric pressures: How representative is the particle concentration in the region sampled of the particle concentration in the total sample? (This same noninvasive configuration can also be used at subatmospheric pressures.)

The background signal from optical scattering by molecules increases with the increasing molecular densities of above atmospheric pressures. Thus, OPCs operating at above atmospheric pressures cannot detect particles as small as OPCs operating at atmospheric pressures because of the higher background signal at the higher pressures. Nonetheless, particles as small as 0.1 μm can be detected at pressures up to 3000 psi (21 MPa), although the higher pressure ranges [above 1000 psi (7 MPa)] operate at flow rates on the order of 15–20 cm^3/min. The noninvasive design of these monitors makes them compatible with measuring particles in toxic and flammable gases as well as the inert gases.

Measurements of the concentrations of particles inside of commercial, high-pressure gas cylinders are not compatible with in situ optical technologies. The common technique for making such measurements is to draw a gas sample from the cylinder through a critical orifice, reducing the gas pressure to levels compatible with optical-based particle counters that operate at or near atmospheric pressure. OPCs and CPCs are the typical particle counters used in this measurement. Both counters typically report high concentrations of particles in gas samples drawn from cylinders whose pressure exceeds about 500 psi (3500 kPa)—even when the gas sample is passed through a high-quality particle filter placed between the cylinder and the orifice! The particles evidently form downstream of both this filter and the expansion orifice. In experiments attempting to understand the particle formation mechanism, Reents et al. [18] used a novel, customized ultrasensitive particle analysis system (USPAS), similar to that described previously in Section II.D, to both count particles and analyze their composition. This analytic instrument vaporized emitted particles in a high-intensity laser beam and mass-analyzed the fragments in a time-of-flight mass spectrometer (TOF/MS) (see Chap. 5). Their USPAS was capable of measuring and analyzing particles as small as 2 nm. Reference 18 identified the lubricants used in cylinder construction as the source of the particles but did not unambiguously clarify the mechanism of the gas-to-particle conversion.

The 1997 NTRS identifies the continuous, in-line detection of particles in high-pressure, process gases at their point of use as a "significant challenge," made more difficult by the need to make such measurements with low-cost sensors compatible with multipoint monitoring in corrosive gases. Suppliers of process gases continue to address these measurement problems. Reference 19 reviews the status and capability of the state of the art in making high-pressure/low-pressure particle measurements in process gases.

4. Particle Measurement in Liquids

Particle counting in liquids can be conducted by either optical scattering or light extinction. Light extinction refers to measurements of reductions in the light collected by a detector aligned with the primary beam but on the opposite side of the particle path from the source—in the forward direction at which the scattering angle is 0°. Light extinction follows the Lambert-Beer law:

$$\frac{I}{I_0} = \exp(-\sigma_e L) \tag{6}$$

where

I = light intensity reaching the detector
I_0 = intensity of the incident light
σ_e = particle extinction coefficient
L = path length of the light beam traversing the particle beam

Light extinction is, loosely, the complement to light scattering. The incident light beam illuminating a region containing particles subdivides into components as follows after passing through the particle beam:

Incident beam → light absorbed + light scattered out of the detector path
+ light transmitted without interaction

Light extinction measures the last component only. In the absence of any scattering center, the light intensity reaching the detector is the same as that of the incident beam. When a particle or a collection of particles is inserted into the beam path between the incident beam and the detector, the light reaching the detector is reduced by that scattered out of the beam and away from the detector and that absorbed by the particles. Light that is forward-scattered by particles in the incident beam can be reduced to negligible contributions by optical designs that allow only parallel, unscattered light to pass through an aperture placed immediately in front of the detector.

Extinction methods are most appropriate for particles larger than 1 μm in diameter. For submicron particles, optical scattering remains the preferred technique, as with the measurement of aerosol particles, and is the optical design that will be discussed here.

A basic difference between the design of a light-scattering counter for particles in liquids and that for particles in air is the relationship between the cross-sectional areas of the light beam and the particle beam. In aerosol instruments the light beam is larger than the aerodynamically focused particle beam and, by suitable design and selection of components, all particles in the particle beam can be made to pass through a nearly uniformly illuminated scattering region, even when the light beam has a Gaussian intensity distribution. This desirable design feature means that similarly sized particles scatter with similar light intensities and are thus similarly counted and classified. In liquids, the lack of a suitable hydrodynamic focusing capability means that the particle flow must be defined by tube walls or other material interfaces. With this configuration, having the light beam larger than the particle beam means that the liquid-wall interface is also illuminated and scatters light to the detector, introducing significant background noise into the measurement. This contribution to background noise increases the size of the smallest particle that can be detected using this type of design.

One way to eliminate this undesirable interference is to focus the light beam into a spot size smaller than the cross-sectional area of the particle flow (Fig. 11, in situ). A penalty thereby incurred is that the intensity of light scattered by a given particle depends on where it passes through the nonuniform light beam. Sophisticated dual detector coincidence techniques exist to reject the signals from particles passing through the edges of the light beam [4]. However, the more commonly used instruments in the industry, called *in situ monitors* by one prominent manufacturer of optical particle counting instruments (*not* the same as the ISPMs discussed earlier as particle counters for insertion into processing equipment operating at subatmospheric pressures), ignore these differences and simply accept the miscounting introduced because of light beam nonuniformity in the scattering region. These monitors overcount large particles

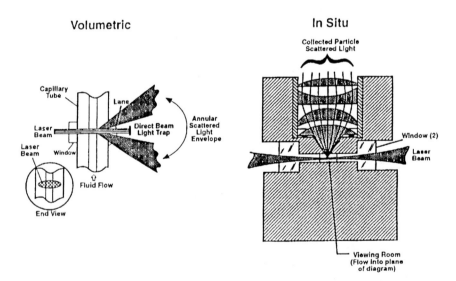

Figure 11 Particle counters for liquids: volumetric designs versus in situ designs. (From Ref. 4.)

and undercount small particles. Since concentrations of small particles are invariably greater than those of the larger sized particles in virtually all liquids, the counting of the large particles that pass through the beam edge as small particles partially compensates for the reduced sampling volume in which actual small particles are counted. Thus, in spite of their neglect of the spatial variation of beam intensity, in situ particle monitors produce surprisingly good agreement with in situ spectrometers when measuring high quality UPW [4]. When measuring water containing significant concentrations of particles larger than 0.2 μm, however, particle monitors exhibit significant counting and sizing errors [20].

Instruments that retain the aerosol counter configuration in which the spot size of the light beam is larger than that of the confined particle beam are still commercially available for counting particles in liquids. They are called *volumetric* counters (Fig. 11)—in principle they sample the entire volume of the particle beam passing through the counter and thus can count a statistically significant number of particles in a given size range in less time than an in situ counter which counts the particles in just a fraction of the particle beam passing through the counter [21]. However, in situ counters can count particles as small as 0.05 μm, while the size detection limit of volumetric samplers is on the order of 0.1–0.2 μm because of the previously mentioned wall-liquid interface contribution to the background signal level.

Scattering intensity also depends upon the difference in the indices of refraction between the scattering center—the particle—and the liquid media [Eqs. (2) and (3)]. Typically, this difference is less in liquid media than in gases so the scattering signal from any given particle is reduced in liquids compared to gases [22]. Nonetheless, commercial in situ OPCs can detect particles as small as 0.05 μm in water and 0.065 μm in corrosive liquids. The volume flow rates at which these sensitivities are achieved are relatively low, the state-of-the-art instrument now operating at a flow rate of 0.25 mL/min. Thus a counter background noise level of 1 count/4 min contributes 1 mL^{-1} to the reported particle concentration, a value of the same order of magnitude as the actual particle concentrations in many ultrapure waters.

Volumetric OPCs that do not have detection limits as low as the in situ instruments but sample at larger volume flow rates (for example, a detection limit of 0.1 μm at a flow rate of 50 ml/min) may represent a better choice for characterizing water with low-particle concentrations and where the limited size sensitivity is not critical [21].

The ISPMs that have proven valuable in many equipment environments have also been used in the liquid baths of wet benches, measuring particle concentration in the near vicinity of product wafers [23]. The data so collected can be used to optimize bath performance and provide improved understanding of the dynamics of the cleaning and rinsing mechanisms involved.

5. Bacteria Measurement in Water

Bacteria constitute a special particulate problem in UPW systems in that unique measurement methods are used for counting them rather than an OPC even though many bacteria are optically detectable. The problem is that the OPC cannot at present distinguish between bacteria and any of the more numerous, other types of particles in the typical water sample. Special methods, sensitive only to bacteria, are needed. All measurements of bacteria are off-line, since no on-line method exists.

The traditional method for measuring bacteria in water is given in ASTM F-1094 [24]. This method consists of passing water through a 0.45 μm membrane filter, which is subsequently placed on an agar medium and incubated for 24 h at 35°C. The number of colony-forming units (cfu) are counted and reported as a density (cfu/ml) based on the volume of water passed through the filter, typically 1 liter. Results depend somewhat upon culturing temperature, time, and nutrient concentration so that all types of bacteria are not detected and repeatable results are not always observed.

Epifluorescence microscopy [25] is also an ASTM method (F-1095; Ref. 26) for measuring bacteria in water. The method requires passing a known volume of the test water through a predarkened membrane filter and staining the filter with acridine orange (N, N, N', N'-Tetramethyl-3, 6-acridine-amine monohydrochloride). Bacteria cells, both viable and nonviable, adsorb the dye and fluoresce under appropriate excitation, making them countable with an epifluorescence microscope. The method can be completed, start to finish, in 1 h. Since this epifluorescence method counts both dead and living bacteria as well as nonculturable bacteria and some other extraneous detritus (organic debris from microorganisms), it typically yields bacteria concentrations higher than the traditional culturing methods (ASTM F-1094). While the epifluorescence method is technique dependent, its speed makes it probably the most commonly used method for monitoring bacteria in water at present.

A modified fluorescent method [27] uses *carboxy fluorescein diacetate (CFDA)* to detect esterase activity in viable cells. CFDA itself is nonfluorescent but is hydrolyzed by cellular esterase activity into a luminescent fluorescein, which emits light when appropriately irradiated. Only live bacteria exhibit esterase activity so the CFDA method does not count dead cells. Osawa et al. [27] postulated that the CFDA method also counts "injured" bacteria—bacteria that are alive and survive some sterilization treatments even though they are unable to form the colonies that the culturing method requires for counting. The CFDA method can thus be used in conjunction with other methods of counting bacteria, each with its own bacteria responsiveness, to identify the sensitivities and effectiveness of various sterilization methods.

Alternative methods of measuring the concentration of bacteria in water have also been developed. One such method is the *ATP* (adenosine triphosphate— $C_{10}H_{16}N_5O_{13}P_3$) monitor [28]. ATP is a product of bacteria, which can be detected by a reaction with lucifern and oxygen in the presence of luciferase, and Mg^{2+} to yield a luminescent product whose intensity is proportional to the bacteria present. To make this measurement, a sample is collected on a filter as in the ASTM methods. This sample then serves as the source of ATP for the bioluminescent reaction.

Still another alternative technique is based on the *polymerase chain reaction (PCR)* [29]. This method detects DNA molecules that are characteristic of bacteria, living or dead. The detection technique depends on separating the double strands of the DNA into single strands and resynthesizing each by introducing a primer molecular chain to form two DNA molecules. This sequence of doubling the initial DNA by separating and replicating each strand can be repeated to achieve large amplification of the starting numbers, simplifying detection by gel electrophoresis. In this method water samples are collected on a filter as with the other methods of measuring bacteria. The filter is then placed in the appropriate solutions and environments for the sequence of steps in the amplification process to be carried out. The method requires about 6 h, including detection following the amplification sequences. In spite of its relatively rapid speed of response by traditional bacteria-measuring standards, this method is not yet widely used.

III. MEASUREMENT OF MOLECULAR CONTAMINANTS

In an operational sense, molecular contaminants are any contaminants not detected by a particle counter, implying that different measurement methods must be used to quantify them. Failure of a molecular cluster (a particle) to be detected by a particle counter implies that the cluster consists of too few molecules for detection, so one property of a detector of molecular contamination must be sensitivity to a smaller number of molecules than are found in a particle and, ideally, sensitivity to a single molecule.

The major environments considered in this section are atmospheric air, low-pressure processing chambers, high-pressure gases, and liquids. Off-line measurement technology for molecular contaminants is more advanced and complete and is discussed first. On-line measurement of molecular contamination, however, is growing in importance and capability, especially in liquids. Some on-line instrumentation is commercially available and more is in development. This section reviews both commercial and developmental on-line instrumentation.

A. Off-Wafer/Off-Line Measurement of Molecular Contaminants

1. Airborne Molecular Contaminants at Atmospheric Pressures

Detection limits in the ppt range are now routinely reported in the analyses of airborne molecular contaminants at atmospheric pressure [30–32]. The technique typically consists of passing an air sample through appropriate adsorbing tubes, or bubbling the air through sorbing solutions or concentrating the analytes from the sample on other surfaces, all of which retain the contaminants for later laboratory analyses [33]. Environmental sampling time can be hours or even days, depending on the analyte concentration. In the off-line analytical laboratory, *gas chromatography/mass spectroscopy (GC/MS)* is generally used to analyze organic contaminants [34], although other measurement methods are also used [30]. *Inductively coupled plasma spectroscopy/mass spectroscopy (ICP/MS)* is the most common technique now used for analyzing inorganics, including most trace metals; *ion chromatography* and, more recently, *capillary electrophoresis* are the methods of choice for analyses of ions collected in solutions.

This type of sampling technology for cleanroom air has been shown to be compatible with remote sample collection followed by laboratory analyses far removed in location from the sampling site. Samples collected from overseas sites, for example, have been successfully analyzed in Japan, using a portable sampling kit developed to preserve sample integrity over long-distance transport [32].

2. Description of Analytical Instrumentation

The gas chromatograph of a *GC/MS* system is a fractionating column which separates species but may not uniquely identify the compounds. A mass spectrometer downstream of the GC column can generally provide unambiguous species identification. Detection limits for common organic contaminants found in a cleanroom range from less than 1 ppm by weight to sub-ppb concentrations. The *ICP/MS* measurement consists of aerosolizing the liquid sample into an argon plasma which vaporizes, dissociates, and ionizes dissolved solids in the sample [35]. These species are then separated and analyzed in a mass spectrometer of either quadrupole or magnetic sector design, the latter having superior mass resolution but greater cost and complexity. Table 1 lists reported detection limits for some trace metals in air when analyzed by ICP/MS.

Ion chromatography (IC), or ion-exchange chromatography, is a liquid chromatographic method that can be used to determine dissolved ionic species in solutions. It is a fractionation method in which there is an interchange of ions between an insoluble stationary phase and ions in a liquid mobile phase.

Table 1 Detection Limits for Trace
Metals in Air, Measured by ICP/MS

Analyte	Detection limit (ng/liter of air)
Aluminum (Al)	0.003
Boron (B)	0.02
Calcium (Ca)	0.1
Chromium (Cr)	0.001
Copper (Cu)	0.003
Gold (Au)	0.002
Iron (Fe)	0.02
Lead (Pb)	0.002
Magnesium (Mg)	0.001
Manganese (Mn)	0.001
Molybdenum (Mo)	0.002
Nickel (Ni)	0.002
Sodium (Na)	0.002
Tin (Sn)	0.001
Zinc (Zn)	0.002

Source: Balazs Analytical Laboratory, 252 Humboldt Court, Sunnyvale, CA 94089-1315.

The technique was first developed as a laboratory instrument in the early 1970s. However, with continued improvement in detector and column sensitivities, preconcentration pumps, and automation, ion chromatography has been used for both off-line and on-line analyses since 1984 [36]. Detection limits for anions and cations are in the ppb to ppt range, depending on the sample matrix. Run time is typically about 20 min. With a wide range of customized columns available, multicomponent analyses can be performed in a single run.

Capillary electrophoresis (CE) is an analytical technique is which ions are separated out of a sample matrix based on their electrophoretic mobility—a liquid phase analog to ion mobility spectroscopy. First introduced in 1990 [37], this technique has gained acceptance as an alternative technique to IC for on-line applications. This is due to its relative simplicity and speed, low cost, and reduced solvent consumption. The sample solution is introduced into one end of a capillary column filled with electrolyte. Ions in the sample solution migrate under an electric field that is imposed across the column and are detected using UV absorption [38]. CE has been used to detect both anions and cations in solution in the low ppb to ppt range in an analysis time of less than 5 min. The measurements reported by CE analyses correlate well with those of IC analyses on the same samples, and the two techniques exhibit similar detection limits in routine analyses. However, preconcentration columns, when used, allow IC to achieve higher sensitivity [39].

3. Off-Line Measurements at Above Atmospheric Pressures

Sampling and measurement methods used at atmospheric pressures are adaptable to high-pressure sources as well. Passing the sample through an absorbent tube for subsequent desorption into an analytical instrument is aided by the high pressure. High pressure also makes it possible to collect high-pressure grab samples that can be discharged directly into an analytical station, such as an atmospheric pressure ionization mass spectrometer *(APIMS)*, without the need for the absorption/desorption cycle. However, the concentrating action of the sorption processes often makes an absorption/desorption cycle worthwhile, especially for the measurement of trace concentrations of contaminants.

4. Off-Line Measurements of Molecular Contaminants in Liquids

The most common liquid environment encountered in semiconductor processing is that of ultrapure water (UPW). Molecular contaminants typically measured in off-line analyses of UPW samples are dissolved silica, total silica, trace metals, anions, cations, and total organic carbon (TOC).

Silica exists in both soluble and insoluble forms in UPW. Total silica is the sum of the two forms. Dissolved silica is usually measured by a spectrophotometric method called the molybdate blue technique. Silicate ions react with ammonium molybdate in acidic solution to form yellow dodecamolybdosilicic acid. A reducing agent such as ascorbic acid reduces the dodecamolybdosilicic acid to molybdenum blue, which is measured by absorbance at a wavelength in the range of 810–860 nm. The detection limit of this method is estimated to be 0.5 ppb SiO_2. Total silica can be measured by ICP/MS, ICP/AES, or the colorimetric method just described, after dissolving all the insoluble silica in the sample with HF [40]. The colorimetric method is recommended for total silica concentrations less that 10 ppb; the less complicated ICP/AES and ICP/MS methods are suitable for concentrations greater than 10 ppb [40]. Insoluble silica is taken to be the difference between the total silica as measured by one of these three total silica methods and the dissolved silica measured by the colorimetric method.

Most *trace metals* ions are measured by ICP/MS, although atomic absorption and atomic emission spectroscopy are also used for certain metals. The masses of calcium, iron, and potassium match those of various argon-based ions formed in the plasma, so graphite furnace atomic absorption spectroscopy (GFAAS) is used for measuring those elements. A cold plasma technique shows promise of making even those three elements amenable to analysis by IC/MS [41]. When measured by ICP/MS, detection limits for many elements in water are now in the low-ppt concentration range (Table 2), and for process chemicals in the sub-ppb concentration range [32]. *Anions and cations* in water and liquids are usually analyzed by ion chromatography [42], but ICP/MS is also used [43]. *Off-line analyses of TOC* in water are often carried out by a high-temperature ($\sim$700 C) oxidation of the water followed by nondispersive infrared detection (NDIR) of the CO_2 thereby generated from the organics in the sample. With suitable modifications this method also applies to the measurement of TOC in selected processing chemicals [44]. An alternative technique for measuring TOC in water, also used for on-line monitoring (Sec. B, below), relies on either chemical oxidation of the organics in a persulfate solution or an ultraviolet photocatalytic oxidation, or a combination of both. Changes in water conductivity

Table 2 Detection Limits for Trace Metals in Ultrapure Water, Measured by ICP/MS

Element	DL (ppb)	Element	DL (ppb)	Element	DL (ppb)
Aluminum (Al)	0.003	Indium (In)	0.001	Samarium (Sm)	0.002
Antimony (Sb)	0.002	Iridium (Ir)	0.002	Scandium (Sc)	0.01
Arsenic (As)	0.005	Iron (Fe)	0.02	Selenium (Se)	0.02
Barium (Ba)	0.001	Lanthanum (La)	0.001	Silicon (Si)	0.5
Beryllium (Be)	0.003	Lead (Pb)	0.003	Silver (Ag)	0.001
Bismuth (Bi)	0.001	Lithium (Li)	0.002	Sodium (Na)	0.007
Boron (B)	0.05	Lutetium (Lu)	0.001	Strontium (Sr)	0.001
Cadmium (Cd)	0.003	Magnesium (Mg)	0.002	Tantalum (Ta)	0.004
Calcium (Ca)	0.2	Manganese (Mn)	0.002	Tellurium (Te)	0.005
Cerium (Ce)	0.001	Mercury (Hg)	0.02	Terbium (Tb)	0.001
Cesium (Cs)	0.001	Molybdenum (Mo)	0.004	Thallium (Tl)	0.006
Chromium (Cr)	0.004	Neodymium (Nd)	0.001	Thorium (Th)	0.003
Cobalt (Co)	0.001	Nickel (Ni)	0.004	Thulium (Tm)	0.001
Copper (Cu)	0.003	Niobium (Nb)	0.001	Tin (Sn)	0.005
Dysprosium (Dy)	0.001	Osmium (Os)	0.002	Titanium (Ti)	0.002
Erbium (Er)	0.001	Palladium (Pd)	0.002	Tungsten (W)	0.005
Europium (Eu)	0.001	Platinum (Pt)	0.009	Uranium (U)	0.002
Gadolinium (Gd)	0.001	Potassium (K)	0.1	Vanadium (V)	0.003
Gallium (Ga)	0.001	Praseodymium (Pr)	0.001	Ytterbium (Yb)	0.001
Geranium (Ge)	0.002	Rhenium (Re)	0.003	Yttrium (Y)	0.001
Gold (Au)	0.003	Rhodium (Rh)	0.001	Zinc (Zn)	0.005
Hafnium (Hf)	0.006	Rubidium (Rb)	0.001	Zirconium (Zr)	0.005
Holmium (Ho)	0.001	Ruthenium (Ru)	0.002		

Source: Balazs Analytical Laboratory, 252 Humboldt Court, Sunnyvale, CA 94089-1315.

is typically the method used to measure the CO_2 generated in such instruments rather than NDIR.

Using ICP/MS, Balazs Analytical Laboratories [45] reports the detection limits and recoveries listed in Table 3 for metals in 49% HF. Table 4 lists similar measurement capabilities from Balazs for metals in 70% nitric acid.

B. On-Line Measurements of Molecular Contaminants

Ideally, on-line analyzers imply real-time or near real-time measurements without the transport or preparatory steps required by off-line measurements. Real-time or near-real-time instruments for monitoring contaminants in processing environments remain limited in number, but their availability is growing in response to the need for improved contamination control in process environments whether at atmospheric pressure, below atmospheric pressure, above atmospheric pressure, or in liquids.

Table 3 Detection Limits and Percent Recoveries for Trace Metals in 49% HF, Measured by ICP/MS

Element	DL (ppb) (ng/g)	Recovery (%)	Element	DL (ppb) (ng/g)	Recovery (%)
Aluminum (Al)	0.1	114	Magnesium (Mg)	0.1	117
Antimony (Sb)	0.01	99	Manganese (Mn)	0.05	94
Arsenic (As)	0.5	*	Molybdenum (Mo)	0.01	91
Barium (Ba)	0.01	109	Nickel (Ni)	0.01	91
Beryllium (Be)	0.01	102	Niobium (Nb)	0.05	93
Boron (B)	0.2	82	Potassium (K)	0.3	112
Cadmium (Cd)	0.01	90	Silver (Ag)	0.05	107
Calcium (Ca)	0.5	118	Sodium (Na)	0.1	98
Chromium (Cr)	0.01	89	Strontium (Sr)	0.01	105
Cobalt (Co)	0.005	88	Tantalum (Ta)	0.05	84
Copper (Cu)	0.01	85	Thallium (Tl)	0.05	97
Gallium (Ga)	0.005	94	Tin (Sn)	0.01	96
Germanium (Ge)	0.01	90	Titanium (Ti)	0.1	*
Gold (Au)	0.5	*	Vanadium (V)	0.01	88
Iron (Fe)	0.3	91	Zinc (Zn)	0.5	80
Lead (Pb)	0.01	114	Zirconium (Zr)	0.05	99
Lithium (Li)	0.005	104			

Source: Balazs Analytical Laboratory, 252 Humboldt Court, Sunnyvale, CA 94089-1315.

Table 4 Detection Limits and Percent Recoveries for Trace Metals in 70% HNO$_3$, Measured by ICP/MS

Element	DL (ppb) (ng/g)	Recovery (%)	Element	DL (ppb) (ng/g)	Recovery (%)
Aluminum (Al)	0.05	97	Lithium (Li)	0.005	94
Antimony (Sb)	0.01	104	Magnesium (Mg)	0.05	100
Arsenic (As)	0.05	109	Manganese (Mn)	0.01	107
Barium (Ba)	0.01	93	Molybdenum (Mo)	0.01	98
Beryllium (Be)	0.01	111	Nickel (Ni)	0.01	108
Boron (B)	0.2	114	Niobium (Nb)	0.05	100
Cadmium (Cd)	0.01	105	Potassium (K)	0.3	95
Calcium (Ca)	0.5	109	Silicon (Si)	10	*
Chromium (Cr)	0.01	98	Silver (Ag)	0.01	129
Cobalt (Co)	0.005	108	Sodium (Na)	0.05	102
Copper (Cu)	0.02	106	Strontium (Sr)	0.01	95
Gallium (Ga)	0.005	100	Thallium (Tl)	0.05	97
Germanium (Ge)	0.01	105	Tin (Sn)	0.01	98
Gold (Au)	0.2	*	Vanadium (V)	0.01	95
Iron (Fe)	0.3	103	Zinc (Zn)	0.05	96
Lead (Pb)	0.01	93	Zirconium (Zr)	0.01	81

Source: Balazs Analytical Laboratory, 252 Humboldt Court, Sunnyvale, CA 94089-1315.

1. On-Line Measurements at Atmospheric Pressures

Many of the traditional off-line methods of measuring molecular contaminants can be adapted to on-line operation. Typically some form of concentration is built into the on-line configuration which slows the instrument response time but often not critically. For example, *ion mobility spectrometry (IMS)* has been demonstrated to be capable of rapid on-line measurements of a wide variety of molecular contaminants in cleanrooms [46,47] and other environments including continuous emissions into the ambient air from industrial sources (Table 5). During the summer of 1997, IMS was evaluated at the Albuquerque (New Mexico) International Airport as a screening device for detecting residues of explosives on airline passengers. It proved to be sensitive, selective, and fast in this role and a promising candidate to replace or supplement the less discriminating metal detectors now in use. In spite of its design's incorporating a concentrator, it was able to sample emissions from passengers in about 12 s from booth entry to exit. Reducing the total sampling time to 3–4 s is now the goal for such equipment.

Table 5 Gases Detectable by Ion Mobility Spectroscopy*

Toluene diamine (TDA)	Organophosphorus compounds
Dinitrotoluee (DNT)	Illicit drugs
Trinitrotoluene (TNT)	Pesticides
Toluene diisocyanate	Phenol
Methylene bis phenyl	Ethyl ether
Isocyanates (MDI)	Pyridine
Vinyl acetate	Piperidine
Tetrahydrofuran (THF)	Hydrogen cyanide (HCN)
Formaldehyde	Hydrogen bromide
Acrylonitrile	Hydrochloric acid (HCl)
Cyclohexanone	Benzyl chloride
Acetone	Nitric acid (HNO_3)
Ketones	Iodine (I_2)
Halogenated compounds	Acetic acid
	Acetonitrile
Nitro-compounds, explosives	Ammonia (NH_3)
	Hydrogen sulfide
Amines	Chlorine dioxide
Esters	Bromine
Chlorine	Hydrogen fluoride (HF)
Sulfur dioxide	Sulfur trioxide
Aldehydes	

*Typical limits of detection (1 ppb).
Source: Molecular Analytics, LLC, 25 Loveton Circle, Sparks, MD 21152-1123.

An IMS ionizes the molecular species to be detected and then separates the ions formed by drifting them in an electric field against the aerodynamic drag of a controlled gas flow. Figure 12 shows a gas sample containing analytes A, B, and C entering the ionization chamber where an easily ionized reactant gas R assists in the ionization of the sample species. The shutter grid periodically opens to admit the now ionized sample gas and the reactant gas into the drift region. The drift velocity of each ionized species depends on its ion mobility and the aerodynamic drag of an external gas flow opposing the ion drift. Quantitation is by an electrometer placed at the far end of the drift region which measures ion current as a function of time following the opening of the shutter grid. The time for one scan is on the order of milliseconds and scans can be repeated rapidly. Sensitivities are in the low ppb range for many cleanroom molecular contaminants. A major advantage of this technique is that it operates at atmospheric pressure unlike most other high-resolution/high-sensitivity analyzers such as mass spectrometers. IMS has been in use for about 25 years but

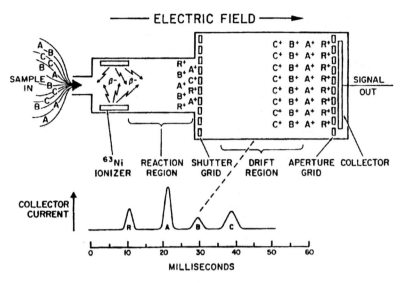

Figure 12   Ion mobility spectrometer. (From Ref. 48.)

mostly in specialty niche applications such as detection of explosives, drugs, and more recently, specialized industrial emissions [49]. Growing recognition of the versatility and capability of IMS, however, suggests a larger market with an increasing number of suppliers and applications.

IMS, along with chemiluminescence and ion chromatography, has been used to monitor molecular contaminants in photolithographic areas [50]. Ammonia, various amines, and other molecular bases degrade chemically amplified photoresists and need to be controlled in such areas. Of these three techniques, Kishkovich [50] preferred the chemiluminescent technique for monitoring DUV facilities, although he reported the detection limit of each of these measurement methods to be less than 1 ppb, well within the required sensitivity for monitoring such photolithographic areas.

Chemiluminescence is a technique that depends on light emission as a product of a chemical reaction—in the application discussed here a chemical reaction between NO and ozone. It is a well-established EPA test method for oxides of nitrogen. Kishkovich's chemiluminescent detector includes a catalytic reaction chamber that converts all nitrogen-containing species in the sample gas, except N_2, into NO. An upstream scrubber removes any basic compounds from a portion of the sample flow stream. This scrubbed portion of the sample is chemiluminescently measured separately from the unscrubbed portion of the sample, allowing the detector to separate the basic species from the other nitrogen-containing species.

In principle, infrared adsorption, mass spectrometry, and gas chromatography methods can all also be configured to operate on-line at atmospheric pressure. Some commercial instrumentation employing these measurement techniques is available. For example, Table 6 lists the published detection limits in the absence of interfering species for a commercial, on-line FTIR spectrometer. However, apparatus cost and, in some cases, size and complexity makes the on-line use of such instrumentation in semiconductor processes the exception rather than the rule, although recognition of the value of having on-line monitoring of processes is growing [51] and improved performance continues to be reported [52].

X-ray fluorescence (XRF) (Chap. 5) is also the basis for on-line monitoring of metals, either as particles or as dissolved species in solution. A slipstream from the process passes through a filter as a method of concentrating metallic particles; an ion exchange membrane concentrates the dissolved metallic species. Detection limits depend upon sampling time and sampling flow rate but can

Table 6 Reported Detection Limits for a Commercial FTIR Cell in the Absence of Interferences

		5 min measurement	1 s measurement
Standard gases			
Carbon monoxide	CO	6 ppb	100 ppb
Carbon dioxide	CO_2	0.8 ppb	15 ppb
Water	H_2O	10 ppb	170 ppb
Sulfur dioxide	SO_2	3 ppb	51 ppb
Nitric oxide	NO	17 ppb	300 ppb
Nitrous oxide	N_2O	7 ppb	120 ppb
Nitrogen dioxide	NO_2	2 ppb	34 ppb
Hydrogen chloride	HCl	7 ppb	120 ppb
Methane	CH_4	3 ppb	50 ppb
Other gases			
Ammonia	NH_3	2 ppb	36 ppb
Hydrofluoric acid	HF	1 ppb	17 ppb
Benzene	C_6H_6	0.7 ppb	12 ppb
Toluene	C_7H_8	5 ppb	85 ppb
Xylene	C_8H_{10}	5 ppb	85 ppb
Formaldehyde	H_2CO	3 ppb	50 ppb

Source: On-line Technologies, Inc., 87 Church St., East Hartford, CT 06108.

be sub ppb in many practical applications. Fluid flow ceases while the XRF measurement is made but the measurement is rapid so that the off-line time is short.

2. On-Line Measurements at Subatmospheric Pressures

Residual gas analyzers (RGAs) are mass spectrometers, usually of quadrupole (QMA) design, that can be incorporated into low-pressure processing equipment for the in-line measurement of gaseous composition within the chamber. These analyzers have been used to monitor molecular contaminants, such as moisture [53], in processing chambers and hence to detect leaks, measure outgassing, adjust reactant concentrations, and optimize process parameters [54–57]. Versions exist that are compatible with corrosive environments. Typical operating pressures are 10^{-4} torr (10^{-2} Pa) or lower, although optional features, such as differential pumping, can raise the operating pressure by 1 to 2 orders of magnitude.

RGAs are typically calibrated with nitrogen at a pressure of 10^{-6} torr (10^{-4} Pa). Sensitivity for other gases is different and has been shown to vary among RGAs [58].

Optimum operation probably varies with application. Users must be aware of both the properties of the RGA and its interactions with the specific environments and detection problems it is being used to monitor [59].

Moisture measurements at pressures between those limiting RGAs and atmospheric pressure have been made by *tunable diode laser absorption spectroscopy (TDLAS)*. McAndrew et al. [56,57] describe TDLAS as complementary to RGA because of its preferred operation at pressures above 10^{-3} torr and hence its superior suitability for process-monitoring in this pressure range which is typical of many processing steps. They report a detection limit for moisture in nitrogen of 100 ppb but note that the unit can be used in "more aggressive" process environments. They used TDLAS as an in situ monitor of moisture to reduce purge times in process equipment, enhancing equipment utilization. This technique has also been used to monitor deposition rates in physical vapor deposition equipment and seems capable of real-time measurement of flux levels in molecular beam epitaxy processes [60].

Improved on-line sensors for measuring trace concentrations of both oxygen and water vapor continue to be developed. For example, the *resonant electron attachment method* "exploits the fact that the cross section for electron dissociative attachment is largest at the target resonance energy, giving a much higher detection sensitivity and enabling lower concentrations to be measured" [61]. Both oxygen and water vapor molecules form O^- ions under electron bombardment, each very efficiently at its own resonant electron energy (6.8 eV for oxygen and 10.5 eV for water vapor). The O^- so formed is then separated by a quadrupole mass analyzer and counted in a channel multiplier. Detection limits for both oxygen and water vapor are calculated to be in the sub-ppt range [61]. However, the QMA limits this method to environments substantially below atmospheric pressure ($<10^{-5}$ torr).

3. On-Line Measurements at Above Atmospheric Pressures

Measurement of molecular contaminants at high pressure is important for assuring the quality of process gases used in semiconductor processing. As in subatmospheric process equipment, a universal molecular contaminant in process gases in the seemingly benign species, water vapor. Even trace quantities of water can induce undesired chemical reactions with other constituents in a process gas and can alter the surface properties of product wafers in process. The state-of-the-art instrument for measuring the concentration of water vapor in high-pressure gases is an *atmospheric pressure ionization mass spectrometer (APIMS)* plumbed into the gas delivery system. Such instruments can detect low-ppt concentrations of water vapor on-line. Unfortunately APIMS are expensive and thus relatively few in number. They are not a routine on-line instrument.

Ion mobility spectrometry (IMS), an atmospheric technique previously described, can be adapted to the detection of molecular contaminants in high-pressure gases. It has been shown to detect sub-ppb concentrations of water, oxygen, carbon dioxide, carbon monoxide, and methane in argon and nitrogen gases [48]. The sensitivity and cost of this analyzer make it an attractive alternative to APIMS.

FTIR spectroscopy can also be used to measure water vapor on-line at low-ppb concentrations in high-pressure gases, including high-purity hydrides (ammonia, silane, phosphine, etc.) [62] and hydrogen chloride and hydrogen bromide [63]. This approach to the measurement of trace water vapor in hydrides is nondestructive and does not decompose the hydrides.

Solid state devices are becoming available for low-cost monitoring of contaminants such as water vapor in high-pressure process gases. A *quartz crystal microbalance (QCM)* with a barium coating on its surface has detected water vapor in nitrogen at concentrations less than 1 ppb and with response times on the order of minutes [64]. This design can operate as an in-line sensor.

Electrolytic sensors, generating a current proportional to the partial pressure of water in a gas, also claim the detection of single-digit ppb concentrations of moisture at above atmospheric pressures [65]. Currents in an electrolytic sensor can be self-generated as in a galvanic cell or the sensing can be based on the magnitude of the current flowing under a constant external bias, the potentiometric method [66]. The sensor described in Ref. 65 is of the latter type.

Its initial response times—the time to first clear indication of a 5 ppb moisture intrusion—was reported to be on the order of 5 min. Instrument response to this 5 ppb moisture challenge reached 50% of final value in about 15 min. *Impedance-based sensors*, using aluminum oxide or other moisture sensitive materials whose electrical impedance varies with the concentration of the analyte in the environment to which the sensor is exposed, operate in-line and span the full range of pressures found in processing environments. Detection limits are in the low-ppb range and response times, on the order of 5–10 s.

Development of improved on-line, in-line moisture sensors in gases remains a goal of considerable importance and activity [67–70].

Oxygen, too, is a contaminant in certain processing environments and specialized sensors are available for measuring its presence. For example, sensors based on the reduction of oxygen in an electrochemical cell can operate over the full spectrum of processing pressures—from subatmospheric to supraatmospheric. Sensitivities range from low-ppb concentrations to as high as 25%. Thermoparamagnetic sensors for oxygen extend the detection range to 100% oxygen atmospheres, although oxygen would not generally be considered a molecular contaminant in such oxygen-rich environments.

4. On-Line Measurements in Liquids

The most common on-line measurements used to monitor molecular contamination in UPW are resistivity/conductivity, TOC, dissolved silica, and nonvolatile residue. Oxygen in UPW is also of interest to some users [71]. In feed waters and spent rinse waters, turbidity, pH and oxidation-reduction potential (ORP) are useful measurements. The resistivity cell, the pH cell and the ORP cell are usually in-line (in situ) monitors; the others are on-line instruments, drawing a slipstream from the UPW system for the measurements.

Commercially available analyzers based on ion chromatography can measure ions on-line [72]. Customized ICP/MS configurations for on-line measurement of metallic contaminants in liquids have been reported [73]. On-line measurements of these species, while not yet widespread, are becoming more common.

The *resistivity* (ρ)/*conductivity* ($\sigma \equiv 1/\rho$) *cell* is typically a concentric electrode configuration made of an appropriate nonreactive metal such as titanium or monel which can be inserted directly into the UPW system for nondestructive measurements of resistivity/conductivity. The resistivity of the water is measured by applying an ac voltage (ac minimizes polarization effects) between the electrodes and measuring the current

$$\rho = \frac{VA}{Iw} \tag{7}$$

where

ρ = resistivity
V = applied voltage
w = electrode separation
I = measured current
A = electrode area

w/A is called the cell constant and for cells used to measure the resistivity of high purity water is typically ~0.1–0.01 cm^{-1}.

Resistivity depends strongly on temperature so an accurate temperature probe, such as a platinum resistor, must also be part of the cell. The resistivity measurement provides a measure of ionic purity. In UPW at 25°C, the contribution of impurity ions to the conductivity is less than that of the H$^+$ and OH$^-$ ions of the water. Since impurity ions do not make a measurable contribution to conductivity, this quality water is called "intrinsic." The resistivity of intrinsic water at 25°C is 18.2 MΩ-cm (conductivity = 0.055 μS/cm). Measurements made by the cell at operating water temperatures are usually converted to equivalent 25°C readings for easy comparison with intrinsic water.

The cell for measuring the *pH* of water is also an in-line device immersed directly in the primary UPW system. It operates by measuring the potential difference between a pH-sensitive electrode and a reference electrode, which is not pH sensitive. The pH-sensitive electrode develops an electrochemical potential directly related to the H$^+$ activity of the water. Traditional pH-sensitive electrodes are thin-walled glass bulbs through which H$^+$ must permeate to make electrical contact with a buffer solution connected to a metal wire on the inside of the bulb. The design of the reference electrode, however, is the greater challenge [74].

An *ORP* cell has similar construction.

TOC, often referred to as total organic carbon, is actually total oxidizable carbon, since all measurement methods rely on the oxidation of the organic carbon constituents, not all of which are necessarily oxidized with equal efficiency or even oxidized at all. TOC analyzers measure the CO$_2$ generated by the oxidation of the organic carbon and deduce the TOC concentration in the liquid from the measured CO$_2$ concentration.

One method of measuring TOC is to acidify the water stream with H$_3$PO$_3$ in order to drive the pH below 3–4 and shift the total inorganic carbon (TIC) equilibrium, all of which is already oxidized, toward CO$_2$ [75]. This CO$_2$ is separated from the water stream by passing through a semipermeable membrane and is swept away in a dry oxygen flow. The carbon remaining in the now oxygen-saturated water sample is all organic carbon, which is oxidized by an UV lamp, again forming CO$_2$. This organic-derived CO$_2$ is also separated from the water stream by another semipermeable membrane. It enters an IR cell, which measures the CO_2 concentration. Appropriate algorithms then deduce the TOC concentration in the water stream that generated that concentration of CO_2. Similar NDIR-based methods have been the primary off-line laboratory method for measuring TOC in water but have not traditionally been used in on-line TOC analyzers.

The most common method for the on-line measurement of TOC in water streams uses the change in electrical conductivity of the water brought about by the addition of the CO_2 produced by an oxidation step. When operated in a batch mode, the analyzer draws a water sample into an oxidation chamber, seals the chamber, measures the water conductivity, and then photooxidizes the organics in the sample by flooding the chamber with UV, perhaps in the presence of a titania photocatalyst. The ions created by the oxidation reaction change the conductivity of the water, which is continually measured throughout the oxidation cycle. The end of the oxidation cycle is determined by stability criteria in the conductivity measurement that are incorporated into the analyzer software. The increase in water conductivity brought about by the oxidation is a measure of the oxidizable carbon in the water sample. In an alternative design, the change in conductivity is measured after a fixed time in the UV reactor without regard to achieving stability. This approach allows for more rapid responses but the calibration becomes species dependent since the oxidation rate varies with species, as does the conductivity of intermediate species formed during the UV oxidation reaction chain. Both of these analyzers measure TOC directly in the water sample, unlike the indirect-measuring designs discussed next.

An alternative TOC analyzer that relies on electrical conductivity measurements features a different configuration—a configuration in which the CO_2 conductivity change is measured indirectly. This analyzer also uses phosphoric acid to reduce the pH of the water sample to low values, assuring that virtually all oxidized carbon present will be present as CO_2. The water stream is then subdivided into two separate streams only one of which is photooxidized by UV, aided at high TOC concentrations (>1 ppm) by the addition of ammonium persulfate. As before the product of the oxidation of the acidified water in both streams is CO_2. The CO_2 in each water line passes through a separate semipermeable membrane into separate ultrapure water lines, one for each stream, isolated from the sample water by the membrane (Fig. 13). It is the conductivity change of these isolated, intrinsic waters that measures total inorganic carbon (TIC) in the nonphotooxidized water line and total carbon (TC) in the photooxidized line. The concentration difference between the two streams is the TOC (= TC minus TIC).

A major advantage of the indirect measurement design is the capability of measuring TOC in conductive and impure water streams, since noncarbon conductive species that raise the background conductivity of the sample water do

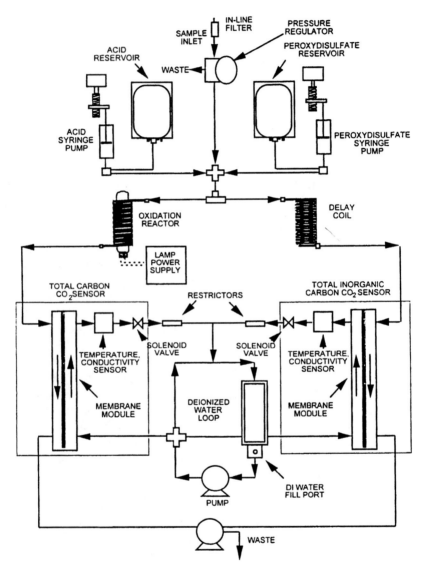

Figure 13   Schematic of one type of contemporary TOC analyzer. (Courtesy of Sievers Instruments, Inc., Boulder, CO.)

not interfere with the indirect measurement of conductivity made in an isolated pure water source that is coupled to the water sample only by the transport of CO_2 through the isolating membrane or other medium. Direct measuring TOC analyzers are limited to measuring the TOC of just high-purity waters (resistivity >0.1–1 MΩ-cm).

GC and GC/MS systems are also available for on-line measurement of some organics. These instruments can identify specific organic compounds rather than just TOC. Analytes can be concentrated by a purge-and-trap sampling arrangement in which organics are first purged from the sample water by a gas and then adsorbed on a suitable surface from that gas. Subsequently, the adsorbed organic species are desorbed into the GC or GC/MS analyzer. Response times of such measurements are longer than that of the TOC analyzers just described, but their speciation capability can be useful in pinpointing sources of TOC spikes or excursions in concentration and for general troubleshooting.

As in the off-line method (Sec. III.A), on-line measurement of *dissolved silica* is by the molybdate blue method. The forms of silica detected by this method include monomeric silica, silicic acid, and most likely some polymeric silica. Colloidal silica is not measured. The on-line instrument draws sample water from a slipstream, using a peristaltic pump to move the sample and the needed reagents through the analyzer. Spectrophotometric measurements, after the addition of ascorbic acid forms the blue-colored solution, are compared with similar measurements made earlier in the measurement cycle on the reagent blank at the same wavelength. Response time for this type of on-line measurement is on the order of 8 min and detection limits are about 1 ppb. Phosphates can be an interferent and should be eliminated prior to the measurement if present in concentrations exceeding 50 ppb.

Measurements of *nonvolatile residue*, traditionally called residue after evaporation (RAE), can now also be made on-line, using a nonvolatile residue monitor (NRM) which solves the problem of measuring this type of molecular contaminants in liquids by adopting an aerosol particle measurement technique. The NRM aspirates a sample water stream into a spray of fine water droplets. The water and volatile components of each droplet readily evaporate in a downstream dryer stage, leaving only the nonvolatile constituents of each droplet behind as a small aerosol particle. The aerosol stream is then analyzed by a CPC (Sec. III.A) to obtain the particle concentration in the aerosol stream. Aerosol particle concentration is converted to nonvolatile residue concentration by calibration curves prepared from solutions of potassium chloride. Figure 14 schematically illustrates the NRM configuration [76].

The NRM detects particles, bacteria, silica, ionic impurities, and nonvolatile or semivolatile organics. It is a sensitive measure of nonvolatile residue and is now the basis of an ASTM standard for the measurement of RAE [77]. However, it does not identify the species dominating or contributing to any

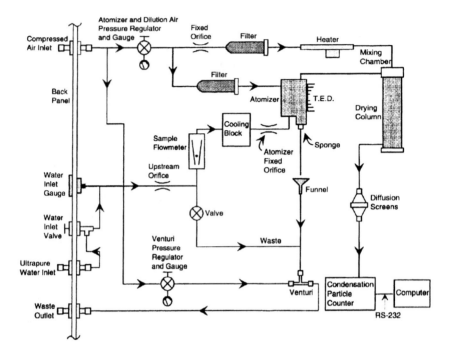

Figure 14   Nonvolatile residue monitor. (Courtesy of D. Blackford, Fluid Measurement Technologies, Inc., Vadnais Heights, MN, and Particle Measurement Systems, Inc., Boulder, CO.)

given measurement. This lack of composition specificity has made it difficult for operators of UPW systems to find a role for the instrument in spite of its acknowledged high sensitivity. No one yet bases a control decision on the NRM reading. It remains a solution looking for a problem.

A potential role for the NRM is to take advantage of the varying temperatures that can be selected to dry the water droplets generated by the atomizer. This feature allows the temperature of the heated air entering the drying column (Fig. 14) to be changed so that species that are completely volatilized at the highest temperature (120°C) may not be volatized when the heated air enters the drying column at a lower temperature, say 45°C. A temperature signature, based on nonvolatile residue measured at two or more temperatures, is postulated to be a means of identifying "semivolatiles" in UPW systems.

Instrumentation exists for the on-line measurement of *ionic contaminants* in water, based on ion chromatography [42,78]. Using components similar to those of the off-line apparatus, real-time configurations can detect ionic concentrations in the ppt range and perform automatically with minimal operator attention. Measurement times, however, can be on the order of an hour. This technique also applies to the measurement of ionic contaminants in other processing liquids.

Ion selective electrodes are a lower cost but somewhat less sensitive method of measuring ionic contaminants on-line. These sensors are devices which measure the voltage difference between an ion-sensitive electrode and a reference electrode which is indifferent to the concentration of the ion being sensed—its potential is independent of the ion concentration. The Nernst equation predicts this voltage difference to vary logarithmically with ion concentration:

$$E = E_0 - \left(\frac{RT}{nF}\right) \ln Q = E_0 - \left(\frac{0.059}{n}\right) \log_{10} Q$$

where

E = measured potential difference between the electrodes

E_0 = standard potential of the half-reaction at the ion-sensitive electrode

R = gas constant 8.314×10^3 J/kmole $\cdot$ K

T = temperature = 298 K

n = number of electrons exchanged in the half reaction

F = Faraday constant = 9.649×10^7 C/kmole

Q = ratio of the product concentrations to the reactant concentrations
   = [product]/[reactant]

ASTM methods, such as D2791 for sodium ($n = 1$) [79], describe the technique. With sub-ppb detection capability, sodium ion selective electrodes can provide an early indication of breakthrough of the ion exchange resins in a UPW system.

Ion-sensitive field effect transistors (ISFETs) are also now commercially available which are faster and more rugged than the traditional ion selective electrodes [80]. ISFETs are field effect devices fabricated without a gate electrode but with a chemically sensitive coating in the gate region that adsorbs the ion(s) of interest. Detection of these ions is by the change they induce in the source to drain current of the ISFET.

Measurement of *dissolved oxygen in UPW* can be made electrochemically. The basic sensor design is one in which the water stream containing the dissolved oxygen passes by an oxygen-permeable membrane. The dissolved oxygen diffuses through the membrane into an electrolyte and is electrochemically reduced at the cathode. The voltage between the anode and cathode is from an external source and is constant in a polarographic cell so that the observed current flow is a measure of the oxygen concentration reacting at the cathode and hence of the dissolved oxygen concentration in the water sample. Galvanic cells having no external potential but electrodes made of suitably selected dissimilar materials can also be used to measure dissolved oxygen in water [81].

Identification of contaminating species in liquids by fast-responding, on-line instrumentation is growing in importance. One promising approach to obtaining such instrumentation is to adapt established off-line measurement methods into on-line configurations. A gas chromatograph/flame ionization detector (GC/FID) combination, configured for direct injection of 0.5 μL water samples, has performed well in on-line operation [82,83]. Response times <30 s and sensitivities below 30 ppb have been reported in water for mixtures of polar organics including acetone and isopropanol. However, no commercial product yet exists.

Ion mobility spectroscopy, already recognized and used in atmospheric sampling, is another candidate for adaptation to on-line sampling of water [84]. This technique is rapid and sensitive and has been demonstrated to be capable of detecting organic contaminants of interest in UPW systems. It is particularly sensitive to ethylene glycol, a contaminant not readily analyzed by the GC/FID approach described in the preceding paragraph.

5. Detection of Metallic Contaminants in Hydrofluoric Acid

Metallic molecular contaminants in liquid-processing chemicals constitute a threat to device performance (Chap. 10). Metallic cations of high-oxidation potential, such as silver, copper, and gold, readily deposit on oxide-free silicon surfaces by displacement plating [85]. This phenomenon can be used to quantify trace concentrations of these metallic ions in hydrofluoric acid solutions by measuring the open circuit voltage of a specially prepared silicon electrode immersed in the solution and a suitable reference electrode [86]. Detection limits at ppt concentrations of silver ions have been reported. However, the technique cannot identify the type of metal producing the detected voltage. No commercial hardware yet exists for making this measurement.

Anions in HF have been measured at the low-ppb levels by using two ion chromatographs in series [72,87]. The first IC separates the anions of interest from the F$^-$ anion matrix, and the second performs the analytical separation. By thus reducing the fluoride interference, chloride, sulfate, nitrate, and phosphate in 24.5% HF become detectable at concentrations of 10 ppb or less in 3–4 min fractionation time.

IV. CONCLUSIONS

Measurement technology must of necessity precede advances in processing technology and so it is with advances in contamination control in semiconductor processing. Happily, at least some of today's contemporary measurement technology for contaminants in wafer environments is adequate for the needs of the semiconductor industry as outlined in the 1997 National Technology Roadmap for Semiconductors (NTRS) (Chap. 2). The usual caveats regarding the uncertainties associated with the NTRS forecasts apply—in most instances the relationship between product yield/reliability and environmental contamination is not known so that the target values specified in the NTRS are estimates at best.

Measurement technology, both off-line and on-line, is adequate for aerosol particle measurement in benign gaseous environments. Particle composition, however, is not readily available on-line and would be useful. Instrumentation for on-line identification of particle composition has been reported in various laboratory experiments but off-the-shelf commercial instrumentation for such determinations is not yet available. Particle counting in reactive gases lags the desired capability, as defined in the NTRS. The working fluids now available in CPCs restrict their use to compatible environments, ruling out CPC measurements in some specialty gases. Particle measurements in liquids, even UPW, rely primarily on OPCs, which are unable to detect particles smaller than about 0.05–0.07 μm. Corrosive liquids and liquids with indices of refraction closer to that of the particles to be detected further limit the detection capability of the present instrumentation.

Off-line analytical technology for measuring molecular contaminants in air and water samples has improved dramatically in sensitivity and sampling methodology over the past decade. Off-line capability exists for measuring trace concentrations of contaminants below most of the aggressive target values contained in the NTRS. On-line measurements, however, are not as advanced. In particular, measurement technology for meeting the NTRS targets for *in-line*, real-time detection of TOC, silica, metals, and nonvolatile residue in UPW at concentrations below 0.01 ppb is not now available. Indeed, resistivity remains one of the few UPW parameters now routinely measured by in-line instrumentation. On-line measurement technology for TOC and nonvolatile residue in UPW is close to or even at the desired sensitivity but does not yet meet the desired response times.

REFERENCES

1. JB Callis, DL Illman, and BR Kowalski. "Process analytical chemistry." Anal Chem 59, 1987; pp. 624A–637A.
2. HP Adler and E Schenker. "Advanced analytical techniques for boiling water reactor chemistry control." Presentation to the 1994 International Conference on the Chemistry in Water Reactors (Nuclear), Paul Scherrer Institut, Wurenlingen und Villigen, CH-5232 Villigen PSI, Switzerland.
3. WC Hinds. Aerosol Technology. John Wiley, New York, 1982.

4. RG Knollenberg and DL Veal. "Optical particle monitors, counters and spectrometers: performance characterization, comparison and use." 1992 SPWCC Proceedings, pp. 197–240 (Balazs Analytical Laboratory, 252 Humboldt Court, Sunnyvale, CA 94089-1315); 1991 Proceedings of the Institute of Environmental Sciences, pp. 751–771 (Institute of Environmental Sciences, 940 East Northwest Highway, Mount Prospect, IL 60056).

5. D Xu, RF Pinizzotto, and JA Sees. "Analysis of contaminants in IC processing chemicals at the sub-part per trillion level." 1997 SPWCC Proceedings, vol. 2, pp. 31–48 (Balazs Analytical Laboratory, 252 Humboldt Court, Sunnyvale, CA 94089-1315).

6. DE Pendlebury and D Pickard. "Examining ways to capture airborne microorganisms." Cleanrooms 11(6), June 1997, pp. 34–40.

7. LA Peach, "Mass spectrometer analyzes fine aerosols." Laser Focus World 33, No. 9, September 1997, p. 29.

8. RL Myers and WL Fite. "Electrical detection of airborne particulates using surface ionization techniques." Environ Sci and Tech 9, 1973, pp. 334–336.

9. RL Myers and WL Fite. "Submicron and centimicron particulate detection using surface ionization." American Laboratory, December 1975, pp. 23–29.

10. FED-STD-209E. "Airborne particulate cleanliness classes in cleanrooms and clean zones." Sept. 11, 1992 (General Services Administration, General Products Commodity Center, Federal Supply Service, 819 Taylor Street, Fort Worth, TX 76102).

11. ISO 14644-1. Cleanrooms and associated controlled environments—part I: classification of air cleanliness, May 1, 1999 (available from the Institute of Environmental Sciences and Technology, 940 East Northwest Highway, Mount Prospect, IL 60056).

12. K Whitby and K Willeke. "Single particle counters." in Aerosol Measurement (DA Lundgren et al., ed.), University of Florida Press, Gainesville, FL, 1979.

13. ML Malczewski, AE Holmer, and HC Demmin. "A sampling system for measurement of nanometer-sized particles in hydrogen." JIEST 41(2), March/April 1998, pp. 23–26.

14. AE Holmer and ML Malczewski. "Design and calibration of a condensation nucleus counter suitable for use in hydrogen service." 1993 Proceedings of the Institute of Environmental Sciences, pp. 309–314 (Institute of Environmental Sciences, 940 East Northwest Highway, Mount Prospect, IL 60056).

15. WT McDermott. "A gas diluter for measuring nanometer-sized particles in oxygen or hydrogen." Journal of the IEST 41(4), July/August 1998, pp. 17–23.

16. PD Kinney, G Bae, DYH Pui, and BDH Liu. "Particle behavior in vacuum systems." Journal of the Institute of Environmental Sciences XXXIX(6), November/December 1996, pp. 41–45.

17. J Mitchell, RG Knollenberg, S Lopez, and D Long. "Fiber-optic coupling for in-situ measurement of particles in vacuum process." 1996 Proceedings of the Institute of Environmental Sciences, pp. 400–405 (Institute of Environmental Sciences, 940 East Northwest Highway, Mount Prospect, IL 60056).

18. WD Reents, HC Wang, and R Udischas. "Monitoring and identification of particles formed during pressure reduction." 1995 Proceedings of the Institute of Environmental Sciences, pp. 30–34 (Institute of Environmental Sciences, 940 East Northwest Highway, Mount Prospect, IL 60056).

19. WT McDermott. "Particle measurement in semiconductor process gases." Solid State Technology 42(8), August 1999, pp. 65–66, 68, 70.
20. B Fardi. "An evaluation of performance of high sensitivity DI water particle counters." 1992 Proceedings of the Institute of Environmental Sciences, pp. 431–437 (Institute of Environmental Sciences, 940 East Northwest Highway, Mount Prospect, IL 60056).
21. M Xu and HC Wang. "Minimum sampling time/volume for liquid-borne particle counters and monitors. J Inst Envir Sci 40(6), December 1997, pp. 29–34.
22. RG Knollenberg. "The importance of media refractive index in evaluating liquid and surface microcontamination measurements." Journal of the Institute of Environmental Sciences, XXX(2), March/April 1987, pp. 50–58.
23. JR Mitchell and BA Knollenberg. "New techniques move in situ particle monitoring closer to the wafer." Semiconductor International 19(10), September 1996, pp. 145–146, 148, 150, 152, 154.
24. ASTM F 1094-87 (Reapproved 1992). Standard test methods for microbiological monitoring of water used in processing electron and microelectronic devices by direct pressure tap sampling valve and by the presterilized plastic bag method (ASTM, 100 Barr Harbor Drive, West Conshohocken, PA 19428).
25. ME Mittelman, PW Johnson, and J McN Sieburth. "Epifluorescence microscopy, a rapid method for enumerating viable and nonviable bacteria in ultrapure water systems." Microcontamination 1(2), 1983, pp. 32–37, 52.
26. ASTM F 1095-88 (Reapproved 1994). Standard test method for rapid enumeration of bacteria in electronics-grade purified water systems by direct-count epifluorescence microscopy (ASTM, 100 Barr Harbor Drive, West Conshohocken, PA 19428).
27. M Osawa et al. "Analysis of bacteria in ultrapure water systems using fluorescent probe and culture method." 1998 SPWCC Proceedings, pp. 121–139. (Balazs Analytical Laboratory, 252 Humboldt Court, Sunnyvale, CA 94089-1315).
28. T Manabe. "ATP Monitor." in Ultraclean Technology Handbook, Vol. 1: Ultrapure Water, T. Ohmi, ed. Marcel Dekker, New York, 1993, pp. 595–600.
29. IL Pepper et al. "A rapid and systematic analytical method for measuring bacterial contaminants in ultrapure water." 1993 SPWCC Proceedings, pp. 50–62 (Balazs Analytical Laboratory, 252 Humboldt Court, Sunnyvale, CA 94089-1315).
30. M Camenzind. "Identification of organic contamination in cleanroom air, on wafers and outgassing from gloves and wafer shippers." 1996 SPWCC Proceedings, pp. 352–372 (Balazs Analytical Laboratory, 252 Humboldt Court, Sunnyvale, CA 94089-1315).
31. J Fucsko, J Mikulsky, and M Balazs. "Keeping pace in contamination monitoring with advanced technology." 1996 SPWCC Proceedings, pp. 313–324 (Balazs Analytical Laboratory, 252 Humboldt Court, Sunnyvale, CA 94089-1315).
32. T Fujimoto, N Takeda, T Tochikazu, and M Sado. "Evaluation of contaminants in the cleanroom atmosphere and on silicon wafer surfaces." 1996 SPWCC Proceedings, pp. 324–351 (Balazs Analytical Laboratory, 252 Humboldt Court, Sunnyvale, CA 94089-1315).
33. MJ Camenzind, H Liang, J Fucsko, and MK Balazs. "How clean is your cleanroom air?" MICRO 13(8), October 1995.

34. M Camenzind and A Kumar. "Organic outgassing from cleanroom materials including HEPA/ULPA filter components: standardized testing proposal," 1997 Proceedings of the Institute of Environmental Sciences, pp. 211–226 (Institute of Environmental Sciences, 940 East Northwest Highway, Mount Prospect, IL 60056).

35. P Gupta et al. "Novel methods for trace metal analysis in process chemicals and DI water and on silicon surfaces," in Proceedings of the Symposium on Contamination Control and Defect Reduction in Semiconductor Manufacturing III, vol. 94-9, May 1994, pp. 200–221 (The Electrochemical Society, 10 South Main Street, Pennington, NJ 08534-2896).

36. MJ Wojtusik, J Berthold, EQ Kaiser, and DK Berg. "Advances in on-line ion chromatography for the 90's." in Advances in Instrumentation and Control 48, part 1, pp. 305–318, ISA, Research Triangle Park, NC 27709, 1993.

37. WR Jones and P Jandik. "Capillary ion analysis." Am Lab 22, p. 51, 1990.

38. P Jandik, WR Jones, A Weston, and PR Brown. "Electrophoretic capillary ion analysis: origins, principles and applications." Liquid Chromatography/Gas Chromatography Magazine 9, p. 634, 1991.

39. P Sun and M Adams. "Demonstrating a contamination-free wafer surface extraction system for use with CE and IC." MICRO 17(4), April 1999, pp. 41–46.

40. T Chu and M Balazs. "Determination of total silica at ppb levels in high-purity water by three different analytical techniques." Ultrapure Water 11(1), February 1994, pp. 56–60.

41. R Mortensen and TJ Gluodenis. "Cold plasma extends trace metal detection capability." Semiconductor International 21(8), July 1998, pp. 261–262, 264, 266.

42. S Heberling, E Kaiser, and J Riviello. "Advances in the application of ion chromatography for the determination of ultratrace ions in semiconductor pure water and chemicals." 1995 SPWCC Proceedings, pp. 207–236 (Balazs Analytical Laboratory, 252 Humboldt Court, Sunnyvale, CA 94089-1315).

43. Z Pourmotamed and P Gupta. "Parts per billion detection of non-metallic elements on silicon surfaces and in semiconductor processing chemicals." 1996 SPWCC Proceedings, pp. 301–311 (Balazs Analytical Laboratory, 252 Humboldt Court, Sunnyvale, CA 94089-1315).

44. MJ Camenzind and MK Balazs. "Analysis of organic impurities in semiconductor processing chemicals." 1992 SPWCC Chemical Proceedings, pp. 143–161 (Balazs Analytical Laboratory, 252 Humboldt Court, Sunnyvale, CA 94089-1315).

45. Balazs Analytical Laboratory, 252 Humboldt Court, Sunnyvale, CA 94089-1315.

46. KR Dean and RA Carpio. "Real-time detection of airborne contaminants in DUV lithographic processing environments." 1995 Proceedings of the Institute of Environmental Sciences, pp. 9–16 (Institute of Environmental Sciences, 940 East Northwest Highway, Mount Prospect, IL 60056).

47. BT Price. "Ion mobility spectrometry—advanced contamination control monitoring." 1996 Proceedings of the Institute of Environmental Sciences, pp. 184–188 (Institute of Environmental Sciences, 940 East Northwest Highway, Mount Prospect, IL 60056).

48. R Stimac and S Ketkar. "Use of ion mobility spectrometry to determine trace level impurities in ultra high purity gases." 1996 Proceedings of the Institute of

Environmental Sciences, pp. 5–12 (Institute of Environmental Sciences, 940 East Northwest Highway, Mount Prospect, IL 60056).

49. AT Bacon, R Getz, and J Reategui. "Ion-mobility spectrometry tackles tough process monitoring." Chemical Engineering Progress, June 1991.

50. OP Kishkovich. "Conducting real-time monitoring of airborne molecular contamination in DUV lithography areas." MICRO 17(6), June 1999, pp. 61–62, 64, 67, 70–71, 74–75, 78–79.

51. T Studt, "Spectrometers edge their way into process monitoring." Research and Development 40(1), January 1998, pp. 36–38.

52. PJ McCann, K Namjou, and I Chao. "Using mid-IR lasers in semiconductor manufacturing." MICRO 17(7), July/August 1999, pp. 93–94, 96, 98–99.

53. A Shapiro et al. "Residual moisture: its role and measurement in semiconductor processing equipment." Micro 94 Proceedings, pp. 70–77 (Canon Communications, Inc., 3340 Ocean Park Blvd., Suite 100, Santa Monica, CA 90405-3216).

54. TM Banks, GR Diamond, and SJ Ruck. "Integrating mass spectrometry data into the fab environment." Semiconductor International 20(6), June 1997, pp. 137–138, 140, 142, 144, 146.

55. C D'Couto and J Liu. "Matching equipment and process through the use of residual gas analysis." MICRO 14(9), October 1996, pp. 103–104, 106, 108, 110, 113, 115.

56. J McAndrew et al. "Increasing equipment uptime through in situ moisture monitoring." Solid State Technology 41(8), August 1998, pp. 61–62, 64–66, 68, 71.

57. J McAndrew. "Progress in in situ contamination control." Semiconductor International 21(5), May 1998, pp. 71–72, 74, 76, 78.

58. MG Rao and C Dong. "Evaluation of low cost residual gas analyzers for ultrahigh vacuum applications." J Vac Sci Technol A 15(3), May/June 1997, pp. 1312–1318.

59. G Brucker. "Getting the most from your RGA." Research and Development 40(7), June 1998, pp. 38–40.

60. M Bortz and T Day. "Diode lasers monitor vapor deposition." Laser Focus World 32(11), November 1996, pp. 95–96, 98, 100, 102, 104.

61. KF Man and S Boumsellek. "Instrument design for sub-ppb oxygenated contaminants detection in semiconductor processing." J Inst Environ Sci 40(4), July/August 1997, pp. 17–21.

62. A Gupta and S Salim. "Measurement of trace levels of moisture in UHP hydride gases by FTIR spectroscopy." Cleanrooms 11(2), February 1997, pp. 28–30.

63. BR Stallard, LH Espinoza, and TM Niemczyk. "Trace water determination in gases by infrared spectroscopy." 1995 Proceedings of the Institute of Environmental Sciences, pp. 1–8 (Institute of Environmental Sciences, 940 East Northwest Highway, Mount Prospect, IL 60056).

64. J Wei, J Pillion, and C Hoang. "In-line moisture monitoring in semiconductor process gases by a reactive-metal-coated quartz crystal microbalance." J Environmental Sciences 40(2), March/April, 1997, pp. 43–48.

65. C Ma, F Shadman, J Mettes, and L Silverman. "Evaluating the trace-moisture measurement capability of coulometric hygrometry." MICRO 13(4), April 1995, pp. 43–49.

66. EM Zdankiewicz. "Gas detection in theory and in practice, part 1: chemical sensing technology." Sensors 14(10), October 1997, pp. 20, 22–24, 26, 28, 30, 32, 33–34, 36, 38.

67. MJ Kelly et al. "Sensors for the detection of moisture in inert and corrosive gases." Ultra Clean Technology 8(6), December 1996, p. 153.

68. KB Pfeiffer et al. "Development of solid state moisture sensors for semiconductor fabrication applications." Micro 94 Proceedings, pp. 87–97 (Canon Communications, Inc., 3340 Ocean Park Blvd., Suite 100, Santa Monica, CA 90405-3216).

69. GH Atkinson. "High sensitivity detection of water via intracavity laser spectroscopy." Micro 94 Proceedings, pp. 98–111 (Canon Communications, Inc., 3340 Ocean Park Blvd., Suite 100, Santa Monica, CA 90405-3216).

70. RK Rowe et al. "FTIR spectroscopy for the determination of water in corrosive gases." Micro 94 Proceedings, pp. 112–120 (Canon Communications, Inc., 3340 Ocean Park Blvd., Suite 100, Santa Monica, CA 90405-3216).

71. JE Martyak, NM Martyak, and J Dietrich. "The effects of vacuum degasification on deionized water." Presentation to the Micro 94 Conference (Canon Communications, Inc., 3340 Ocean Park Blvd., Suite 100, Santa Monica, CA 90405-3216).

72. ML Wu and JG Chen. "Measuring trace ionic impurities in ultrapure acids and bases with ion chromatography." MICRO 15(9), October 1997, pp. 65–67, 70, 74, 76.

73. LW Shive, K Ruth, and P Schmidt. "Using ICP-MS for in-line monitoring of metallics in silicon wafer-cleaning baths." MICRO 17(2), February 1999, pp. 27–31.

74. S West et al. "Combination electrodes with platinum redox reference systems for on-line pH measurements." Ultrapure Water 15(10), December 1998, pp. 45–50, 52.

75. CD Jolly and EL Jeffers. "Ultrapure water total organic carbon analyzer—advanced component development." SAE Paper 911436, SP-874, 1991 (SAE, 400 Commonwealth Drive, Warrendale, PA 15096-0001).

76. DB Blackford, T Kerrick, and G Schurmann. "The measurement of nonvolatile residue in high-purity water and clean liquids." Ultrapure Water 11(5), July/August 1994, pp. 57–63.

77. ASTM D 5544-99. Standard test method for on-line measurement of residue after evaporation of high-purity water (ASTM, 100 Barr Harbor Drive, West Conshohocken, PA 19428).

78. A Meuter et al. "The use of on-line ion chromatography for the submicron ultrapure water plant." 1992 SPWCC Water Proceedings, pp. 83–107 (Balazs Analytical Laboratory, 252 Humboldt Court, Sunnyvale, CA 94089-1315).

79. ASTM D 2791-93 (Reapproved 1997). "Standard test methods for continuous determination of sodium in water" (ASTM, 100 Barr Harbor Drive, West Conshohocken, PA 19428).

80. BC Syrett. "Reference manual for on-line monitoring of water chemistry and corrosion." TR-104928, RP 8044, March 1995 (EPRI, 3412 Hillview Ave., Palo Alto, CA 94304).

81. M Lewandowski and B Langie. "Galvanic sensors reduce maintenance and startup time. Research & Development Magazine 41(5), April 1999, pp. 27–28, 30.

82. GC Frye et al. "Development and evaluation of on-line detection techniques for polar organics in ultrapure water." J Inst Environ Sci. 39(6), September/October 1996, pp. 30–37.

83. DS Blair, CD Mowry, PJ Rodacy, and SD Reber. "Automated gas chromatography." U.S. Patent 5,922,106, July 13, 1999 (to Sandia National Laboratories).
84. P Rodacy et al. "Trace identification of organic molecules in ultrapure water using ion mobility spectroscopy." 1996 Proceedings of the Institute of Environmental Sciences, pp. 363–367 (Institute of Environmental Sciences, 940 East Northwest Highway, Mount Prospect, IL 60056).
85. L Mouche, F Tardif, and J Derrien. "Mechanisms of metallic impurity deposition on silicon substrates dipped in cleaning solution." J Electrochem Soc 142(7), July 1995, pp. 2395–2401.
86. OMR Chyan et al. "A new potentiometric sensor for the detection of trace metallic contaminants in hydrofluoric acid." J Electrochem Soc 143(10), October 1996, pp. L235–L237.
87. JG Chen and ML Wu. "Using two-dimensional ion chromatography to measure contaminants in ultrapure chemicals." MICRO 15(1), January 1997, pp. 31, 32, 34, 36–37.

4

On-Wafer Measurement of Particles

Rodolfo E. Díaz
Arizona State University, Tempe, Arizona

Brent M. Nebeker and E. Dan Hirleman
Purdue University, West Lafayette, Indiana

Optical systems that perform on-wafer measurement of particles have become one of the most important process control tools available to semiconductor manufacturers and the IC industry. They give the process owner information necessary to determine whether or not a specific process is under control, and so are ultimately significant contributors to yield-increasing strategies. It follows that the functional requirements of these systems depend on which process in the production cycle they support. Thus, during process development, a system with high throughput is not as important as a system with high-detection sensitivity and classification capability. On the other hand, a system employed to monitor a production line must have a throughput compatible with that line, without sacrificing sensitivity to those defects that are yield limiting [1]. The manner in which this trade-off between speed and sensitivity is accomplished is what distinguishes different commercial inspection systems from one another. However, regardless of the individual approaches, all such systems are based on the same physical principles. They are all concerned with the detection and classification of wavelength- or subwavelength-sized defects using light scattering in an environment that includes a large scattering surface (the wafer) and possibly other light-scattering features intentionally incorporated onto the wafer surface (e.g., circuit patterns). In this chapter we focus on the physical foundations underlying the operation and design of a generic on-wafer inspection system.

   The results presented here are primarily based on work that has been performed over the last five years at Arizona State University. Most of this work has been funded through the Consortium for Metrology of Semiconductor Nano-defects, a university/industry research partnership dedicated to precompetitive research and development related to in-line detection and characterization of defects in the semiconductor wafer manufacturing and processing industries.

I. BRIEF DISCUSSION OF GENERAL METHODS

The methods available to measure defects on wafers are generally classified as off-line or in-line depending on the process step they serve. Off-line tools address multiple-process requirements and typically involve single-use monitor wafers. These off-line measurement methods include x-ray fluorescence, resistivity measurements, and most optical unpatterned wafer defect detection systems. In-line methods fall generally in the category of metrology instruments, and measure such parameters as film thickness, pattern-to-pattern overlay, and critical dimensioning. When an in-line measurement instrument is integrated into the process tool itself, the system is referred to as an on-line system. When the coupling between the measurement instrument and the process tool is so tight that it can dynamically adjust the tool's performance in real time, the system is referred to as an in situ sensor [2]. As in Chapter 3, the methods available to gauge the number and type of defects on wafers range from the chemical (in situ residual gas analysis) to the optical (film thickness sensors and laser-scattering detectors) and beyond (x-ray and electron beam scattering). Our emphasis in this chapter will be strictly on the optical methods. It is acknowledged that ever-shrinking device sizes, and the increasing size of wafers, may eventually demand the higher resolution achievable with electron beam and ultraviolet-based systems.

The problem of the optical detection and classification of particles on wafers shares many of the characteristics of the particle measurement problem in air and fluids discussed in Chapter 3. In particular, optical particle counting and methods analogous to particle spectroscopy for particles on a surface are at present the premier approaches for fast and effective wafer inspection. The most significant differences between particle measurements in space and measurements on a surface lie in the topology and the environment. With regard to topology, the on-wafer problem is strictly a two-dimensional problem. However, the reduction in complexity in going from three-dimensional volume sampling to two-dimensional area scanning is more than offset by the requirement to inspect the entire wafer at rates of the order of 50 to 100 wafers per hour. With regard to the environment, the differences are 1) the inherent assumption that the process to be inspected is almost under control, and therefore the number of surface particles per square millimeter is assumed to be small, and 2) the fact that on-wafer particles are always in the vicinity of a large surface that affects the optical interaction. Thus, on-wafer particle measurement is more than just a Mie scattering problem.

Optical detection and classification methods for on-surface measurements can be categorized in various ways: If the emphasis is placed on the mode of acquisition of the light scattered by the defect, the natural separation is *bright-field* versus *dark-field* systems. Roughly, bright-field systems are high-speed microscopes, using the same aperture to collect both reflected and scattered light. Dark-field systems avoid the reflected (also referred to as *specular*) light and concentrate on the scattered light. Dark-field systems may illuminate and collect the scattered light from a variety of angles of incidence. Whereas bright-field systems can detect and classify simultaneously, directly measuring the size of any defect their optics can resolve, dark-field systems can detect defects much smaller than the resolution spot size. Defect classification in the case of dark-field systems is then based on the angular distribution and strength of the scattered light.

Another natural separation of optical methods is based on the way collected data are processed. An *imaging system* illuminates the area to be inspected and then reconstructs an image onto a camera [TDI (time delay integrator) or CCD (charge coupled device)]. A *scanning system* "paints" the surface with a laser beam and collects the light with a few (usually one) detectors [PMT (photo multiplier tube) or photodiode]. The assumption of a few particles per square millimeter allows the use of scanning laser beam spots measured in millimeters, which consequently can operate with very high wafer throughput. At the same time, focusing the laser beam down to a few micrometers allows the detailed inspection of isolated particles for characterization of individual signatures. For a very thorough discussion of the details, and pros and cons of each of these systems, the reader is directed to Ref. [1].

Regardless of the specific implementation, all optical on-wafer detection and classification methods rely on the ability to discriminate the scattering properties of the defect from the scattering properties of the background surface. That background may be smooth or rough, and need not be a bare surface but in fact may be a patterned wafer surface (such as a DRAM array). The scattering defect may not be an actual particle but a subsurface void in the bulk silicon [also known as a *COP* (crystal originated pit), typically octahedral in shape and about 100 nm across]. And in general, the nature of the contaminants may range from smooth dielectric objects (PSL calibration spheres) to rough and inhomogeneous process byproducts (such as metal or grit residues from CMP processing). Therefore, we are interested in the optical scattering from arbitrarily shaped, inhomogeneous, possibly anisotropic, and usually dispersive objects in the presence of a dielectric (also possibly lossy) surface.

Because we are always interested in the smallest detectable object (below the resolution limit of the optics), the focus is on the dark-field approach in its most general form. The fundamental question is: Can we, by collecting as much of the scattered light as possible, determine, from the distribution of that light, the size, location, and composition of the scatterers? Or, rephrasing the question into its most practical form: How should an inspection system be arranged to extract enough information from the scattered light to allow the user to detect, size, and classify the defect on the wafer? To the degree that the scattering problem can be analyzed and the results translated into useful inspection guidelines, to that degree the science of on-wafer inspection is advanced. The balance of this chapter is dedicated to the development of the theoretical approach to this problem and its experimental confirmation.

II. BACKGROUND OF THE THEORETICAL APPROACH

The ability to predict the scattering behavior of particles on surfaces is critical to the development of the inspection approaches that will be required to detect yield-limiting defects in the next generation of semiconductor wafer devices. Sensitivity to scatterers below 100 nm in size, and as small as 30 nm, is the goal to be met within the next 6 years. Because real semiconductor surfaces are never perfect, smooth, infinite half-spaces, sensitivity to scatterers in nonideal, clustered environments is the crucial issue for on-line applications of these techniques. In particular, the detection and classification of particles in the presence of a circuit pattern on the wafer, or the discrimination of one defective via within an array of vias, are timely problems whose solutions have immediate applications.* In order to solve these problems, the inspection tools must be designed to account for all the scattering phenomena expected to occur. Rigorous and practical computation methods to quantify those phenomena are required.

The rigor of the computation method determines the confidence level that we give to its solutions and the degree to which we trust it as we push it beyond the envelope of our experimental experience. The practicality of the computational method determines how widespread its use becomes. A practical method is a method that can be utilized by university researchers as well as by industry instrument developers. Rigor is usually associated with the exactness with which the computational method satisfies the physical equations governing the phenomena, in this case Maxwell's equations. Practicality is a combination of ease of conceptual understanding, simplicity in modeling of arbitrary configurations, and speed of execution.

Various authors have addressed the problem of light scattering by a particle on a penetrable surface. The most successful analytic solution of this problem

---

*A via is a cylindrical conductive path or interconnection between different conductive layers inside a substrate. At several points during the manufacturing of an integrated circuit, these vias appear as cylindrical holes or metal rods connecting the surface of the substrate to one or more subsurface features.

(that is, with a minimum of simplifying assumptions) is that of Assi [3]. In his approach, the Mie formalism is extended to include a planar surface. Repeated transformations from the spherical vector wave functions representing the scattering from the sphere, to a plane wave spectrum whose interaction with the planar surface can be expressed in closed form, allow the solution of the problem (Fig. 1).

Inhomogeneous particles could be modeled as concentrically layered spheres. However, the limitation of the method to spherical, or nearly spherical particles, leaves most realistic defects outside of its domain of rigorous validity. Therefore, a method which allows for the modeling of irregular and inhomogeneous scatterers is preferred.

The method we use is the natural extension of the Purcell-Pennypacker method as developed by Draine and Flatau [4–5]. It is implemented at ASU in the computer code DDSURF. The discrete dipole approximation (DDA) is also known in the electrical engineering literature as the volume integral equation method in the frequency domain, and it can be traced to its two-dimensional version in the work of Richmond [7]. It is formally based on the equivalence principle and the Stratton-Chu formulation [8]. This formulation states that the total electromagnetic fields outside a bounded region of space containing sources is completely and uniquely described by the tangential values of those fields on the surface of the bounding region. For the purpose of any calculation, the structure of the sources inside the volume is unimportant, and they can be replaced by equivalent sources on the surface of the volume given by

$$\vec{K}_e = \hat{n} \times \vec{H}, \quad \vec{K}_m = -\hat{n} \times \vec{E} \tag{1}$$

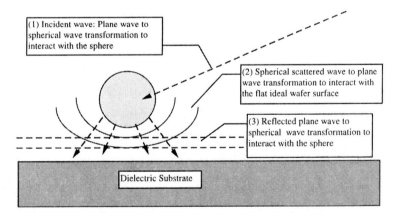

(1) Incident wave: Plane wave to spherical wave transformation to interact with the sphere

(2) Spherical scattered wave to plane wave transformation to interact with the flat ideal wafer surface

(3) Reflected plane wave to spherical wave transformation to interact with the sphere

Dielectric Substrate

Figure 1 The Mie-Weyl formalism can be extended to include the interaction with a dielectric substrate through transformations from the plane wave spectrum to the spherical wave spectrum.

where $\hat{n}$ is the surface normal, $\vec{E}$ and $\vec{H}$ are the total electric and magnetic fields, and $\vec{K}_e$ and $\vec{K}_m$ are the equivalent electric and magnetic surface current densities.

If we now imagine an inhomogeneous dielectric body under electromagnetic illumination, dissected into small subregions, it follows that the fields radiated by any subregion onto the rest of the body can be obtained from the equivalent sources on the surface of that subregion. If we further imagine doing this with all the subregions except one, we end up replacing the volume of the body with an array of sources, all of which are radiating *in free space* [9]. This radiation goes in every direction of space. In particular, it impinges on the single dielectric subregion that has not been removed (see Fig. 2).

Such an isolated material subregion, illuminated by the total field produced by all the other sources, will also scatter radiation in all directions of space. However, if the isolated subregion is assumed to be a sphere of radius *electrically small* compared to the internal (and the external) wavelength, λ, of the radiation impinging upon it (that is, $r \ll \lambda/\varepsilon^{0.5}$, where ε is the highest of the relative dielectric constants of the subregion or the medium), its scattering properties are particularly simple. It radiates like an elementary dipole. This is the key concept of the DDA method. The assumption of smallness guarantees that the total electric field over the subregion sphere is nearly uniform or at worst slowly varying, and this allows the use of the electrostatic case to derive the strength of the induced dipole moment. This is given by

$$\vec{p} = \alpha \vec{E}_{\text{tot}} \tag{2}$$

where $\vec{p}$ is the induced vector dipole moment, $\vec{E}_{\text{tot}}$ is the total vector electric field incident on the subregion (that is, the total from all other subregion sources plus

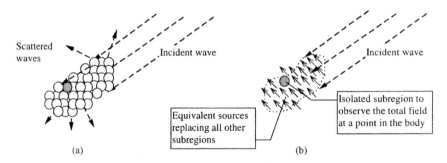

Figure 2 In the discrete-dipole approximation, the body of the scattering object is discretized into spherical subregions (a), which in turn are replaced by equivalent sources (b), to determine the total field at every point of the body.

the original illuminating field), and α is the polarizability, which in the Clausius Mossotti approximation takes the form:

$$\alpha = \frac{3 \, dV \varepsilon_0 (\varepsilon - 1)}{\varepsilon + 2} \tag{3}$$

where dV is the differential volume occupied by the subregion, ε_0 is the permittivity of free space, and ε is the relative dielectric constant of the subregion. In practice, the body is usually dissected into elementary cubes, cut along the axes of a Cartesian coordinate system. Then the volume of the spherical region is taken to be equal to the elementary volume $dx * dy * dz$ (with dx usually $= dy = dz$). Since the volume of the elementary spherical region is $4\pi r^3/3$, this results in the spherical regions with radii r greater than $dx/2$ that slightly overlap their nearest neighbors. This overlap does not significantly affect the validity of the results until the permittivity ε becomes very high (of the order of 50 or higher).

The dipole field radiated by the illuminated (isolated) subregion exists everywhere in space, including, in particular, every other subregion in the body. Because in the frequency domain we have a steady state solution, this field must have been part of the total field that illuminated every other subregion, and contributed to their total internal field and therefore to the magnitude of the sources with which we replaced those subregions in the first place. That is to say, if we repeat the process of leaving an isolated subregion of the body intact while replacing every other subregion with their equivalent sources, across the entire dielectric body, we quickly realize that the $\vec{K}_m$ and $\vec{K}_e$ of every subregion must all be summarized by the elementary dipole moments of Eq. (2), evaluated at each particular subregion. The entire system of dipole sources interacting with each other as a result of being excited by an external illuminating field must form a self-consistent system of equations of the type in Eqs. (2) and (3), in which every dipole in turn reacts to the total field impinging upon it. Because every dipole is "coupled" to every other dipole in the body, the DDA method is also known as the coupled dipole method (CDM). The elegance of the DDA method comes from the inherent simplicity of this formulation.

Using the DDA method, the problem in the presence of an adjacent substrate is not very different from that of a particle in free space. The field radiated by every dipole must simply include the effect of the dielectric half-space representing the substrate. This can be done by replacing the free-space Green function* with the Green function for the case of an arbitrary dipole at a height

---

*The Green function, or Green's function, is defined by P. M. Morse and H. Feshbach, *Methods of Theoretical Physics, Part I*, McGraw-Hill, 1953, p. 791, as follows: "To obtain the field caused by a distributed source calculate the effects of each elementary portion of the source and add them

h over a penetrable half-space [10]. It is customary to separate the resulting Green function into a "direct" component identical with the free-space term, and a "reflected" component which is the rest of the solution. The term *reflected* must be taken with a grain of salt, because in the near field it includes not only reflected waves but also all the surface waves and evanescent fields* excited on the substrate by the near field of the dipole. In the far field, the reflected term tends asymptotically to the reflected field that would be expected from Fresnel's equations.[†]

It can be shown that the solution to the system of simultaneous equations connecting N subregions of the body to each other through their respective polarizabilities, in the presence of the original illuminating field, rigorously satisfies Maxwell's equations as long as the subregions are electrically small enough. For further discussions on the foundations of the DDA method and comparisons with other computational scattering methods, the reader is directed to Refs. 11 and 12.

III. THE DDA METHOD FOR SCATTERING FROM PARTICLES ON DIELECTRIC SUBSTRATES

For every dipole in the body, let

$$\overline{P}_i = \alpha_i \overline{E}_{\text{tot},i} \tag{4}$$

so that at a dipole i the dipole moment $\overline{P}_i$ is related to the total electric field present at the dipole $\overline{E}_{\text{tot},i}$, by the dipole polarizability α_i. The total electric field at each dipole is the summation of the field incident from the beam, the field present by direct interaction between the dipoles, and the dipole fields "reflected" from the surface:

$$\overline{E}_{\text{tot},i} = \overline{E}_{\text{inc},i} + \overline{E}_{\text{direct},i} + \overline{E}_{\text{reflected},i} \tag{5}$$

---

all. If $G(\mathbf{r} \mid \mathbf{r}_0)$ is the field at the observer's point $\mathbf{r}$ caused by a unit point source at the position $\mathbf{r}_0$, then the field at $\mathbf{r}$ caused by a source distribution $\rho(\mathbf{r}_0)$ is the integral of $G\rho$ over the whole range of $\mathbf{r}_0$ occupied by the source." The function G is called the Green function.

*Surface waves are electromagnetic waves that remain bound, or attached, to a dielectric boundary and travel along it at a speed slower than the speed of light. Evanescent fields are electromagnetic fields that exist in the immediate neighborhood of a material discontinuity, incapable of carrying power away from the discontinuity but which nevertheless can interact with objects near that discontinuity.

[†]The equations governing the transmission and reflection of plane electromagnetic waves from plane dielectric boundaries (cf. J. A. Stratton, Electromagnetic Theory, McGraw-Hill, 1941, pp. 492ff).

Note that in this formulation, a dipole never sees its own self-field, and no questions about singularities in the free space Green function arise. Substituting (5) into (4) gives

$$(\alpha_i)^{-1}\overline{P}_i - \overline{E}_{\text{direct},i} - \overline{E}_{\text{reflected},i} = \overline{E}_{\text{inc},i} \tag{6}$$

The purpose of formulating Eq. (6) as shown is to place all the unknown variables on the left-hand side and isolate the only known input, the incident field, on the right side. The $\overline{E}_{\text{direct},i}$ field is the free-space interaction between the dipoles through the free-space dyadic Green function $\overline{G}_{ij}$.

$$\overline{E}_{\text{direct},i} = \frac{k_0^2}{\varepsilon_0} \sum_{j \neq i} \overline{G}_{ij} \cdot \overline{P}_j \tag{7}$$

The interaction term due to the surface is written

$$\overline{E}_{\text{reflected},i} = \sum_{j=1}^{N} \left(\overline{S}_{ij} + \frac{k_2^2}{\varepsilon_0} \frac{k_1^2 - k_2^2}{k_1^2 + k_2^2} \overline{G}_{ij}^I \right) \cdot \overline{P}_j \tag{8}$$

where k_1 and k_2 are the wave numbers for the particle and the surface, respectively, and the image dyadic Green function is defined as

$$\overline{G}_{ij}^I \equiv -\overline{G}_{ij} \cdot \overline{I}_R \quad \text{with} \quad \overline{I}_R = \mathbf{e}_x\mathbf{e}_x + \mathbf{e}_y\mathbf{e}_y - \mathbf{e}_z\mathbf{e}_z \tag{9}$$

where $\mathbf{e}_x$, $\mathbf{e}_y$, $\mathbf{e}_z$ are the unit vectors along the coordinate axes. The Sommerfeld integrals S_{ij}, in Eq. (8), are computed numerically using a procedure due to Lager and Lytle [13].

Now, to obtain the simultaneous solution of this set of equations for N dipoles we represent them as a matrix equation:

$$(\overline{B} + \overline{A} + \overline{R})\overline{P} = \overline{E}_{\text{inc}} \tag{10}$$

where $\overline{B}$ is the diagonal matrix containing the inverse of the polarizabilities of every subregion (which in the most extreme case are all different from each other and anisotropic):

$$\overline{B} = \text{diag}\,(a_{1x}^{-1}, a_{1y}^{-1}, a_{1z}^{-1}, \ldots a_{Nx}^{-1}, a_{Ny}^{-1}, a_{Nz}^{-1}) \tag{11}$$

Matrices $\overline{A}$ and $\overline{R}$ contain the direct and the reflected interactions between N arbitrarily polarized elements in three dimensions and therefore consist of N^2 3×3 submatrices.

$$\overline{A} = \begin{bmatrix} \overline{A}_{11} & \cdots & \overline{A}_{1N} \\ \vdots & & \vdots \\ \overline{A}_{N1} & \cdots & \overline{A}_{NN} \end{bmatrix} \qquad \overline{R} = \begin{bmatrix} \overline{R}_{11} & \cdots & \overline{R}_{1N} \\ \vdots & & \vdots \\ \overline{R}_{N1} & \cdots & \overline{R}_{NN} \end{bmatrix} \tag{12}$$

The submatrices are defined in the appendix of Schmehl et al. [14]. The matrix equation thus connects N 1×3 vector dipole moments in every subregion of the object to the N 1×3 incident electric field vectors in every subregion of the object.

$$\overline{P} = [\overline{P}_1 \quad \cdots \quad \overline{P}_N]^T \quad \text{and} \quad \overline{E}_{\text{inc}} = [\overline{E}_{\text{inc},1} \quad \cdots \quad \overline{E}_{\text{inc},N}]^T \quad (13)$$

where the exponent T means the transpose. Since the goal is to determine the dipole moments induced by the incident field, Eq. (10) is solved by casting it in the form:

$$\overline{P} - (\overline{M})\overline{E}_{\text{inc}} \quad (14)$$

where the matrix $\overline{M}$ is the inverse of the matrix $(\overline{B} + \overline{A} + \overline{R})$. And now a known matrix operates on a given incident field to obtain the unknown dipole moments. Because of the smallness requirement placed on each subregion, the number of subregions required to represent a realistic particle easily exceeds 1000, making the direct inversion of the matrix impractical. Iterative methods are therefore needed to solve Eq. (10). The success of such methods will depend on the number of elements, the symmetry of the overall matrix, and on how "well-conditioned" the matrix is. Tenuous particles in which the internal and scattered fields are fractions of the external field lead to well-conditioned matrices. Particles with regions of high index of refraction (or even metallic regions) lead to "ill-conditioning" problems, because the fields they scatter and their internal fields can be comparable to the external driving fields.

Complex conjugate gradient (CCG) methods have been used to perform this iteration efficiently in the case of particles in free space [15]. In such methods an initial guess for $\overline{P}$ is inserted into Eq. (10) and successively improved by evaluating the degree of satisfaction of the equation. The disagreement in the form of a residual field is used to drive the next iteration step toward a closer approximation and, eventually, the exact solution. In theory, the conjugate gradient method should converge in less than $3N$ steps. Practical iteration methods set convergence criteria as a tolerance level (typically $<10^{-3}$) for the mean ratio of the residual field to the incident field strength, and then seek to reach this level in at most 100 iterations. Once the matrix equation is self-consistent to better than this level, the problem is said to be solved. Because the evaluation of Eq. (10) involves matrix products, and matrix products can be done efficiently using the fast Fourier transform (FFT) method, iteration methods are frequently coupled with an FFT algorithm. If every dipole is located on a regular Cartesian grid, it is easy to see that the interaction terms connecting any two dipoles reoccur all over the matrix at least N times and as many as $8N$ times. This repetition by symmetry distinguishes the special kind of matrices called circulant matrices and it makes them particularly well suited for FFT implementation. For particles in free space, this threefold symmetry leads to the use of a three-dimensional FFT.

For the problem of particles on a surface, one axis (the z axis) is lost from the symmetry but two still remain, and so a two-dimensional FFT can still be used. However, the presence of the surface introduces another difficulty. In the absence of a surface, the matrices $\overline{A}$ and $\overline{B}$ of Eq. (10) are highly symmetric and amenable to inversion by CCG. The surface introduces matrix $\overline{R}$, which is highly unsymmetric. The result is that the CCG method becomes slow to converge as the index of refraction of the subregions increases. It has been found that the quasiminimal residual method (QMR) is a robust alternative to CCG iteration [16]. In numerical experiments with a 500 nm silicon sphere of complex refractive index 3.88-j0.23, illuminated with 632.8 nm light, analyzed in free space, it is found that both QMR and CCG converge, with QMR reaching a given level of accuracy 2 to 10 times faster than CCG. However, when the particle is on a silicon surface, CCG diverged, while QMR still converged to a solution. The CPU time to a given accuracy on the silicon surface with QMR was 2 to 4 times slower than for the particle in free space, also using QMR. Clearly, high-refractive-index surfaces make the matrix harder to invert.

In the cited example, the electrical size of the silicon sphere was 3 wavelengths in diameter. This is typical of the size of problem addressable with DDSURF at present: approximately 3 electrical wavelengths at indices of refraction of the order of 4. For lower index (lower contrast), the object can be larger. For higher index (the case of metals), the object must be smaller.

Once the problem is solved, we have a vector describing the internal dipole moments at every subregion inside the object. The total scattered field is the vector sum of the far-field radiation from each of these dipoles. In the presence of the surface these fields include a direct part and a reflected part. These could be obtained from the Green function of the dipole in the presence of the dielectric half-space by taking the limit as the observation point goes to infinity. However, this is not necessary because by the reaction theorem [9] the field transmitted by a dipole to a unit detector at infinity is identical to the field received by that dipole from a unit source at infinity. The fields from a unit source at infinity are simply plane waves interacting with the surface through the well-known Fresnel coefficients. Thus,

$$\overline{E}_{\text{sca}}(r) = k_0^2 \frac{\exp{(ik_0 r)}}{4\pi r} \sum_{j=1}^{N} \tag{15}$$

$$\times \left\{ \begin{array}{l} \exp{(-ik_{\text{sca}}r_j)}[(\overline{P}_j \cdot \mathbf{e}_1)\mathbf{e}_1 + (\overline{P}_j \cdot \mathbf{e}_2)\mathbf{e}_2] + \\ \exp{(-ik_{\text{sca}}r_j)}[R^{\text{TM}}(\overline{P}_j \cdot \mathbf{e}_1)\mathbf{e}_1 + R^{\text{TE}}(\overline{P}_j \cdot \mathbf{e}_2)\mathbf{e}_2] \end{array} \right\}$$

with R^{TM} and R^{TE} being the Fresnel coefficients for TM polarization (also known as p—parallel to the plane of incidence) and TE polarization (also known as s—perpendicular to the plane of incidence) respectively, and e_1 and e_2 are the unit vectors in the θ and φ directions of spherical coordinates.

The far-field irradiance is defined as

$$I_{sca} = \overline{E}_{sca} \overline{E}_{sca}^*$$ (16)

From this we define the differential scattering cross section as the power scattered into a unit solid angle in a given direction of space:

$$\frac{dC_{sca}}{d\Omega} = \lim_{\Omega \to 0} \left(\frac{C_{sca}}{\Omega} \right) \approx \frac{I_{sca} A}{I_{inc}(A/r^2)} = \frac{r^2 I_{sca}}{I_{inc}}$$ (17)

with A as the area of the detector, and Ω the solid angle spanned by that detector.

The differential scattering cross section is the signature of the scatterer. Its pattern of maxima and minima as a function of θ and φ over the two-dimensional surface of the far-field hemisphere constitute the sum total of the information available to us for detecting, sizing, and classifying the defect under inspection. Since in general the far-field scattered from an object is the Fourier transform of the currents induced on that object, it is intuitively clear that there is enough information in the far-field differential scattering cross section to reconstruct the object. This is true even though, due to the lack of phase information, we cannot perform a rigorous inverse Fourier transform. It is true because we have available extra information; namely, we can determine the dependence of the differential scattering cross section on angle of incidence and polarization of the incident light.

The importance of these extra degrees of freedom cannot be overemphasized. The scattering from a particle on a dielectric surface is strongly dependent on these parameters because the Fresnel reflection coefficients for the surface are strong functions of the same. In particular, for p polarization, from normal incidence to above the Brewster angle, the total electric field on the surface of the wafer is a maximum at the surface and tends to twice the strength of the incident field. For s polarization, the surface field tends to zero. Therefore, p polarization strongly excites small particles on a surface. However, s polarization becomes useful as an adjunct to p when the surface is covered by a film. In that case it can be shown that the combination of s and p polarizations into circular polarization is relatively insensitive to the interference induced by the film [1]. s polarization is also useful by itself because it excites the transverse dimensions of the object more strongly than p polarization.

Thus, the real question is not whether we can reconstruct the object with all this information. Rather, what is the minimum amount of information required to reconstruct the important parameters of the object? And, second, how do those requirements change in the presence of realistic substrates and other scatterers?

IV. APPLICATION EXAMPLES

To emphasize the difference between the scattering phenomena of particles on a surface and particles in free space, it suffices to consider the case of PSL spheres on a silicon wafer. This case is extensively used in the industry as the basis for the calibration of wafer scanners. In fact, when scanning instruments report the size of a defect it is not a real physical size but the "PSL equivalent" size, that is, the size of a PSL sphere that would scatter the same amount of light. The classic work in this area is that of Liu et al. [17]. They describe the system used by the older Tencor Surfscan-4000, and then proceed to compare the instrument's reported particle sizes and cross sections with independently determined values. The reader is directed to this excellent article for all the details. Here we concentrate on the scattering cross section as reported by the instrument in Figure 5 of Ref. 17, reproduced here as Figure 3a.

The figure shows the measurement results for PSL spheres on a silicon wafer ranging in size from 0.26 to 2 μm under 633 nm illumination. In addition to the measurements, six calculated results using a full-wave axisymmetric solution to Maxwell's equation implemented on a Cray supercomputer are reported, plus two curves representing two approximate models based on combining Mie theory with the Fresnel coefficients. (No more supercomputer calculations were

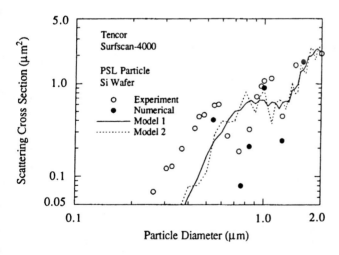

Figure 3a Measured scattering cross section for PSL spheres (open circles) illuminated near normal incidence with 633 nm light, as reported by Liu et al. [17]. The resonant nature of the scattering cross section is evident. Black circles show the results of six fullwave calculations confirming the results. The solid and dashed curves represent two simplified models proposed in the reference.

made because of the large amount of time each calculation took.) The most significant features of the results reported are: 1) The measured data show high peaks and deep nulls in the scattering cross section that cannot be replicated by the approximate models, the typical error being as high as a factor of 4 near the peaks and a factor of 7 near the nulls. 2) By contrast, the six super-computer results appear to be typically within 10% of the measured data. The significance of the observations above should not be underestimated. Most computational electromagnetics codes used in the electrical engineering community are expected to agree with experimental results within fractions of a decibel.* Without such stringent requirements, the computational methods become of little use in trying to extrapolate beyond experimental knowledge to obtain new results or design new inspection approaches. The failure of the simple Mie-based models of Liu et al. [17] to correctly describe the resonance properties of a particle on a surface stem from the omission of the effect of the surface on the *near field* of the particle. The surface effectively makes the particle appear almost twice as large (because of its image), moving the first resonance from the simple models' prediction of 1.0 μm diameter down to about 0.5 μm. This can be readily confirmed today using DDSURF. Figure 3b shows our results for this case.

These calculations were not performed on a supercomputer but rather on an Alpha workstation running at 500 MHz. The average calculation, which makes no assumption of axial symmetry, takes less than a minute, and the results are very close to the supercomputer results and the measured data shown in Figure 3a.

In Figure 3b we also report the results for fictitious PSL cubes. An effective diameter was assigned to them based on their volume. These data highlight the role that the particle's shape plays on the scattering cross section and therefore on the size an inspection tool would assign it. Next we consider one of the most challenging and practical applications of an on-wafer inspection system: the contaminated patterned wafer.

Whereas an unpatterned wafer surface scatters most of the energy imping-ing upon it in the forward specular direction, the patterns on the wafer (circuit traces) scatter energy in all directions of space. Thus the pattern on a wafer con-tributes its own signature to the map of the differential scattering cross section. Out of this background we need to detect the presence of contaminants. When the pattern on the wafer is repetitive and large compared to the illuminating beam, optical Fourier filtering techniques can be used to subtract out the pat-tern. However, when the pattern is irregular or consists of isolated features, the subtraction will have to be done by software that recognizes the characteristic

---

*One decibel of error would be equivalent to a 25% error in the evaluation of the scattering cross section.

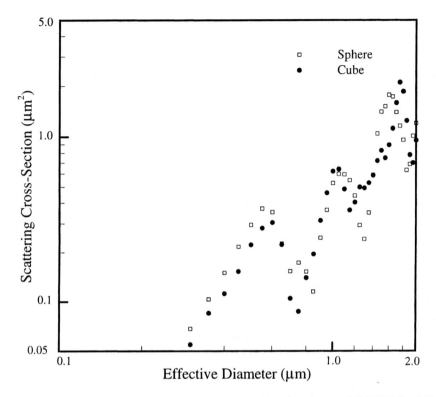

Figure 3b Scattering cross section as calculated by the program DDSURF for PSL spheres (black dots) illuminated near normal incidence with 633 nm light, for the range examined by Liu et al. [17]. The resonant nature of the scattering cross section is evident. Fictitious PSL cubes (open squares) of comparable volumes are also included to demonstrate the effect shape has on the cross section. The effective diameter is equal to the actual diameter for the spheres and to a volume equivalent diameter for the cubes.

scattering of the features and can subtract them from the total signature, in order to locate and identify the scatterer.

Consider the situation of Figure 4. A PSL sphere models a contaminant in the neighborhood of a rectangular silicon dioxide line feature (a), in the neighborhood of a line pair (b), and in the neighborhood of a corner feature (c). The contaminant sphere is 482 nm in diameter, while the features are approximately 270 nm in thickness. The substrate is silicon and the illuminating wavelength is 632.8 nm incident at 45 degrees from normal.

The differential scattering cross section for each of these cases is displayed in Figures 5 to 7, as it would be seen on a plane inclined 45 degrees to the normal, perpendicular to the specular reflected beam, and centered on that beam. We

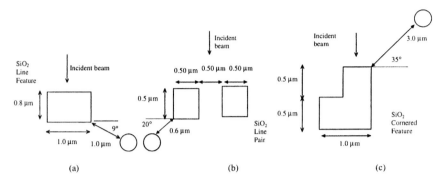

Figure 4 Simulation of the contaminant-near-a-pattern-feature problem. A 482 nm PSL sphere simulates the contaminant near (a) a line feature, (b) a line pair feature, (c) a corner feature.

examine this plane because it is well known that particles on a surface scatter most strongly in the forward direction, close to the specularly reflected field. This is particularly true as their dimensions grow, and is understandable from the observation that the strong forward scatter expected from Mie theory must reflect off the surface and travel in the same direction as the reflected incident beam. Figure 5a shows the irradiance on that plane for the line feature alone. Figure 5b shows the effect of inserting the contaminant. Similarly Figures 6a and 7a show the scattering from the line pair and corner features alone, while Figures 6b and 7b show the effect of the contaminant. The periodicity of the fringes that develop on the observation plane is a strong function of the distance between the particle and the feature. The magnitude of the fringes is a function of the strength of the particle scattering signature, which is itself a function of particle size and composition [18].

Similar experiments have led to the observation that the apparent size of a standard PSL calibration sphere is affected by its proximity to a pattern line. When the PSL sphere is located on top of a line, near one of its edges, it has the largest apparent size. This is to be expected based on electrostatic considerations: At the top edge of a feature, the electric field is a maximum and thus the induced dipole moment on the sphere is a maximum.

The capability to model dielectric objects of arbitrary composition and shape above the substrate translates directly into the modeling of subsurface defects. The Sommerfeld integral terms in the equations are essentially the same as in the above surface case except the dipoles are in the optically denser medium. The Clausius Mossotti definition of the polarizability [Eq. (3)] becomes

$$\alpha = \frac{3\,dV\,\varepsilon_0(\varepsilon_{particle} - \varepsilon_{medium})}{\varepsilon_{particle} + 2\varepsilon_{medium}} \tag{18}$$

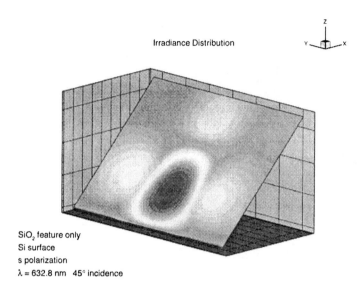

Irradiance Distribution

SiO₂ feature only
Si surface
s polarization
λ = 632.8 nm   45° incidence

(a)

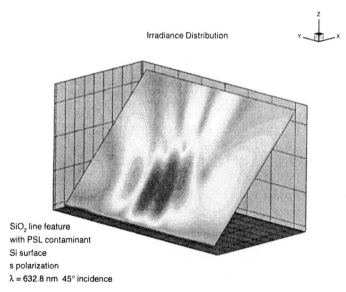

Irradiance Distribution

SiO₂ line feature
with PSL contaminant
Si surface
s polarization
λ = 632.8 nm   45° incidence

(b)

Figure 5 Irradiance distribution on the observation plane perpendicular to the specular beam for the case of Figure 4a: (a) for the line feature alone; (b) with the contaminant 1 μm away.

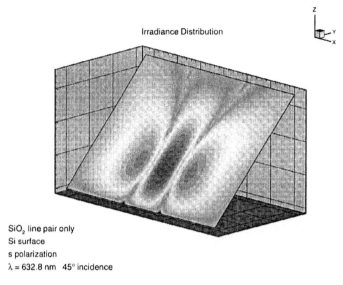

(a)

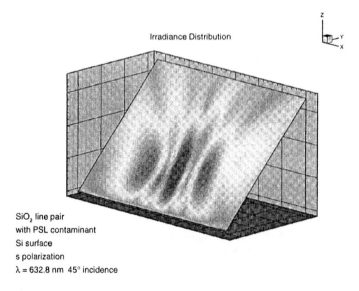

(b)

Figure 6 Irradiance distribution on the observation plane perpendicular to the specular beam for the case of Figure 4b: (a) for the line pair alone; (b) with the contaminant 0.6 μm away.

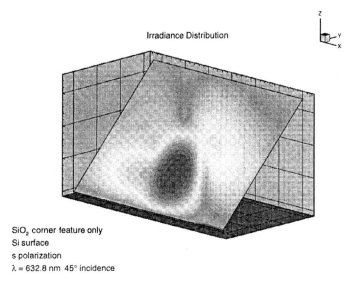

SiO$_2$ corner feature only
Si surface
s polarization
λ = 632.8 nm 45° incidence

(a)

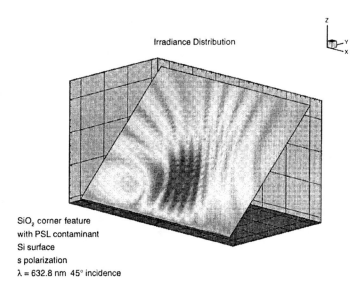

SiO$_2$ corner feature
with PSL contaminant
Si surface
s polarization
λ = 632.8 nm 45° incidence

(b)

Figure 7 Irradiance distribution on the observation plane perpendicular to the specular beam for the case of Figure 4c: (a) for the corner feature alone; (b) with the contaminant 3 μm away.

where it is clear that if the subsurface defect has a lower dielectric constant than the substrate medium, its dipole moment will be negative. Regardless of the sign of the dipole moment, any amount of contrast between a defect region and the surrounding space leads to scattered radiation. Thus, both a contaminant on the surface and a void (COP) near the surface will appear as scattering objects in a wafer scanner. This is one of the critical issues in unpatterned wafer inspection. Whereas contaminants above a certain size are clearly yield-limiting defects, COPs buried a certain depth below the surface may not be. To discriminate between the two types of scatterers is paramount in the quality assurance of the bare semiconductor wafer.

A rigorous simulation of the electromagnetic scattering problem can help in developing strategies to effect this discrimination. Because the COP, or bulk void, problem is similar to the via problem, much can be gained from inspecting the scattering from the relatively simple cylindrical geometry of a via. Consider the geometry of Figure 8, a cylindrical hole in silicon dioxide ($n = 1.46$), 600 nm in diameter, 732 nm deep, illuminated with 488 nm light at 70 degrees from normal.

Figure 9 shows the differential scattering cross section in the principal plane for both s (a) and p (b) polarization. The scattering from this large subsurface void is stronger toward the negative angles, that is, forward scatter, just as it would be for a large particle on top of the surface. However, as will be seen below, small voids mostly scatter in the direction of the surface normal.

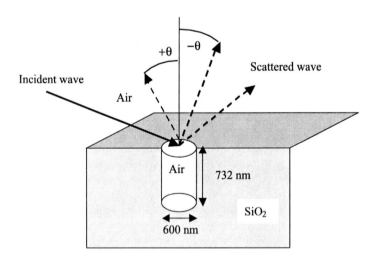

Figure 8 A cylindrical hole in SiO$_2$ serves as a test base for understanding the scattering from subsurface voids and vias.

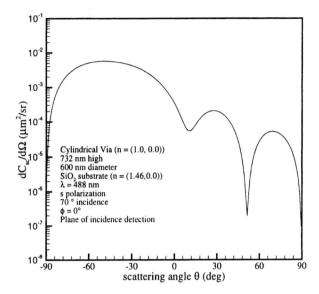

(a)

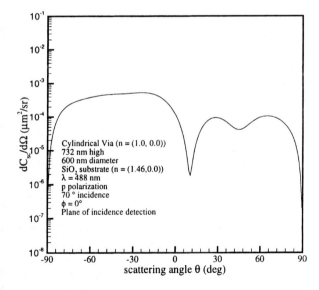

(b)

Figure 9 Differential scattering cross section in the principal plane of the large cylindrical hole of Figure 7, illuminated at 70 degrees off the normal: (a) *s* polarization, (b) *p* polarization. Negative angles are in the forward scatter direction.

Note that for this case the s-polarized scatter is stronger than the p-polarized scatter, which is opposite to the behavior exhibited by particles above the surface. This is very significant and is a clue that can be used to distinguish particles from subsurface voids. However, the physics of the scattering mechanism must be understood, to derive useful inspection guidelines. As we mentioned before, the Fresnel coefficients play a large part in all the phenomena associated with a surface. Figure 10 shows the Fresnel reflection coefficients for a silicon dioxide surface (a) and for a silicon surface (b), at 488 nm. For a given index of refraction, both s and p polarizations reflect equally strongly at normal incidence to the surface. However as the angle of incidence becomes increasingly shallow, the s polarization reflects more strongly while p polarization reflects more weakly. At the Brewster angle the p-polarization reflection goes to zero and the beam completely penetrates the surface.

Associated with the strength of the reflection there is a phase shift in the reflected wave. For s-polarization, the strong reflected wave flips 180 degrees relative to the incident beam, resulting in cancellation of incident and reflected wave near the surface. For p polarization, the only component of the electric field that flips is the tangential one, leaving the normal component unchanged and resulting in a field that is mostly vertical (perpendicular to the surface) and of magnitude roughly equal to the magnitude of the incident beam plus that of the (generally weak) reflected beam. p polarization produces a stronger vertical field at the surface of the substrate than was present in the incident beam while s polarization produces a weaker horizontal field. However, once we cross the Brewster angle, the p polarization also experiences a 180 degree flip and its reflection coefficient climbs rapidly. The result is that past the Brewster angle the p-polarized field on the surface is weaker than the field in the incident beam until it vanishes at grazing incidence.

From the standpoint of the electric currents (polarization currents in the case of dielectrics) associated with the total field interacting with the surface, we can also develop some physical guidelines: The electric currents are there to support the tangential components of the fields. Because, off the normal, p polarization tends to be mostly vertical, it carries with it a small surface current whose vector direction is aligned with the plane of incidence. On the other hand, s polarization, being all transverse to the plane of incidence, is also wholly tangential to the surface and therefore carries with it a large surface current perpendicular to the plane of incidence.

Therefore, particles above the surface get polarized strongly by the vertical field of p-polarized waves, and generally their scattering cross section is greater with p than with s polarization. However, a void on and under a surface is involved in an entirely different mechanism. The mouth of the void, which breaks the surface, interrupts the surface currents. The body of the void under the surface interacts with the part of the incident wave that penetrates through the

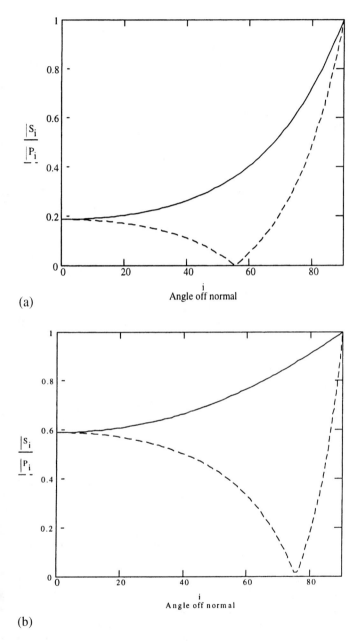

(a)

(b)

Figure 10 Absolute value of the Fresnel reflection coefficients as a function of angle of incidence for a plane wave impinging on a half-space of (a) silicon dioxide, (b) silicon. The solid line is for s polarization, the dashed line for p.

surface. This means that for a large void, as the via in Figure 9, the interruption of large s-polarized surface currents becomes a major contributor to the scattering. This effect is exacerbated by the fact that in the example the incident beam is beyond, but close, to the Brewster angle for SiO_2. Thus the p-polarized wave is highly penetrating, and it tends to cross any air-dielectric boundary with very little scattering. The net result is that, even though s polarization does not penetrate the surface as well as p, it actually scatters more strongly from the large void than p polarization. The difference is of the order of a factor of 10.

If, however, we had a small void, say a 65 nm cylindrical hole, in a high index of refraction material, like silicon, and we illuminated it at the same 70 degree incidence (which is close to but not yet at the Brewster angle), we get entirely different results. Figure 11 shows that most of the (weak) scattering is nearly perpendicular to the surface and that p polarization is slightly stronger than s.

In this case, since we are not yet at the Brewster angle and the mouth of the void is small, the p-polarization scattering gets a slight edge over the s polarization. It is interesting to note in comparing Figures 11a, a 50 nm deep hole, and 11b, a 100 nm deep hole, that the scattering is nearly independent of the depth of the hole. This is true for both polarizations for very different reasons. For p polarization, the proximity to the Brewster angle again makes the radiation very penetrating. Even after refraction at the surface, its angle of incidence is close to the Brewster angle when measured relative to the vertical sides of the hole. As a result, the p-polarized wave passes through the hole almost unscattered. The s-polarized wave reflects so strongly at the surface that it has almost no energy left to penetrate down to the base of the hole and scatter from there.

The above examples serve to illustrate the wealth of insight a rigorous and practical computational code like DDSURF can bring to the problem of defect scattering in the presence of a surface. Many other applications are being considered at present. For instance, taking into account the roughness of a real wafer surface is not impossible. All that would need to be done for that case is to place "bumps," of the same material as the substrate and of the appropriate size, randomly around the scattering scene we wish to study. It can be argued from electromagnetic theory that a bumpy surface will tend to enhance surface wave effects. Thus, in addition to the well-known surface spatial spectrum noise around the specularly reflected beam [19], we expect to find enhanced near-field coupling between particles on the surface. The Fresnel coefficient discussion above suggests that if the scattering from the roughness of the surface needs to be avoided at all cost, we will end up illuminating our wafer with the highly penetrating p polarization near the Brewster angle. Then the surface, bumps included, becomes nearly transparent, and all that scatters is the particle

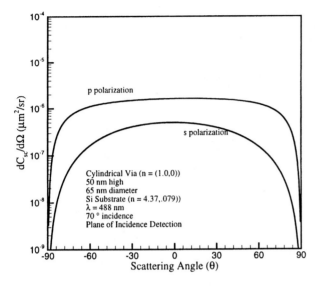

(a)

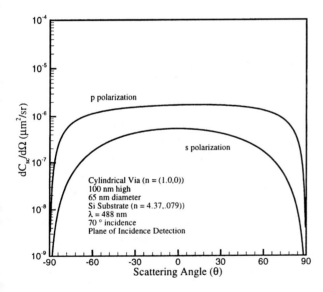

(b)

Figure 11 Differential scattering cross section from a small cylindrical hole (65 nm diameter) in silicon, illuminated at 70 degrees off normal: (a) for 50 nm depth; (b) for 100 nm depth.

under test. Configurations in which the surface has film layers also occur in real applications and involve more complex Green functions, which can support tightly bound surface waves. The capability to model these configurations is being added to DDSURF.

In summary, in this section we have demonstrated the application of a computational electromagnetics method to the simulation of the scattering from practical defects on a realistic surface. Through numerical experimentation of this type, a variety of strategies can be explored to obtain guidelines for the design of the next generation of on-wafer inspection equipment. These guidelines can range from the optimal angle and polarization of illumination for a particular type of defect to a catalog of the salient scattering signatures of typical defects of interest. These guidelines, then, provide the answer to our original question: What is the minimum amount of information required to perform the functions of detection, sizing, and classification of defects on wafer surfaces? The final task left to do is to support the computational method with experimental results. This is the subject of the next section.

V. ANGLE RESOLVED SCATTERING MEASUREMENTS

In order to obtain experimental confirmation of the numerical model, it is imperative to characterize the features under test as accurately as possible before measuring their scattering cross section. A combination of optical microscope, scanning electron microscope (SEM), and atomic force microscope (AFM) is used for this purpose. The optical microscope is primarily used to view sample areas and ensure that undesired particle contaminants are not present prior to testing.

In conjunction with the Semiconductor Research Corporation (SRC), ASU designed a special test wafer with a variety of "typical" silicon dioxide pattern features, onto which "contaminant" PSL spheres were deposited. The Joel 840 SEM was used to verify the particle shapes and relative size. Figure 12a is an example of one structure shape located on the ASU/SRC die. This micrograph is for one of the lines with a defect near the feature. The SEM micrographs allow verification of the sample, but do not provide the accurate data required for modeling since sample charging leads to some ambiguity in exact feature size, especially for smaller-sized particles and features. Figure 12b shows the same feature measured with the AFM. From this image, the feature size was determined to be 1×2 μm with a thickness of 0.28 μm. This thickness was also verified on a Digital Instrument NanoScope E. These values were used as input parameters for modeling.

Figure 13 shows the setup for the ASU scatterometer. The equipment consists of a laser (HeNe or Argon Ion), polarizer (for selection of s or p polar-

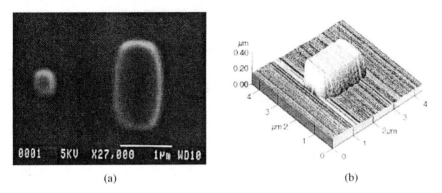

(a)   (b)

Figure 12 (a) SEM and AFM image of feature and contaminant PSL sphere on the ASU/SRC die; (b) AFM image of feature alone.

ization), a beam splitter to provide a beam power reference, and focusing optics. Spot sizes as small as 6.5 μm may be achieved, giving the ASU scatterometer the capability to measure the differential scattering cross section from individual contaminants and features. The sample is mounted to an xy translation stage which allows movements as small as 1 μm.

The conventional approach with this setup is to position a single (PMT or photodiode) detector in the far field and scan it over θ, φ, to measure the differential scattering cross section. In this approach there is a trade-off among 1) maximizing the detector aperture (which increases gain but also sensitivity to

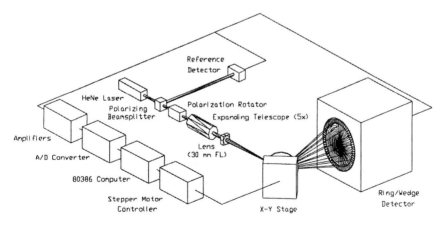

Figure 13 ASU scatterometer setup for measurement of differential scattering cross section.

ambient noise), 2) reducing the angular resolution by integrating many angles in one measurement, and 3) the effect of these two parameters on scanning time. An alternative to this conventional detector arrangement is one that combines the gain advantage of integrating the signal over various angles, while preserving the scan speed and the angular resolution necessary to recognize scattering signatures. This unique detector is known as a ring-wedge detector (Fig. 14). Since the signature of scatterers is displayed in the θ, φ dependence of the scattered light, the detector is divided into 64 sections of rings and wedges. Of these, the 32 rings measure angle resolved scatter in the θ direction, while the 32 wedges measure angle resolved scatter in the φ direction. The area of the rings increases as a function of the radius. The wedge area is approximately the same for all wedges. The figure shows a sketch of the detector and its typical mounting position. It is usually located 8.2 mm from the sample, thus providing a range of mean scattering angles from approximately 0.6 to 62.7 degrees off the specular beam. The center of the detector serves as a beam dump for the strong reflected specular beam. Even so, the proximity of the specular beam saturates the first nine rings, and they are not used during measurements.

To compare experiment with theory, the scattered irradiance calculated from DDSURF was integrated over the appropriate regions of space to simulate the signature that the ring wedge detector would see. Figure 15 shows a comparison of predicted vs. measured integrated differential scattering cross sections as read by a ring wedge detector for an isolated 482 nm PSL sphere on a Si surface. Figure 15a is for the ring region; Figure 15b is for the wedge region. The numerical predictions correspond well with the experimental measurements. The error bars on the experimental values are a 3σ distribution of the results found by repeating the experiments five times.

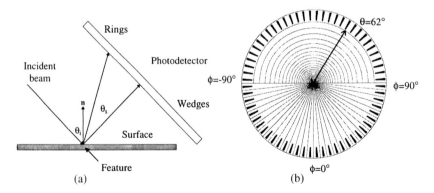

Figure 14   Location (a) and front view (b) of ring-wedge photodetector.

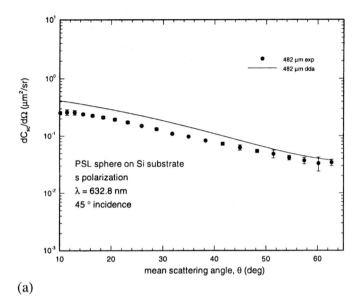

(a)

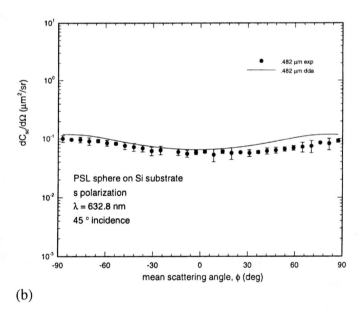

(b)

Figure 15 Comparison of calculated (line) and experimental (dots) differential scattering cross section as measured by the ring-wedge detector for a 482 nm PSL sphere on Si, with 632.8 nm illumination at 45-degree incidence: (a) rings; (b) wedges.

Next we show the same comparison for the cases shown earlier in Figures 4 to 7. Ring data are presented first. Figures 16a, 17a, and 18a show the ring data for the differential scattering cross section for the line feature, line pair feature, and cornered feature, respectively, when no contaminant is present. Figures 16b, 17b, and 18b are the corresponding differential scattering cross sections when a contaminating particle is introduced. Good agreement is found between the experimental results and the numerical predictions.

The corresponding wedge data is presented in Figures 19 to 21. As with the ring data, good agreement is found between the experimental results and the numerical predictions by DDSURF. Most of the results here are very close to, or lie within, the experimental error. Introduction of the particle contaminant has a significant effect on the $dC_{sca}/d\Omega$ variation in the wedge data in the form of a "bump" in the signature, dependent on the position of the contaminant.

VI. BRIEF LIST OF INDUSTRY RESOURCES

This section is a brief and incomplete list of commercially available on-wafer inspection equipment, circa 1999. Most manufacturers consider their particular implementation of the principles discussed in this chapter to be proprietary information. Therefore, few technical details can be gleaned from their product literature. One notable exception is Candela Instruments, which gives a very detailed description of their particular approach (see www.candela-inst.com).

The manufacturers we are most familiar with are:

- *ADE*. With reported capability to detect 60 nm particles on epitaxial Si.
- *Applied Materials*. Supply both optical and SEM tools. Their images, obtained by "fusing" the information from various detector types, highlight defect characteristics that a single instrument could not detect. They report the use of bright-field and dark-field techniques.
- *Candela Instruments*. Applied to magnetic disk inspection, uses both *p*- and *s*-polarized light, collected at specular and scattered directions, to detect film thickness, carbon wear, and particle contamination.
- *Inspex*. Imaging as well as scanning equipment. Their imaging systems boast 20 nm resolution for defect classification. The scattering systems use oblique dark-field illumination and image subtraction to detect 100 nm defects at 60 wafers per hour.
- *KLA-Tencor*. Utilize proprietary bright-field and double dark-field technology. The use of Nomarsky objectives in the bright-field optics provides an added phase discrimination capability.

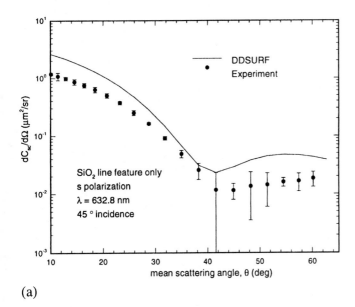

(a)

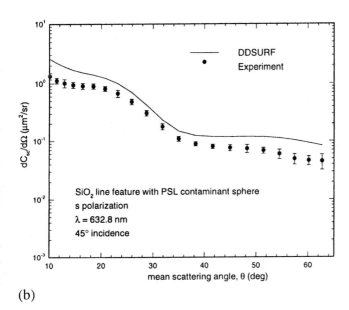

(b)

Figure 16 Calculated (line) and measured (dots) ring data for the configuration of Figure 4a: (a) line feature only; (b) line feature with contaminant.

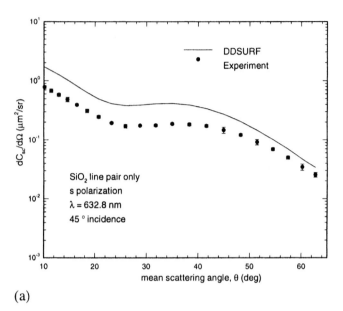

(a)

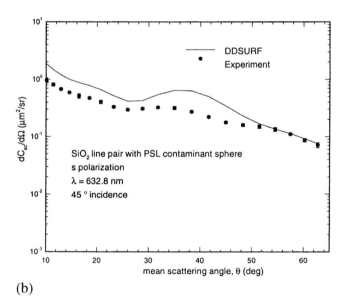

(b)

Figure 17 Calculated (line) and measured (dots) ring data for the configuration of Figure 4b: (a) line pair only; (b) line pair with contaminant.

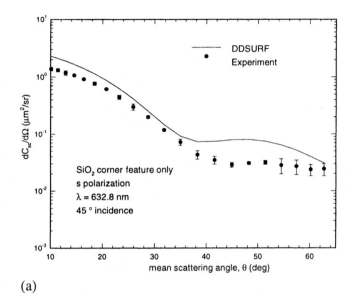

(a)

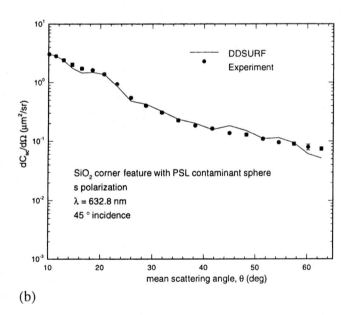

(b)

Figure 18 Calculated (line) and measured (dots) ring data for the configuration of Figure 4c: (a) corner feature only; (b) corner feature with contaminant.

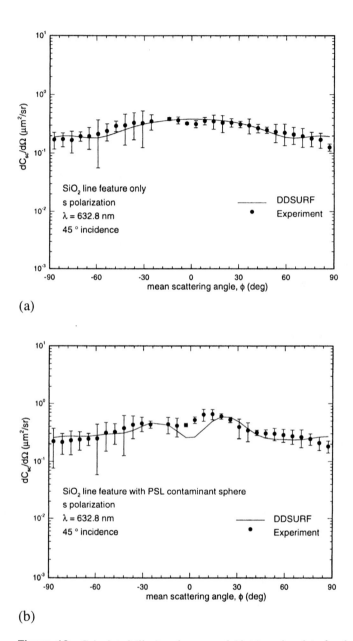

(a)

(b)

Figure 19 Calculated (line) and measured (dots) wedge data for the configuration of Figure 4a: (a) line feature only; (b) line feature with contaminant.

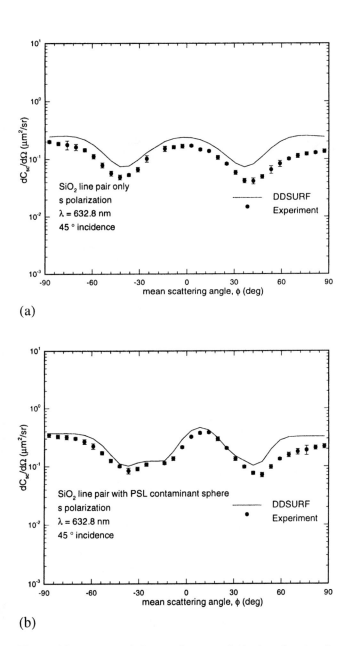

(a)

(b)

Figure 20 Calculated (line) and measured (dots) wedge data for the configuration of Figure 4b: (a) line pair only; (b) line pair with contaminant.

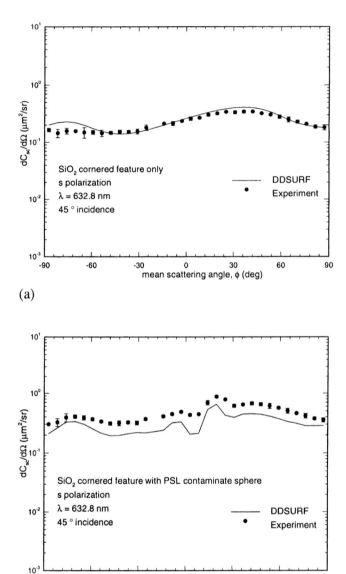

(a)

(b)

Figure 21   Calculated (line) and measured (dots) wedge data for the configuration of Figure 4c: (a) corner feature only; (b) corner feature with contaminant.

VII. CONCLUSION

In this chapter we have reviewed the foundations of on-wafer particle detection technology. Applied as a quality assurance tool at various points along the semiconductor wafer and device production cycle, this technology is a significant contributor to the growth of the industry. The inspection problem it addresses is in essence the same as the one addressed in every other chapter of this book. In the drive toward contaminant-free manufacturing, contaminants must be detected, sized, and identified, with reliability and speed. Only then can their source be identified and reduced or eliminated. Because of the ever-shrinking size of semiconductor devices (defect detection goals as low as 30 nm), this inspection problem will always be pushing the resolution edge of the present inspection tools. Economics will drive the inspection tool manufacturers toward improving the sensitivity of those present systems until the laws of physics win out and we are forced to switch to illumination sources capable of smaller wavelengths.

When the size of the objects to be detected falls below the resolution limits of conventional optics, we are forced to resort to inspecting the scattered light. It turns out that even in the presence of a large scattering background (the wafer surface), there is enough information in the scattering signature of objects to locate, size, and identify them. The task at hand is to determine the most efficient way to use that information in order to attain realistic production inspection goals in the 50 to 100 wafers per hour range. We have shown that the scattering phenomena can be accurately modeled for defects of arbitrary shape and compositions, on or under the surface of the wafer. The results of this modeling tool (DDSURF) have been experimentally verified by differential scattering cross-sectional measurements made with single particles on patterned and unpatterned surfaces.

REFERENCES

1. S. Stokowski and M. Vaez-Iravani, "Wafer Inspection Technology Challenges for ULSI Manufacturing," Characterization and Metrology for ULSI Technology, 1998 International Conference, AIP 1-56396-753-7/98.
2. D. S. Perloff, "Gauging the Future: The Long Term Business Outlook for Metrology and Wafer Inspection Equipment," Characterization and Metrology for ULSI Technology: 1998 International Conference, AIP 1-56396-753-7/98.
3. F. I. Assi, "Electromagnetic Wave Scattering by a Sphere on a Layered Substrate," Master of Science Thesis, Department of Electrical and Computer Engineering, University of Arizona, Tucson, 1990.
4. E. M. Purcell and C. R. Pennypacker, "Scattering and Absorption of Light by Nonspherical Dielectric Grain," Astrophysical Journal 186, pp. 705–714, 1973.

5. B. T. Draine, "The Discrete Dipole Approximation and Its Application to Interstellar Graphite Grains," Astrophysical Journal 333, pp. 848–872, 1988.
6. B. T. Draine and P. J. Flatau, "The Discrete Dipole Approximation and for Scattering Calculations," J Opt Soc Am A 11, pp. 1491–1499, 1994.
7. J. H. Richmond, "Scattering by a Dielectric Cylinder of Arbitrary Cross-Sectional Shape," IEEE Trans Antennas and Propagation, AP-13, pp. 335–341, 1965.
8. J. A. Stratton and L. J. Chu, "Diffraction Theory of Electromagnetic Waves," Phys Rev 56, pp. 99–107, 1939.
9. V. H. Rumsey, "Reaction Concept in Electromagnetic Theory," Phys Rev, 94(6), pp. 1483–1491, 1954.
10. M. A. Taubenblatt and T. K. Tran, "Calculation of Light Scattering from Particles and Structures by the Coupled Dipole Method," J. Opt. Soc. Am. A, 10, pp. 912–919, 1993.
11. T. Wriedt and U. Comber, "Comparison of Computational Scattering Methods," J Quant Spectrosc Radiat Transfer 60(3), pp. 411–423, 1998.
12. A. Lakhtakia, "Macroscopic Theory of the Coupled Dipole Approximation," Optics Com 79(1,2), pp. 1–5, 1990.
13. D. L. Lager and R. J. Lytle, "Fortran Subroutines for the Numerical Evaluation of Sommerfeld Integrals unter Anterem," Rep. UCRL-51821, Lawrence Livermore Laboratory, Livermore, CA, 1975.
14. R. Schmehl, B. M. Nebeker, and E. D. Hirleman, "Discrete-Dipole Approximation for Scattering by Features on Surfaces by Means of a Two-Dimensional Fast Fourier Transform Technique," J Opt Soc Am A, 14(11), pp. 3026–3036, 1997.
15. J. J. Goodman, B. T. Draine, and P. J. Flatau, "Application of Fast-Fourier-Transform Techniques to the Discrete-Dipole Approximation," Optics Lett 16(15), pp. 1198–1200, 1991.
16. B. M. Nebeker, G. W. Starr, and E. D. Hirleman, "Evaluation of Iteration Methods Used When Modeling Scattering from Features on Surfaces Using the Discrete-Dipole Approximation," J Quant Spectrosc Radiat Transfer 60(3), pp. 493–500, 1998.
17. B. Y. H. Liu, S. Chae, and G. Bae, "Sizing Accuracy, Counting Efficiency, Lower Detection Limit and Repeatability of a Wafer Surface Scanner for Ideal and Real-World Particles," J Electrochem Soc 140(5), pp. 1403–1409, 1993.
18. T. L. Warner and E. D. Hirleman, "Toward Classification of Particle Properties Using Light Scattering Techniques," Journal of the Institute of Environmental Sciences, XL(3), pp. 15–21, 1997.
19. J. C. Stover, Optical Scattering Measurement and Analysis, McGraw-Hill, New York, 1990, Chaps. 2 and 4.

5

On-Wafer Measurement of Molecular Contaminants

Victor K. F. Chia
ChemTrace, Fremont, California

Michael J. Edgell
Charles Evans & Associates, Redwood City, California

I. INTRODUCTION

To keep pace with the increasing demands of high-technology industries, equipment manufacturers must strive to improve product throughput and reduce contamination. The emphasis on contamination-free manufacturing (CFM) continues within the manufacturing environment through the use of cleanrooms and the practice of contamination-free procedures. In a typical production process, analytical tools are required to monitor the process in order to optimize product yields. The production yield enhancement cycle illustrated in Figure 1 involves the interaction of analytical tools to inspect the quality of incoming material, to monitor the production activity, and to qualify the outgoing material. These analytical tools may be used in-line, on-line, or off-line for material inspections to ensure CFM and for failure analyses to determine the original cause of the contaminant that is causing a particular failure. It is important to understand, through modeling and simulation, which parameters can improve product yield. It is also important to systematically identify these parameters within the process system and to control, and preferably reduce/eliminate, these parameters by identifying their root cause. To verify that CFM is being maintained in a production process these parameters must be monitored using a statistical process control (SPC) sample.

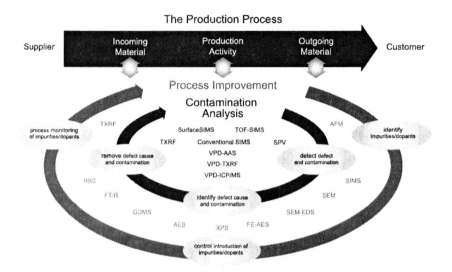

Figure 1   Typical product yield enhancement cycle in industry. Interaction between the production process and analytical monitoring tools is key to optimizing the production yield.

In this chapter we focus on common analytical techniques (acronyms for measurement methods are listed in Table 1) used to characterize on-wafer measurement of molecular contaminants. A molecular contaminant, as defined in Chapter 1, is any contaminant not classified as a particle, such as a single molecule, a small collection of molecules, or a film. Reducing molecular contamination (and particles) that may occur inadvertently during the production process is important because it can be detrimental to overall product performance. The key is to know what concentrations of which species are tolerable for a product or process.

Common contaminants can be considered as particles, native oxides, organic films, mobile ions, and elemental metals. The appropriate analytical techniques to characterize them are summarized in Figure 2. Accurate characterization of molecular contaminants requires the correct measurement technique, together with careful experimental design and data analysis to eliminate random and systematic errors. When selecting the analytical technique for a specific application, it is important to recognize and differentiate the limitations of the technique from the limitations of the instrument itself. Very often a correct analytical technique is used but inconclusive results are obtained because of instrument limitations from an outdated detector/source or data reduction system.

Table 1 Acronyms of Measurement Methods

Acronym	Measurement method
AES	Auger Electron Spectroscopy
FE-AES	Field Emission AES
AFM	Atomic Force Microscopy
EDS	Electron Dispersive X-Ray Spectrometry
ESCA	Electron Spectroscopy for Chemical Analysis
FTIR	Fourier Transform Infrared Spectroscopy
GCMS	Gas Chromatography Mass Spectrometry
HIBS	Heavy Ion Backscattering
IMS	Ion Mobility Spectrometry
SEM	Secondary Electron Microscopy
FE-SEM	Field Emission SEM
SIMS	Secondary Ion Mass Spectrometry
TEM	Transmission Electron Microscopy
TOF-SIMS	Time-of-Flight SIMS
TXRF	Total Reflection X-Ray Fluorescence
VPD	Vapor Phase Decomposition
XPS	X-Ray Photoelectron Spectroscopy

The analytical needs for on-wafer measurements of molecular contaminants are shown in Table 2, the 1997 National Technology Roadmap for Semiconductors. The time scale is from 1997 to 2012 for technology generation from 250 to 50 nm. The requirement for particle analysis size decreases from 75 to 15 nm. FE-AES using a thermal field emitter can provide spatial resolution of 15 nm routinely during elemental analysis of particles. TOF-SIMS is limited to 100 nm spatial resolution. The contamination limit for surface Al, Ti, and

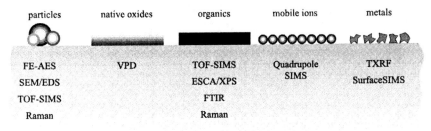

Silicon Wafer

Figure 2 Illustration of molecular contaminants on a silicon wafer and appropriate analytical techniques to monitor them.

Table 2 Metrology Technology Requirements from the 1997 National Technology Roadmap for Semiconductors

Year of first product shipment technology generation	1997 250 nm	1999 180 nm	2001 150 nm	2003 130 nm	2006 100 nm	2009 70 nm	2012 50 nm
Particle analysis area (on patterned wafers) (nm)	75	60	50	45	35	25	15
Surface detection limits (Al, Ti, Zn)/Ni, Fe, Cu, Na, Ca) (atoms/cm^2)	5×10^9 5×10^8	2.5×10^9 4×10^8	2×10^9 3×10^8	1.5×10^9 2×10^8	1×10^9 1×10^8	5×10^8 $\leq 10^8$	$\leq 5 \times 10^8$ $\leq 10^8$
Dopant concentration precision (integrated dose)	5%	5%	4%	4%	3%	2%	2%

Zn decreases from 5×10^9 atoms/cm^2 to $\leq 5 \times 10^8$ atoms/cm^2. In addition, the contamination limit for surface Ni, Fe, Cu, Na, and Ca decreases from 5×10^8 atoms/cm^2 to $\leq 10^8$ atoms/cm^2. SurfaceSIMS with a detection limit in the 10^8–10^9 atoms/cm^2 region is well positioned to provide this analytical need. Dopant concentration (integrated dose) precision requirements of 5 to 2% can be routinely achieved by SIMS, AES, and XPS (sometimes called ESCA). SIMS is the standard method for implant dose matching with a precision of $\leq 1\%$. The accuracy of the dopant concentration depends on the reference material used for quantification.

Section I provides an overview of many analytical techniques. In Section II we discuss the main considerations for selecting an appropriate analytical technique. Sections III to VI review common surface analytical techniques with good detection sensitivity to characterize molecular contaminants. SIMS (Dynamic SIMS, SurfaceSIMS, and TOF-SIMS), TXRF, HIBS, and VPD are reviewed. Sections VII and IX discuss analytical issues of surface metal and surface organic contamination, respectively.

II. FUNDAMENTALS OF ANALYTICAL TECHNIQUES

A wide range of analytical techniques is used to characterize molecular contaminants. Many of the techniques involve a primary beam of electrons, photons, or ions to probe the sample. The primary beam acts as an excitation source and interacts with the material in some way. Figure 3 illustrates the general principles of selected techniques. In some techniques, such as FTIR and TEM, changes

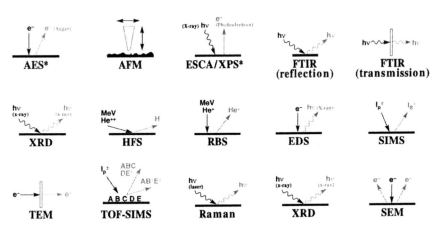

Figure 3 Schematic diagrams of the principles of common analytical techniques showing the excitation source and detected species. AES and ESCA/XPS require ancillary ion sputtering for depth profiling.

induced in the beam (energy, intensity, and angular distribution) are monitored after the interaction. The analytical information is then derived from the observation of these changes. In other techniques the information used for analysis comes from electrons, photons, or ions that are ejected from the sample during primary beam bombardment. In most cases several related quantum processes are occurring more or less simultaneously and the analytical technique is focused on one particular aspect. For example, XPS analyzes the photoelectrons that are emitted from the sample after excitation by x-rays. In AES, the incident electron ejects a K-shell electron, for example. In accordance with quantum theory an electron of higher energy, such as an L-shell electron, fills the K-shell vacancy. Energy loss by this L-shell electron during this process result in the ejection of another L-shell electron that is called an Auger electron.

III. CHOOSING THE APPROPRIATE TECHNIQUE

The selection of an analytical technique for on-wafer measurement of molecular contaminants depends on the specific application. We must first ask ourselves:

- What concentrations of which species are tolerable for a product or process?

Only when we can answer this question can we begin the selection process. Another consideration is:

- What is the cost of the analysis and what is the turnaround time?

In general, the cost of an analysis increases with lower detection sensitivity, improved depth resolution, and improved spatial resolution (i.e., analysis of a small area). This is because the analytical instruments that can provide these specifications are generally more expensive. If you don't need the best detection sensitivity, for example, then don't request it. To illustrate this point, Dynamic SIMS is widely used in the semiconductor industry because of its excellent detection sensitivity, in the parts-per-million/billion range. However, it is rarely used in the hard drive industry because the contamination level requirements are less restrictive. Other relevant questions are:

- Is the contamination elemental or organic?
- At what depth is the contamination?

To address detection sensitivities, Figure 4 shows selected analytical techniques on a plot of detection ranges (in units of ppm or atom/cm^3) vs. analytical area. The families of microscopic techniques, such as AFM, SEM, and TEM, are placed outside of this plot because they exhibit only one of these parameters. These techniques have excellent spatial resolution but provide no information

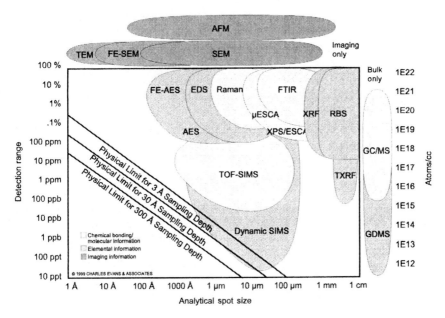

Figure 4 Plot showing the detection range and analytical area of selected techniques. The ideal technique is one with the lowest detection sensitivity and the smallest analytical area. An inhibitor is the physical limitation of insufficient atoms available to measure.

on the concentration of contaminant present on surfaces. The techniques shown inside the plot provide information relating to detection sensitivity and spatial resolution. The best/ideal technique is one that can provide information with the lowest detection limit from the smallest analysis area. Unfortunately, there is a physical limitation due to insufficient atoms available for detection. Sampling depth is therefore an important parameter of detection range and analytical area.

In the past few years, significant improvements have been made by instrument manufacturers to improve specifications. For example, FE-AES can analyze particles as small as 10 nm in diameter. This is a significant improvement to 0.2 μm analysis area with conventional AES in an imaging mode. Also, modern micro-XPS instruments can operate with 10 μm lateral resolution compared to 600 μm to 2 mm by conventional XPS instruments. If the application need is for small area analysis, there is no better technique than FE-AES. However, if organic or chemical state information is required from a small area, techniques such as XPS and TOF-SIMS are the primary choice, despite their limitation to submicrometer area analysis. In general, there is no definitive analytical technique for all applications because more than one technique is often required to determine the original source of the contaminant. Each analytical technique shown in Figure 4 plays a key role in characterizing molecular contaminants. Surface sensitivity and depth of analysis must also be considered. Applications of AES and XPS, as surface sensitive techniques, are well known because they have been established for well over 20 years. Their detection sensitivities are typically in the order of parts-per-thousand or atom % range. However, to evaluate the performance of contamination removal technology, this detection sensitivity is often too poor to be useful. In this chapter, we focus our attention on surface analytical techniques that provide detection sensitivities 2 orders of magnitude or more better (i.e., lower) than AES and XPS. Figure 5 compares the detection sensitivities of several surface analytical techniques. Detection limits are shown in units of atoms/cm^3 and, for comparison, in areal density units of atoms/cm^2. If the application is to monitor a 256 MB DRAM process line for elemental metals below 10^{10} atoms/cm^2, then analytical techniques such as SurfaceSIMS, TXRF, and TOF-SIMS are required.

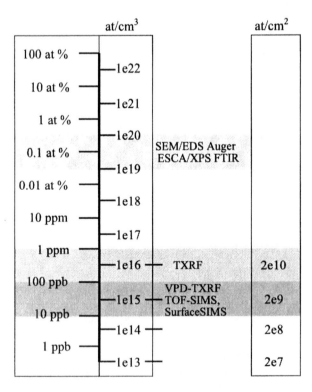

Figure 5 Detection limits of selected surface analytical techniques.

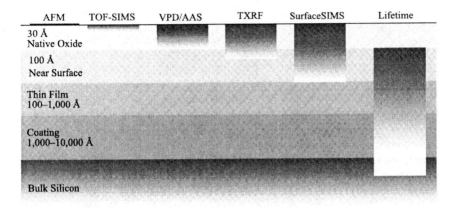

Figure 6 Sampling depths of surface analysis techniques.

Figure 6 shows the analysis depth of AFM, TOF-SIMS, VPD-AAS, TXRF, and SurfaceSIMS. The sampling depth of TXRF is typically 3–5 nm, while SurfaceSIMS can depth profile to 20–50 nm. TXRF and SurfaceSIMS measurements almost always agree to within a factor of 2 when carefully prepared control samples are analyzed, and indicate that no overall relative calibration errors exist between the techniques [1]. However, discrepancies can occur from a number of physical causes that are related to their sampling depths. For example, SurfaceSIMS analysis of an ion implanted silicon wafer can given higher concentrations for tungsten than TXRF. This is because the tungsten has been energetically driven into the wafer during the ion implantation process (i.e., coimplanted with the primary species of interest) beyond the sampling depth of TXRF. The TXRF measurement provides only a partial sampling of the total tungsten present at the near surface of the wafer.

IV. VAPOR PHASE DECOMPOSITION

Vapor phase decomposition (VPD) is a wet chemistry process used specifically to collect molecular contaminants in native oxide films on silicon. Figure 7 shows the steps involved in the process. In principle, the VPD process preconcentrates the surface impurities on the wafer surface using a HF droplet. The first step is to decompose the native oxide on the silicon wafer using HF (hydrofluoric acid) vapor. This is typically carried out in a Lucite box. The decomposition step leaves the surface in a highly chemophilic state. A HF droplet, typically 100 μl in volume, is then pipetted onto the surface and scanned in various directions and patterns to gather the contaminants from the entire wafer surface.

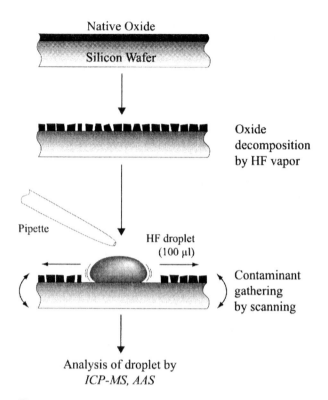

Figure 7 Sample preparation steps of vapor phase decomposition (VPD).

The chemophilic condition of the surface ensures the droplet adheres to the wafer surface during the scanning process.

After VPD sample preparation, a trace element analytical technique is required to analyze and quantify the contaminants in the droplet. Common analytical techniques are AAS and ICPMS. Alternatively, the droplet may be left to dry and the residue analyzed by TXRF. The droplet residue must be localized in a well-defined area that is smaller than the TXRF detector aperture. An improvement in TXRF detection sensitivity is observed that corresponds to the ratio of wafer surface and TXRF detection areas. A serious drawback of VPD-TXRF is the thickness of the residue, proportional to the concentration of the collected impurities, because it can result in adsorption of the x-ray and lead to inaccurate quantification. A description of VPD-TXRF is given by Neumann and Eichinger [2].

Although VPD has many virtues, there are several caveats to the procedure. Complete scanning of the droplet on the wafer surface is difficult to confirm.

Automated scanning systems exist and typically consist of the systematic movement of the scanning collection solution droplet across the entire wafer surface. For manual scanning, the analyst must be skillful in order to achieve proper collection and complete wafer surface coverage. With either scanning method inherent problems exist. They include sample preparation precision, total surface coverage, wafer-edge contamination, and actual sample loss. With the advent of 300 mm wafers, manual scanning of the wafer surface will become even more difficult. There are now several commercially available scanners in the marketplace offering a variety of configurations ranging from semiautomated to fully automated.

The collection efficiency is dependent on the solubility and electrochemical potential of the metallic impurity in the solvent with respect to silicon (-0.86 V). The recovery rates of selected elements are shown in Table 3. Careful recovery experiments are required for accurate quantification. Iron is a common contaminant, a component of stainless steel, introduced during processing of VLSI or ULSI devices. For the purpose of performing VPD, it has been generally assumed that the surface Fe chemistry is an oxide when Fe is deposited from a SC1 cleaning solution. With this assumption, the most appropriate chemistry to remove this surface Fe would be dilute HF. However, studies by Pirooz [3] showed that the presence of boron, for example, in p^+ wafers, probably results in surface Fe forming Fe-B pairs which can affect the VPD recovery rate of surface Fe. The results of the Pirooz study are shown in Table 4. The important part of the results, for our discussion here, is the poor recovery of surface Fe by the last HF step on the p^+ wafer surface compared to the p^- wafer surface, a factor $8\times$ difference. SurfaceSIMS would have obtained a more correct surface Fe value with a detection limit superior to TXRF.

Table 3 Estimated Recovery of Trace Metals from Bare Silicon Wafers Analyzed by VPD-ICPMS

Element	Recovery (%)
Al	98
Cr	98
Fe	110
Mo	87
Na	104
Ti	95
V	98
W	78
Zn	104

Table 4   Surface Iron Concentration Levels Measured by TXRF Remaining on p^- and p^+ Silicon Wafers After Cleaning Process Steps (in 10^{10} atoms/cm^2) [3]

	SC1	SC1 + H_2SO_4 + HNO_3	SC1 + H_2SO_4 + HNO_3 + HF
p^- wafer	60	30	10
p^+ wafer	180	320	80

Another difficulty associated with VPD is the possibility of inadvertently adding contamination from the HF droplet itself. The VPD process requires a HF droplet to be deposited onto the wafer surface; the droplet may be considered a "contaminant" and must be of ultrahigh purity so as not to contribute impurities to the wafer surface. Additional problems include the VPD collection, which assumes that the surface contamination is distributed homogeneously on the wafer. In reality this distribution can vary by 1 order of magnitude from the edge to the center of the wafer. Particles, if present, are also indiscriminately collected during the VPD process and will lead to incorrect results. Furthermore, VPD is limited to applications for unpatterned wafers because it is difficult to scan the droplet over a specific device region. This is consistent with the basic principle of VPD in that it requires a large collection surface area to enhance its detection sensitivity.

V.   SECONDARY ION MASS SPECTROMETRY

A.   Overview

The term SIMS encompasses two broad categories of analysis differentiated primarily by the amount of the analyzed sample that is removed during the analysis measurement:

1. *Dynamic SIMS* (high-current density sputtering: many atomic layers of the sample are consumed)
2. *Static SIMS* (low-current density sputtering: less than a monolayer of sample consumption)

Figure 8 introduces common terminology used to describe different aspects of SIMS. The historical relationships that exist among the different general categories of SIMS, types of SIMS instrumentation, and specific subsets of SIMS analytical techniques are also shown.

B.   Dynamics SIMS

Dynamic SIMS is commonly used for in-depth monitoring of molecular con-taminants in very thin (≤ 100 Å) films. Applications of SIMS include mon-

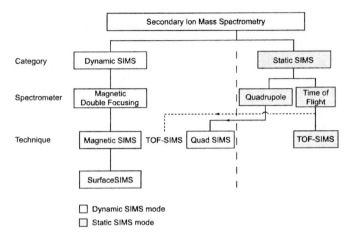

Figure 8   Secondary ion mass spectrometry family tree.

itoring Na, K, Li, Fe, Ni, Cr in SiO_2 films, and elemental and atmospheric contaminants in metal oxide and silicide thin films. Excellent general treatises on SIMS are available [4,5]. SIMS is an ion-sputtering depth-profiling technique that erodes the sample surface, typically from depths of about 5 nm to >100 μm, at rates of <1 nm/s to >40 nm/s. The two most common primary ion-bombarding species used in Dynamic SIMS are O_2^+ and Cs^+. The selection of the primary ion depends on whether the molecular contaminant to be analyzed favors the formation of positive or negative secondary ions (electropositive, EP, or electronegative, EN). Oxygen bombardment increases the yield of positive ions (e.g., Groups I–III and transition metals), while cesium bombardment increases the yield of negative ions (e.g., Groups IV–VII). In general, the selection rule is that the ion yield of positive ions is governed by the ion potential (IP) of the element, and the ion yield of negative ions is governed by the electron affinity (EA).

The most common ion mass separation systems used in Dynamic SIMS instruments are based on either double-focusing magnetic mass analyzers (Magnetic Sector SIMS) or quadrupole mass filter analyzers (Quadrupole SIMS). The strength of a magnetic sector instrument lies with its adjustable mass resolving power (M/Δm as high as 8000), good detection sensitivity due to inherently high transmission, and fast sputtering rate. Magnetic sector instruments can distinguish most interferences that may result from combinations of molecular contamination with a variety of matrix surfaces. For example, Magnetic SIMS can easily resolve ^{56}Fe from $^{28}Si_2$; both species have the same nominal mass of 56. High-mass resolution is crucial for achieving the best detection limit for iron in silicon-type materials.

With Quadrupole SIMS, mass separation is achieved with a quadrupole mass filter, comprised of four cylindrical electrodes and associated RF circuitry. Mass selection requires proper programming of the electrode DC potentials and radio frequency potential. At predetermined potentials, an ion of desired mass can pass through the system, while unwanted masses undergo an oscillating trajectory of increasing amplitude perpendicular to the major axis. Unwanted masses ultimately collide with one of the electrodes or the vacuum wall of the spectrometer. The advantage of this spectrometer for molecular contaminant characterization is the ease of independent control of the primary beam energy and incident angle. This instrument feature allows very thin films (≤ 10 nm) to be measured with minimal "knock-on" or broadening artifact from the SIMS bombardment. The necessity to use the lowest primary beam energy (≤ 500 eV) for near-surface measurement has led to the application of ULE SIMS[TM] (ultra-low-energy SIMS). In addition, the data point acquisition density with time is higher with quadrupole instruments than magnetic sector instruments. This is because of the fast peak switching times associated with the quadrupole instrument design.

C. SurfaceSIMS

SurfaceSIMS has emerged in recent years as an important application of Dynamic SIMS for process monitoring of molecular contaminants. It is frequently used to evaluate the performance of contamination removal technology such as postchemical mechanical polishing (CMP) cleaning, and to certify equipment. The strength of SurfaceSIMS lies in its excellent detection sensitivity, typically 10^9 atoms/cm^2, and depth-related information, to distinguish surface and energetic molecular contaminants. Charles Evans & Associates (California) developed the practical application of SurfaceSIMS, in an attempt to provide reproducible and accurate quantification of Na, K, Al, and Fe on silicon wafer surfaces. Continued development of new analytical procedures has led to measurement of a wide range of surface impurity elements, with detection limits of 10^8 atoms/cm^2 for specific elements [6].

SurfaceSIMS was developed to overcome the changing sputter ion yield, associated with the transient region at the near surface, during the initial sputtering process. SurfaceSIMS measurements are made using an oxygen primary ion beam on a magnetic sector mass spectrometer with the utilization of oxygen flooding. In principle, with oxygen flooding, the silicon surface is saturated with adsorbed oxygen and the secondary ions are emitted from a fully oxidized surface. The ion yields are therefore independent of the depth to which the primary oxygen ion is implanted into the sample during the initial sputtering process. As a result, accurate quantification can be made from the sample surface, which is not possible using conventional SIMS.

The effectiveness of SurfaceSIMS to overcome the low-ion yields of the surface transient region is illustrated in Figure 9. This figure shows depth profiles of aluminum on silicon acquired by conventional SIMS and SurfaceSIMS. A steady state of the surface composition with sample atoms is observed only by SurfaceSIMS. This is indicated by the constant silicon signal. The results by conventional SIMS show dramatic changes in ion yield of silicon to a depth of about 15 nm. In this so-called "transient" region, where the ion yield is changing, it is not possible to quantify the molecular contaminants that are present on the sample surface.

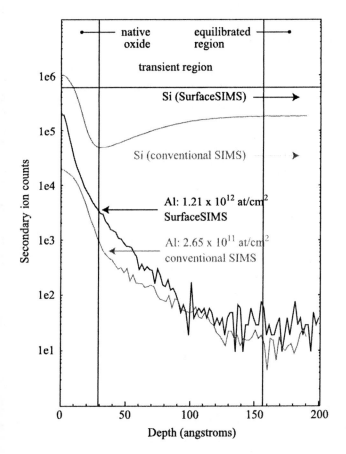

Figure 9 A comparison of SurfaceSIMS and conventional SIMS depth profiles. Oxygen flooding with SurfaceSIMS effectively eliminates the near surface transient yield variations observed by conventional SIMS.

Table 5 SurfaceSIMS Detection Limits of Selected Elements
on Silicon

Element	Detection limit (10^{10} at/cm^2)	Element	Detection limit (10^{10} at/cm^2)
Li	0.001	Ca	0.05
B	0.1	Cr	0.01
C	100	Fe	0.05
Na	0.01	Ni	1
Mg	0.1	Cu	1
Al	0.05	Mo	0.1
P	1	Ta	1
K	0.01	W	0.2

Table 5 shows the detection limits for selected elements on silicon. Surface-SIMS complements TXRF by detecting the low-Z elements such as Li, Na, K, and Al, and by providing valuable depth distribution of molecular contaminants. In-depth profiling of molecular contaminants is necessary to identify contamination that has been energetically or thermally driven into the substrate.

An ASTM spoke-wheel SIMS round robin study using silicon substrates with different surface Na, K, and Al levels was completed in 1996 [7], following the ASTM Test Method F 1617 [8]. Ten laboratories participated. SIMS results are compared to VPD-AAS determinations in Table 6. The SIMS results are mean values of the laboratories and it is interesting to note that they agree

Table 6 Summary of VPD-AAS and SIMS Results for
Round-Robin Samples (in 10^{10} atoms/cm^2)

Sample	Technique	Al	Na	K
A	VPD-AAS	3, 3	15, 10	8, 7
	SIMS	3.5	11	7.8
B	VPD-AAS	8, 7	29, 32	23, 22
	SIMS	9.4	34	24
C	VPD-AAS	25, 22	115, 121	92, 92
	SIMS	29	112	82
E (blank)	VPD-AAS	ND, ND	4, 0.6	2, 0.1
	SIMS	0.8	0.7	0.4

Note: Two measurements were made by VPD-AAS. SIMS results are an average of 7 measurements. ND = not detected.

well with the VPD-AAS results. This agreement for carefully prepared control samples is very encouraging since there are no NIST standard reference materials for surface metal contamination. ASTM Consensus Reference Materials for Na, Al, and K on silicon were used. The significance of this round robin study was threefold:

- It is clearly possible to make acceptable Na, Al, and K quantitative measurements on the surface of silicon wafers in conventional SIMS laboratories (which are most often not cleanrooms) as long as the laboratory follows the ASTM procedure.
- It is possible to make surface Al measurements down to the $10^{10}-10^{11}$ atoms/cm^2 range with both magnetic sector (high-mass resolution) and quadrupole (low-mass resolution) SIMS instruments. This hold true as long as the quadrupole instrument uses an energy filter in the secondary ion mass detection scheme to remove the C_2H_3 and $^{11}B^{16}O$. However, to detect surface Al below 10^{10} atoms/cm^2 range, a magnetic sector mass spectrometer is required.
- Although the areal density values in atoms/cm^2 are accurate by SurfaceSIMS, the actual depth profile acquired during SurfaceSIMS may not be accurate. This is because cascade mixing, impurity diffusion, and thermodynamically driven segregation affect the depth location of the contamination in the near-surface region. For example, Cu atoms segregate quickly to a SiO$_2$/Si interface. So during depth profiling by SurfaceSIMS, where an oxygen-rich altered layer is formed just beneath the analytical surface, Cu will segregate to beneath the sputtering surface resulting in a "snow plow" effect of Cu. The depth profile therefore shows a distribution of Cu extending for many nanometers beyond where the Cu was originally present.

D. Time-of-Flight Secondary Ion Mass Spectrometry

Time-of-flight secondary ion mass spectrometry (TOF-SIMS) is the premier technique of Static SIMS. When operating in the Static SIMS mode, very little material (typically less than a monolayer) is removed from the sample surface during the course of an analysis measurement. Applications of TOF-SIMS in the semiconductor industry are currently less than for Dynamic SIMS, but are expanding rapidly.

Traditional Static SIMS in the early 1960s used quadrupole mass filter instruments. However, in the late 1980s, time-of-flight SIMS (TOF SIMS) was introduced whereby mass separation is achieved by measuring the time of flight from the sample surface to the detector. TOF-SIMS rapidly acquired the reputation as the most sensitive Static SIMS instrumental configuration. Modern

TOF-SIMS instruments have excellent sensitivity, high working mass range and mass resolution, and imaging capability. TOF-SIMS applications include finger-printing of organic contamination on surfaces, small particle characterization, and survey analysis of surface metal contamination.

The basic concept of TOF-SIMS is relatively simple. A pulsed primary ion beam bombards the sample surface, causing the emission of pulses of atomic and molecular secondary ions and neutrals. The secondary ions are then elec-trostatically accelerated into a field-free drift region with a nominal kinetic en-ergy. Since lighter ions will have higher velocities than the heavier ones, they will arrive at the detector at the end of the drift region earlier than the higher masses. The mass separation is therefore obtained by measuring the flight time from the sample surface to the detector [9]. Since different ion types arrive sequentially at the detector, the operating conditions may be chosen such that virtually all of the secondary ions produced by each primary ion pulse of a given polarity are detected and recorded in the mass spectrum. Efficient col-

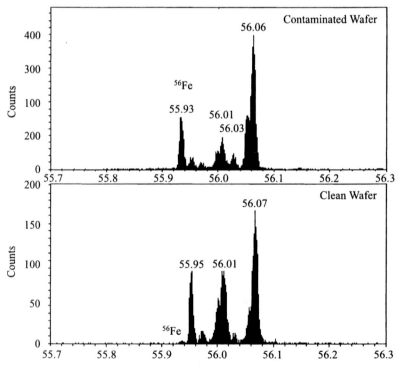

Figure 10 TOF-SIMS mass spectrum of 56 amu. The high-mass-resolution capability of TOF-SIMS clearly resolves eight mass peaks.

lection and detection of secondary ions, together with the short primary ion pulses, minimizes the ion beam-induced damage of the sample and results in a very surface sensitive technique for molecular contaminants (organic or metallic).

TOF-SIMS has several advantages for the measurement of organic contamination. It has very good sensitivity for low levels of contaminants and for low volatility or thermally unstable contaminants that can be difficult to measure by desorption techniques. The high-mass resolution capability of the mass spectrometer can provide unambiguous identification of chemical species. Figure 10 shows the mass spectrum of 56 amu (atomic mass unit). Eight mass peaks are clearly resolved with $^{28}Si_2$ and ^{56}Fe as the two most prominent peaks. The relative integrated counts of Fe, when normalized to the integrated counts of Si_2, provide a representation of the relative Fe concentration on the wafer surface [10]. It can be seen from Figure 11 that the normalized Fe surface concentration determined by TOF-SIMS correlates rather well with the TXRF measurements within the experimental efforts. Semiquantitative analysis of metal impurity concentrations on silicon wafers thus appears to be feasible under conditions where the complete native oxide layer is analyzed.

TOF-SIMS can compare the levels of contamination from different regions on the same wafer since the ions detected are coming from specific, identifiable

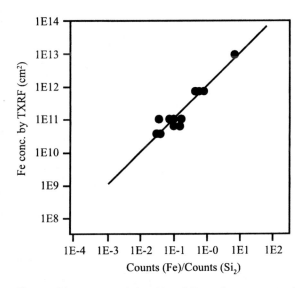

Figure 11 Linear relationship of Fe surface concentration determined by TXRF and normalized Fe intensity (Fe/Si_2) by TOF-SIMS. The approximate analysis depth for the TOF-SIMS is 5 nm.

regions of a wafer [11]. Figure 12 shows TOF-SIMS spectra of a wafer that has been stored in a wafer box at 80°C. Using a 200 mm stage, the whole wafer was analyzed at the center and edge. The results indicate traces of polyethylene glycol at the wafer edge from contact with a wafer box.

Quantification of organics is a more complicated issue, due to the very large number of diverse organic materials that may be encountered in an industrial setting. However, it has been shown that TOF-SIMS is inherently quantitative if suitable steps are taken to control the variables involved, such as substrate homogeneity, primary ion beam energy, and raster size. Recent work also points to a way for indirect quantitative analysis of nonvolatile and semivolatile residues by TOF-SIMS utilizing a SAW (surface acoustical wave) device [12]. The SAW acts as a microbalance and can quantify the amount of material that deposits on its surface. The device containing the organic contamination can then be directly analyzed by TOF-SIMS to chemically characterize the residue on the SAW.

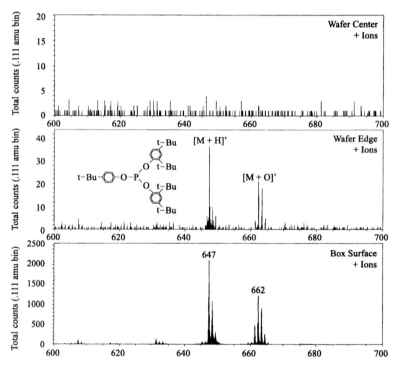

Figure 12   Whole wafer analysis by TOF-SIMS of a wafer stored at 80°C in a wafer box. Wafer edge contamination is clearly observed from contact with the wafer box.

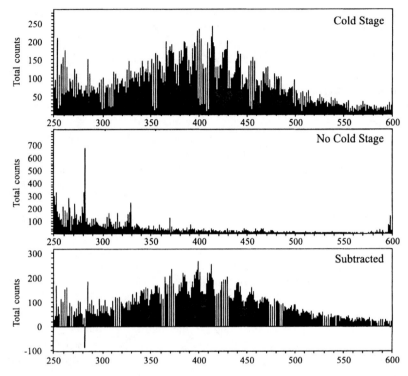

Figure 13 TOF-SIMS analysis results of a wafer surface using a cold stage and without using a cold stage.

The most important limitation of TOF-SIMS, other than limited quantification at this time, is the effect of the UHV conditions of the instrument. The standard UHV conditions of the TOF-SIMS analysis mean that more volatile compounds present on wafers are often removed from the sample before it is measured [13]. Figure 13 shows that by using a cold stage during the analysis this problem can be mitigated.

VI. TOTAL REFLECTION X-RAY FLUORESCENCE

Total reflection x-ray fluorescence (TXRF) is a well-established surface analysis technique. It is a survey technique yielding results for multiple elements in a single measurement with advantages of being an accurate, sensitive, and nondestructive measurement. TXRF can detect medium- and high-Z elements (sulfur to uranium) at very low concentration levels on silicon wafers. TXRF is sensitive to particulate, surface, or near-surface (≤ 5 nm) molecular contaminants, particularly transition metal contaminants. Routine detection limits on silicon wafers with a Technos TREX 630T (300 mm wafer handling capability), using a W rotating anode x-ray source, is approximately 10^{10} atoms/cm^2 or better. TXRF has proved to be particularly effective in the determination of molecular contaminants after cleaning processes. A comprehensive description of this technique is given in ASTM document, F 1526-94 [14]. The strengths and weaknesses of the TXRF technique are summarized in Table 7.

In the TXRF technique, a well-collimated x-ray beam impinges upon an optically flat substrate at an angle less than the critical angle for total external reflection. Typical operating angles, measured with respect to the surface, are between 0.08 and 0.11 degree. The incident beam penetrates the silicon wafer approximately 3–5 nm, causing the trace impurities present to fluoresce with a characteristic energy. A Si(Li) detector obtains the spectra of these fluorescent x-rays. The detection area is large relative to other techniques discussed here, typically 10 mm, which improves sensitivity at some sacrifice of lateral resolution. The surface concentrations of the impurities may subsequently be quantified using the appropriate calibration. Quantification for TXRF is straightforward, as long as the distribution of contamination is known. The fluorescence signal from impurity atoms depends on whether the impurity is present as a thin film on the sample surface, a thick residue, embedded in the matrix, or a subsurface layer. Particles ≥ 100 nm in diameter on the sample surface respond as a residue. Incorrect choice of the quantification factor can cause an error of up to 5× in quantified data.

Precision of TXRF measurement varies with source power, beam intensity, and areal density of the element measured. A properly adjusted TXRF instrument can measure matrix elements with approximately 1% variation. For trace elements near the detection limit, 1σ relative standard deviation is limited by Poisson statistics and may be greater than 100%. To achieve long-term reproducibility of results, a statistical process control (SPC) sample should be measured at frequent intervals to monitor instrument stability. One sigma variation of 15–20% for measurements made over a period of months to years is expected

Table 7 Advantages and Disadvantages of TXRF

Strengths	Weaknesses
Survey analysis	Cannot detect low-Z elements (e.g., Li, Na, K, Al)
Quantitative	No depth distribution information
Nondestructive	10 mm lateral resolution
Full wafers	Require polished surface
(200 mm and 300 mm)	

for a sample at the 10^{13} atoms/cm^2 level, where Poisson statistics have little effect. At this time, there are no worldwide standards for TXRF, and each laboratory must find suitable reference materials [15]. Residues, microdroplets, spin coating, or contaminated baths are commonly used to prepare reference wafers. Calibration is usually by VPD-AAS, RBS, HIBS, and/or round-robin studies. ISO Working Group TC209/WG2 has been created to determine appropriate methods, procedures, and controls for TXRF measurements and calibration.

In 1989, Penka and Hub [16] reported how the detection limit for a particular x-ray series (e.g., K_α) typically increases with decreasing Z. For constant instrumental parameters, the interference-free detection limits vary over 2 orders of magnitude and are a function of the atomic number of the element. Table 8 shows the practical detection limits for TXRF. Often spurious peaks of Fe K_α and Ni K_α appear in TXRF spectra [17]. The intensity of the spurious peak varies with the changes in incident azimuth angle. The origin of this phenomenon is impurity specific and exists along the path of the x-rays. This phenomenon can influence the accuracy of trace level analysis of impurities. The incident azimuth must be selected so as to be off the Bragg condition of the primary beam. Best detection limits are achieved for measurement of transition metals on polished silicon wafers. Lower detection limits can be achieved with synchrotron TXRF but this is not commercially viable.

1.   Synchrotron TXRF (SR-TXRF)

TXRF utilizing a synchrotron source has the potential to lower detection limits for selected transition metals by 2 or 3 orders of magnitude. A detection limit

Table 8   TXRF Practical
Detection Limits
(in 10^{10} at/cm^2)

S	20
Cl	10
Ca	10
Ti	3
Mn	0.4
Fe	0.4
Ni	0.4
Cu	0.3
Zn	0.5
Sn	8
As	3
Au	5
Br	4

for Ni of 3×10^8 atoms/cm^2 has been reported [18]. The synchrotron source has a higher brightness than conventional sources, almost complete linear polarization, and produces an x-ray flux that is several orders of magnitude higher than that of conventional sources used in TXRF. This results in improved signal-to-background ratio. The tunability of the synchrotron source can be used to extend the analytical range and maximize excitation for a given element. Excitation of a given x-ray line or peak is possible only if the beam has a higher energy; excitation is most efficient if the energy gap between exciting and excited x-rays is small. It is possible to tune the primary beam to maximize the excited fluorescence signals. However, the consequences of the increased photon flux to the sample are that major improvement in detectors will be necessary and that our basic understanding of background-forming mechanisms needs to be increased [19]. While SR-TXRF holds great promise for achieving lower detection limits by improving signal-to-background ratio, the general unavailability of synchrotron sources may be a bar to commercial viability of the technique.

VII. HEAVY ION BACKSCATTERING SPECTROMETRY

Heavy ion backscattering spectrometry (HIBS) is an attractive technique because it is inherently quantitative and does not suffer from matrix effects [20]. In theory, HIBS should offer superior sensitivity for heavy elements compared to TXRF and avoid the quantification challenges in SurfaceSIMS. In 1995 an evaluation [21] was completed of the HIBS analytical capability at Sandia National Laboratories, New Mexico. The HIBS experiments focused on answering a series of questions. The conclusion drawn from the experimental results follow.

1. Detection Limit of HIBS

The practical HIBS detection limits are about 5×10^{10} atoms/cm^2 for a single transition metal, 2×10^{10} atoms/cm^2 for Br, 1×10^{10} atoms/cm^2 for Sn, and 3×10^9 atoms/cm^2 for heavy elements such as Pb. It should be noted that there are few wafers where it would be actually possible to reach these detection limits because of the presence of interfering elements such as Br and As.

2. HIBS Mass Resolution for Multiple Elements

The HIBS mass resolution was far worse than expected, not being able to resolve Cr, Fe, Ni, and Zn, unless the element is present at high concentration. Typically, the concentrations of all transition metals are summed and listed under one element such as Fe or Zn. Mass resolution is even worse for heavier elements, so if an unknown heavy element is detected it is always called Pb, but it could be Au, W, Ta, Pt, etc.

3. HIBS as a Calibration Method for TXRF

Excellent agreement was obtained between the HIBS results and TXRF results when the sample contained a single impurity. For example, for some samples TXRF and HIBS agreed to within 2% for Cr at 8×10^{12} atoms/cm^2. For Cr at 3×10^{11} atoms/cm^2, HIBS and TXRF differed by 20%, but it should be noted that, at 3×10^{11} atoms/cm^2 Cr, the HIBS spectrum is quite noisy and this may be the source of the difference here. HIBS and TXRF differed by 35% for Sn at 5×10^{11} atoms/cm^2. Part of this difference may be due to the spectral interferences in TXRF with Sn and K.

4. HIBS to Determine TXRF Accuracy and Sensitivity for Sn

HIBS detected Sn on several of the samples, but only one of them had sufficient Sn to be detected by TXRF. For this wafer, HIBS was 36% higher than TXRF. HIBS detected Sn at 8×10^{10} atoms/cm^2 on a wafer that showed no Sn by TXRF, so the stated detection limit of 9×10^{10} atoms/cm^2 for TXRF appears to be about right. HIBS also detected Sn at 2.5×10^{11} atoms/cm^2 on a Fe-Spin sample that is used as a reference material for SurfaceSIMS.

5. HIBS Utility for Wafer Backside Measurement

HIBS was performed on the backside of two wafers. On one wafer HIBS gave much higher W concentrations than TXRF. It is not clear if this difference is due to calibration problems (unlikely), or due to a TXRF or HIBS artifact associated with the wafer roughness. Good agreement was obtained for Br on the backside of the second wafer. Br was the only element that HIBS could detect that TXRF also detected. This would tend to support a good calibration for TXRF-Mo, and an artifact affecting the measurement of W on the first sample.

6. HIBS Capability for Measuring Near the Edge of a Wafer

HIBS was able to measure to within 2 mm of the edge of a 150 mm wafer with no artifacts being observed. When the beam was centered on the edge of the wafer (one-half of the beam being off the wafer), then transition metal signals along with Sn and Pb were detected from the metal wafer holder. In the case of a 200 mm wafer, the goniometer could only get to within 2 mm of the edge and could go no further.

7. Reproducibility of HIBS

The samples containing 3×10^{12} atoms/cm^2 Cr, 3×10^{11} atoms/cm^2 Br, and 1×10^{10} atoms/cm^2 Sn were measured on 3 days of experiment. The 3σ RSDs of 6, 58, and 43% were calculated for Cr, Br, and Sn, respectively. Raw counts for Cr, Br, and Sn were approximately 4700, 630, and 90, respectively, and the background beneath the Cr, Br, and Sn peaks were about 580, 170, and 110. This indicates good day-to-day reproducibility for the measurement, at least over a 3-day time frame.

In summary, HIBS works equally well on polished wafer surfaces, or on rough backsides. Whole wafers up to 200 mm can be measured, as can wafer pieces. The analysis area is a square 4×4 mm. Wafers can be analyzed to within 2 mm of the edge, with no adverse effect. Analysis of wafer backsides may be one area where HIBS could have a unique use.

The Sandia HIBS system cannot seriously compete with TXRF for the analysis of transition metals due to its lack of sensitivity and severe problems with mass resolution. It could be useful for experiments involving specific heavy elements such as Sn, W, and Pb, where the purpose of the experiment is focused on that element, and another technique such as TXRF is used to look for other elements. Even though HIBS is more sensitive than TXRF for heavy elements, SurfaceSIMS can also measure these heavy elements with similar sensitivity and far superior mass resolution. HIBS does not have a useful depth resolution to distinguish a molecular contaminant in the top 3 mm regime from a contaminant, say 8 mm, deeper. The lack of mass resolution for HIBS means that it cannot positively identify heavy elements that are detected, and there are often interference between ubiquitous elements, such as Br and As, and the element of interest, such as Sn. The conclusion is that HIBS (at Sandia) is not a viable survey or problem-solving technique but is useful as a calibration tool for generating standards.

VIII. ANALYTICAL ISSUES FOR SURFACE METAL CONTAMINATION

In a production cycle, yield loss is mainly due to particles and metal contamination. Reliability issues are still primarily affected by metal contamination. Notorious impurity elements from chemicals and processing that can be deleterious to silicon devices are Fe, Ni, Cu, Ca, and Al. Consequences that may arise from molecular contamination are:

- Surface metals introduced during ion implantation affect the electrical properties of the ultrashallow source drain and results in a shift in junction depth.

- Transition metal residues remaining on the surface of a wafer after chemical cleaning result in grown-in defects and affect the GOI (gate oxide integrity). Usually there is a correlation between surface metal concentration and OISF (oxidation-induced stacking faults).
- Elemental metal contamination degrades device performance. For example, excess reverse-bias currents in *pn* junctions and storage-time degradation in MOS circuits can be caused by electrically active stacking faults.
- Metals segregation into the gate oxide during growth degrades breakdown voltage of the junction, resulting in a large current flow from the drain to the substrate.

The 1997 NTRS analytical requirements for critical contaminants are $\leq 5 \times 10^9$ atoms/cm^2 (refer to Table 2). These concentration levels are typically 1 order of magnitude below the contamination limit requirements. Table 9 shows the critical surface metal densities for starting materials and surface preparation with technology generation. VPD-AAS is commonly used today for Fe, but many practitioners do not claim a 2.5×10^9 atoms/cm^2 detection capability. There is interference for VPD-ICPMS from ArN, ArO, and ArOH resulting in a detection limit of 3×10^{10} atoms/cm^2 with a 200 mm wafer. In practice, TXRF can detect 4×10^9 atoms/cm^2. TOF-SIMS can theoretically detect Fe below 2.5×10^9 atoms/cm^2 but calibration procedures have not been adequately developed. Today, SurfaceSIMS can be used for Fe down to 8×10^8 atoms/cm^2.

SurfaceSIMS and VPD-ICPMS (200 mm wafer) can detect Ni at 5×10^9 atoms/cm^2. TOF-SIMS quantification is still an open issue. The detection of Cu at 5×10^9 atoms/cm^2 is questionable for VPD-ICPMS. Presently, VPD-ICPMS can detect 1×10^{10} atoms/cm^2 with a 20 mm wafer, SurfaceSIMS can detect 5×10^9 atoms/cm^2, and TXRF can detect 3×10^9 atoms/cm^2 with the new generation detector. Subsequently, there is a technology gap for detecting Cu at 2×10^9 atoms/cm^2 and lower. Calcium is important because of its effect on gate oxide reliability. However, a Cl-containing oxidation removes the Ca and its negative effects on the gate oxide. VPD-ICPMS can detect Ca at the 5×10^9 atoms/cm^2 level. Consistent measurements at 1×10^9 atoms/cm^2 are very difficult. SurfaceSIMS can detect Ca at 5×10^8 atoms/cm^2.

Detection of surface Al down to about 4×10^9 atoms/cm^2 can be achieved by VPD-ICPMS, 150 mm wafer with a collection sample volume of 0.5 mL, operating in a single-element mode. SurfaceSIMS can detect 5×10^8 atoms/cm^2. The VPD approach provides one average number (this can be a disadvantage) for the entire wafer and may include dissolved particulates. The SurfaceSIMS has the advantage of providing lateral distribution information, avoiding the contribution from particles, and is independent of the impurity chemistry, unlike the VPD technologies.

Table 9 Surface Metal Densities for Starting Materials and Surface Preparation from the 1997 National Technology Roadmap for Semiconductors*

Year of first product shipment	1997	1999	2001	2003	2006	2009	2012
technology generation	250 nm	180 nm	150 nm	130 nm	100 nm	70 nm	50 nm
Starting materials (atoms/cm^2)							
Critical surface metals	$\leq 2.5 \times 10^{10}$	$\leq 1.3 \times 10^{10}$	$\leq 1 \times 10^{10}$	$\leq 7.5 \times 10^9$	$\leq 5 \times 10^9$	$\leq 2.5 \times 10^9$	$\leq 2.5 \times 10^9$
Analytical tools	1, 2, 3	1, 2, 3	1, 2, 3	1, 2, 3	1, 2, 3	2	2
Critical surface metals	$\leq 5 \times 10^9$	$\leq 4 \times 10^9$	$\leq 3 \times 10^9$	$\leq 2 \times 10^9$	$\leq 1 \times 10^9$	$\leq 10^9$	$\leq 10^9$
Surface preparation (atoms/cm^2)							
Analytical tools	1, 2, 3	1, 2	1, 2	1, 2	2	2	2
Other surface metals (atoms/cm^2)	$\leq 10^{11}$	$\leq 10^{11}$	$\leq 10^{11}$	$\leq 10^{11}$	$\leq 10^{11}$	$\leq 10^{11}$	$\leq 10^{11}$
Analytical tools	1, 2, 3	1, 2, 3	1, 2, 3	1, 2, 3	1, 2, 3	1, 2, 3	1, 2, 3

*Table also shows analytical tools capable of measuring metals to required detection levels.
Critical surface metals: Fe, Ca, Co, Cu, Cr, K, Na, and Ni.
Other surface metals: Al, Ti, V, and Zn.
Analytical tools: 1 = TXRF, 2 = SurfaceSIMS, 3 = VPD-ICPMS.

IX. ANALYTICAL ISSUES FOR ORGANIC CONTAMINATION

Organic contamination has not in the past been treated with the same scrutiny as metal contamination. The effect of organic contamination on device processing remains ambiguous, although some examples and recent studies of the effects are as follows:

- Significant amounts (10^{15} atoms/cm^2) of hydrocarbon can cause serious degradation of MOS (metal-oxide-semiconductor) devices grown on hydrogen passivated, HF cleaned <100> Si [22].
- Gate oxides exposed to the cleanroom ambient perform poorer than oxides without such exposure [23]. This was a startling observation since cleanrooms were supposed to be clean. Cleanrooms have been designed to remove particulate contamination, with less thought given to the outgassing properties of the materials used to construct the rooms. Wall materials, window materials, floor and ceiling materials, and air filters all have organic components. Even delivery systems for liquids and gases, bottles, cleanroom garments, and especially the plastic boxes used to hold wafers are all potential sources of volatile contaminants.
- Degradation of gate oxide integrity [24] has been detected by electrical measurements when wafers were intentionally contaminated with the antioxidant butylated hydroxytoluene (BHT) in the presence of a nitrogen atmosphere.
- Room temperature oxide growth rate of HF-etched (hydrophobic) Si wafers is retarded by the presence of surface organics [25].
- Unbonded areas or bubbles at the interface have been observed during Si wafer bonding [26]. The cause was attributed to hydrocarbon contamination originating in storage in plastic boxes and from exposure to controlled amounts or organic vapors.

Published studies to date suggest the form of the organic is not as important as the total carbon atoms/cm^2. The analytical needs to detect carbon at 10^{13} atoms/cm^2 can be met today using, for example, XPS well below the NTRS target concentration of 10^{14} atoms/cm^2. In selecting the analytical technique for organic contamination measurement, it is important to note the differences between the outgassing products and what is ultimately deposited on wafers. At the same time, identifying a contaminant as simply containing certain elements is not sufficient, because specific compounds or classes of compounds will adhere to surfaces preferentially and will produce different effects depending on subsequent processing. In addition, more specific information can be valuable for tracing contaminants to their sources in order to reduce or eliminate them.

There are a number of analytical techniques available ranging from traditional wet chemical analysis to advanced surface analysis. The most commonly
used techniques to evaluate organic contamination on semiconductor surfaces are

- GS-MS
- IMS
- TOF-SIMS
- XPS
- FTIR

These techniques can be classified into two groups. The first, covering
GC-MS and IMS, includes methods that detect contaminants after desorption
from a surface. These techniques take advantage of the volatility of the organic
by measuring the material outgassing from a surface under controlled conditions.
The desorption is typically induced thermally. IMS has the advantage of very
high sensitivity that makes it possible to detect the small amounts of material that
can affect device performance. A practical advantage of GC-MS is its presence
in many laboratories.

In the second group, which includes TOF-SIMS, XPS, and FTIR, the
surface of the material is analyzed directly. The advantage of this approach is a
higher sensitivity to less volatile polar or high molecular weight contaminants.
This can also be important when reactions occur on a surface, for example,
when salts form from volatile organic amines and anions such as sulfates. A
disadvantage of direct analysis is that there is no separation step, so spectra
obtained from the wafer contain data from all of the compounds detected.

X. CONCLUSION

In this chapter we reviewed the capabilities of several analytical techniques to
characterize molecular contaminants. The techniques were surface sensitive with
extremely low detection capability. We discussed the important parameters that
need to be taken into consideration when making a decision on which technique
to use. Individual analytical techniques were described to provide a better understanding of the technique, before entering into a more in-depth discussion
of analytical issues relating to metal and organic contamination measurements.
Beyond these factors, availability and cost are also important to consider when
selecting a technique for a particular application.

In general, no single analytical technique can provide the answer to a contamination problem. Each situation will dictate the choice of technique that is
most appropriate. For the analysis of metal contamination, the analytical techniques are well established and entrenched in product yield improvement pro-

grams. In the analysis of organic contamination, the choice can be complicated because the effects of different contaminants are only beginning to be understood.

REFERENCES

1. SP Smith and J Metz. Materials Res Soc Proc 447, pp. 305–310, 1997.
2. C Neumann and P Eichinger. Spectrochimica Acta 46B, p. 1369, 1991.
3. S Pirooz, LW Shive, and DI Golland. First International Symposium on Ultra Clean Processing of Silicon Surfaces, (UCPSS-92), Leuven, Belgium, 1992.
4. A Benninghoven, FG Rudenauer, and HW Werner. SIMS-Basic Concepts, Instrument Aspects, Applications and Trends. John Wiley, 1987.
5. VKF Chia. "Secondary Ion Mass Spectrometry," In: Materials and Process Characterization of Ion Implantation (M. I. Current and C. B. Yarling, eds.), Ion Beam Press, 1997, p. 163.
6. SP Smith, L Wang, JE Erickson, and VKF Chia. Mat Res Symp Proc 386, p. 157, 1995.
7. RS Hockett and AC Diebold. Cleaning technology in semiconductor device manufacturing (R. E. Novak and J. Ruzyllo, eds.), ECS Proceedings, vol. 95-20, 1996, pp. 500–507.
8. ASTM Document F 1617-98. Test method for measuring surface sodium, aluminum, potassium, and iron on silicon and epi substrates by secondary ion mass spectrometry, 10.05, 1998.
9. BW Schueler. Microsc Microanal Microstruct 3, p. 119, 1995.
10. B Schueler. Microcontamination '93 Conference Proceedings. pp. 783–791, 1993.
11. P Lindley, F Radicati, I Mowat, L McCaig, and M Kendall. Proceedings of secondary ion mass spectrometry (SIMS XI). John Wiley, 1998, pp. 175–178.
12. Private communication with P. M. Lindley and G. Strossman, Charles Evans & Associates, Redwood City, CA 94063.
13. G Strossman, P Lindley, and W Bowers. Proceedings of secondary ion mass spectrometry (SIMS XI). John Wiley, 1998, pp. 699–702.
14. ASTM Document F 1526-94. Standard test method for measuring surface metal contamination on silicon wafers by total reflection x-ray fluorescence spectroscopy. 1.6, 1994.
15. RS Hockett. Analytical Sciences 11, p. 511, 1995.
16. V Penka and W Hub. Analytical Chem, vol. 333, pp. 586–589, 1989.
17. Y Kenji, O Shinji, Y Atsushi, and H Jimpei. Analytical Sciences, 11, p. 1025, June 1995.
18. SS Laderman, A Fischer-Colbrie, A Shimazaki, K Miyazaki, S Brennan, N Takaura, P Pianetta, and JB Kortright. Advances in x-ray fluorescence analysis, vol. 26, 1995, pp. 91–106.
19. RS Hockett. Advances in x-ray analysis, vol. 38, 1995, pp. 687–690.
20. JA Knapp, JC Banks, and DK Brice. MRS proceedings, vol. 354, pp. 389–398, 1995.

21. S Baumann and RS Hockett. Presented at the MRS Symposium on the Science and Technology of Semiconductor Surface Preparation, San Francisco, March 31 to April 4, 1997.

22. SR Kasi, M Liehr, PA Thirty, and M Offenberg. Appl Phys Lett 59, p. 108, 1991.

23. T Iwamoto, T Miyake, and T Ohmi. Proceedings of SPIE, vol. 2875, 1996, pp. 207–215.

24. K Saga and T Hattori. Appl Phys Lett 71, p. 3670, 1997.

25. A Licciardello, O Puglisi, and S Pignataro. Appl Phys Lett 48, p. 41, 1986.

26. K Mitani, V Lehmann, R Stengl, D Feifoo, UM Goesele, and H Massoud. Jpn J Appl Phys, 4, p. 615, 1991.

6

Transport and Deposition of Aerosol Particles

Anthony S. Geller and Daniel J. Rader
Sandia National Laboratories, Albuquerque, New Mexico

This chapter reviews the theoretical models available to describe particle transport in typical semiconductor processing environments. Recent or classic references have been provided wherever possible for additional background. In all of the discussion that follows, particle concentrations are assumed to be low enough so that the influence of the particle on fluid transport can be neglected; particle-particle interactions are also neglected. Under this assumption, the fluid and thermal fields are calculated first (in the absence of particles), and then used as input for subsequent particle transport calculations. The theoretical underpinnings for both the Lagrangian approach (where individual particle trajectories are calculated) and the Eulerian approach (where the particle concentration field is modeled as a continuum) are presented. The strength of the Lagrangian formulation is in predicting particle transport resulting from external forces including particle inertia; but the current implementation cannot describe the chaotic effect of particle Brownian motion (i.e., particle diffusion) on particle transport. On the other hand, the Eulerian formulation can describe particle transport resulting from applied forces and particle diffusion, but the current implementation cannot account for particle inertia.

This chapter begins with a discussion of noncontinuum effects which play a key role in the transport of small particles at low pressure. Next follows a description of the Lagrangian particle transport equation, which basically describes how a particle responds when external forces are applied to it. Brief summaries of some of these forces (which are likely to be important inside a semiconductor process tool) are also included. The concept of particle relaxation time is then introduced, which leads to a discussion of particle deposition velocity. The chapter includes a description of the Eulerian particle transport equations, for predicting diffusion contributions to particle deposition and concludes with a detailed analysis of particle transport in a parallel plate reactor as an illustration of the utility of these models.

I. NONCONTINUUM CONSIDERATIONS

In the following discussion, frequent mention will be made of the continuum and free molecular regimes. These terms are used here to distinguish between the two limiting cases characterizing the nature of the particle/gas interaction.[1] In the continuum limit (large particles or high gas pressures), the gas surrounding the particle appears as a continuous fluid and traditional continuum fluid dynamics apply—such as the Navier-Stokes equations for fluid motion. In the free molecular limit (small particles or low gas pressures), however, the discrete nature of the gas becomes important and individual molecule/particle collisions must be considered. Discrimination between these two regimes is made by comparing the particle diameter to the gas mean free path (which is defined as the average distance a molecule travels between collisions with other gas molecules); a dimensionless parameter known as the Knudsen number is commonly used for these comparisons:

$$\text{Kn} = \frac{2\lambda}{d_p} \tag{1}$$

where

λ = gas mean free path (cm) = $\mu/(\phi\rho\bar{c})$
d_p = particle diameter (cm)
μ = gas viscosity (g cm^{-1} s^{-1})
ρ = gas density (g/cm^3) = PM/RT (for ideal gas)
$\bar{c}$ = mean thermal velocity of the gas molecules (cm/s) = $(8RT/\pi M)^{1/2}$
R = universal gas constant (8.31451 · 10^7 g cm^2 s^{-2} K^{-1} mol^{-1})
M = gas molecular weight (g/mol)
P = gas pressure (dyne/cm^2 or g cm^{-1} s^{-2}) (1 atm = 760 torr = 1.01325 · 10^6 dyne/cm^2)
T = gas temperature (K)

---

[1]The flowfield entraining the particle can also be either molecular or continuum in nature. In this case, discrimination between the two flow regimes is made by comparing some characteristic length associated with the reactor geometry to the gas mean free path. For example, a Knudsen number for the flowfield could be defined by replacing the particle diameter in Eq. (1) with a characteristic reactor length scale. For small particles, it is frequently the case that the flow regime can be considered continuum while the fluid-particle interaction is characterized as free-molecular.

Also, ϕ is a dimensionless parameter that depends on the kinetic-theory model used to define the gas mean free path: in this work the value $\phi = 0.491$ has been adopted [1]. At atmospheric pressures, the mean free path is typically less than 0.1 μm. Gas mean free path is inversely proportional to pressure at constant temperature; for example, the mean free path in air is 0.674 μm at 76 torr, 6.74 μm at 7.6 torr, and 67.4 μm at 760 mtorr at 296 K. Thus, for low-pressure applications, the Knudsen number for submicron particles can be large.

A large Knudsen number (say, >10) corresponds to the free-molecular regime, while a small Knudsen number (say, <0.1) corresponds to the continuum regime. Typically, verified theoretical expressions are available in the literature for the forces acting on particles in both the continuum and free-molecular limits. Unfortunately, theoretical force expressions are difficult to formulate in the transition regime that lies between the continuum and free-molcular regimes (particle size of the order of the mean free path, Kn = 1). Instead, interpolating or correlating functions are used which go to both the continuum and free-molecular expressions in the limit and match experimental data (if available) in between.

II. LAGRANGIAN PARTICLE EQUATION OF MOTION

In the Lagrangian approach to particle transport, particle Brownian motion is neglected and individual particle trajectories (position and velocity as a function of time) are determined by integrating the following system of ordinary differential vector equations:

$$\frac{d\mathbf{x}_p}{dt} = \mathbf{V}_p \tag{2}$$

$$m_p \frac{d\mathbf{V}_p}{dt} = \mathbf{F}_D + \mathbf{F}_G + \mathbf{F}_T + \mathbf{F}_E + \sum \mathbf{F}_i \tag{3}$$

where $\mathbf{x}_p$ is the particle position vector, $\mathbf{V}_p$ is the particle velocity vector, m_p is the particle mass, and $\mathbf{F}_D$, $\mathbf{F}_G$, $\mathbf{F}_T$, $\mathbf{F}_E$, and $\mathbf{F}_i$ are the fluid-drag, gravitational, thermophoretic, electric, and any additional forces (diffusiophoresis, wall drag, etc.) acting on the particle. The forces explicitly listed in Eq. (3) are those of greatest interest in analyzing parallel-plate tools (Sec. VI); additional forces can be added linearly as needed. A review of these forces follows. A brief review of the forces used in this report follows (for greater detail see Refs. 2–4). Because of the low pressures (typically less than 100 torr) and small particle sizes (diameters typically less than 1 μm) of interest in semiconductor processing, particles are usually sufficiently smaller than the gas mean free path, λ, so that the free-molecular limit of the various force expressions apply. The free-molecule assumption is well justified for particle Knudsen numbers (Kn = $2\lambda/d_p$) larger than 10, and is acceptable (accurate within about 20%) for Knudsen numbers as small as 2. For Knudsen numbers less than 2, force expressions that extend into the transition or continuum regime should be used.

A. Fluid-Particle Drag Force

A particle moving at a velocity different from the surrounding gas will experience a gas resistance or fluid-drag force. A great deal of research has been devoted to describing fluid drag; only a brief review of this body of literature is reported here. For a rigid sphere of diameter d_p moving at constant velocity $\mathbf{V}_p$ through a fluid with local velocity $\mathbf{U}$ and viscosity μ, the drag force is given by (e.g., see Friedlander [5, p. 105])

$$\mathbf{F}_D = -\frac{3\pi \mu d_p}{C(\mathrm{Kn})}(\mathbf{V}_p - \mathbf{U}) \cdot C_D(\mathrm{Re}_p)\frac{\mathrm{Re}_p}{24} \tag{4}$$

where $C_D(\mathrm{Re}_p) \cdot \mathrm{Re}_p/24$ is a non-Stokesian correction for fluid inertial effects and $C(\mathrm{Kn})$ is a slip correction factor for noncontinuum effects. A fairly simple correlating equation for the non-Stokesian correction has been suggested by Turton and Levenspiel [6]:

$$C_D(\mathrm{Re}_p)\frac{\mathrm{Re}_p}{24} = 1 + 0.173\mathrm{Re}_p^{0.657} + \frac{0.01721 \cdot \mathrm{Re}_p}{1 + 16,300 \cdot \mathrm{Re}_p^{-1.09}} \tag{5}$$

where the particle Reynolds number is given as

$$\mathrm{Re}_p = \frac{\rho d_p |\mathbf{V}_p - \mathbf{U}|}{\mu} \tag{6}$$

and

$$|\mathbf{V}_p - \mathbf{U}| = [(u_p - u)^2 + (v_p - v)^2 + (w_p - w)^2]^{1/2} \tag{7}$$

and u, v, and w are the x, y, and z components of velocity for the fluid (no subscript) and the particle (subscript p). This correlation applies for particle Reynolds number up to about 200,000. Note that only moderate particle Reynolds numbers (say 10 at most) are expected for most semiconductor process applications, so that the far right terms in Eq. (5) are small compared to unity. In fact, for submicron particles in typical processing environments, the slip-corrected Stokes drag law [Eq. 4] with $C_D(\mathrm{Re}_p) \cdot \mathrm{Re}_p/24 = 1$ can generally be used with negligible error.

It has already been explained that, for small particles or at low gas pressures (large value of Kn) the continuum approximation eventually breaks down and the molecular nature of the gas must be considered. Cunningham [7] was the first to propose that a correction factor (called the Cunningham or slip correction factor), $C(Kn)$, be included in Eq. (4) to account for noncontinuum effects. The functional form first suggested by Knudsen and Weber [8] is in common use:

$$C(Kn) = 1 + Kn \left[\alpha + \beta \cdot \exp\left(-\frac{\gamma}{Kn}\right) \right] \tag{8}$$

where α, β, and γ are parameters that depend on the nature of the gas-particle interaction at the particle surface, and so are affected by both gas composition and particle surface roughness. For example, Ishida [9] used a Millikan apparatus to determine the coefficient α for oil-drops in nine common gases. His results were recently reevaluated by Rader [10] using modern, more accurate values for the electric charge and gas properties; the corrected values for α are given in Table 1.

Based on theoretical considerations [11] and experimental data [12], Rader [10] recommends that the expression $\alpha + \beta = 1.647$ can be used to accurately calculate β for most gas and particle-surface combinations; β values calculated by this formula using Ishida's measured values for α are given in Table 1. The choice of the third constant, γ, is more difficult as there are no theoretical techniques for estimating it, and as complete data for empirically determining it are limited. Rader [10] fit the available data for air, carbon dioxide, and helium and found γ values of 0.78, 0.92, and 2.0, respectively. For other gases, Rader [10] suggests $\gamma = 0.85$. Note that the correlation for the slip correction factor given by Eq. (8) is not very sensitive to the value of γ used. In fact, only small errors typically result if the slip factor is calculated using the fitted constants for oil-droplets in air ($\alpha = 1.207$, $\beta = 0.440$, and $\gamma = 0.78$) for different gases or

Table 1 Gas Properties. Density, Viscosity, Molecular Weight, Mean Free Path, and Slip Correction Constants (for oil drops) for Nine Common Gases at $T_0 = 296.15$ K and $P_0 = 760$ torr

Gases	μ_0 (poise)	M (g/mol)	ρ_0 (g/cm^3)	λ_0 (μm)	α (–)	β (–)	γ (–)
Air	183.47(-6)	28.966	1.192(-3)	0.0674	1.207	0.440	0.78
Ar	224.80(-6)	39.948	1.645(-3)	0.0703	1.227	0.420	—
He	197.11(-6)	4.003	0.165(-3)	0.1943	1.277	0.370	2.00
H$_2$	88.61(-6)	2.016	0.826(-3)	0.1240	1.141	0.506	—
CH$_4$	110.75(-6)	16.043	0.661(-3)	0.0545	1.154	0.493	—
C$_2$H$_6$	93.37(-6)	30.069	1.251(-3)	0.0333	1.254	0.393	—
i-C$_4$H$_{10}$	75.06(-6)	58.123	2.406(-3)	0.0193	1.186	0.461	—
N$_2$O	147.88(-6)	44.013	1.818(-3)	0.0493	1.207	0.440	—
CO$_2$	148.12(-6)	44.010	1.823(-3)	0.0438	1.150	0.497	0.92

Source: Ref. 10.

for particles of different surface roughness. A plot of the $C(Kn)$ against Kn is given in Fig. 1 using these constants. Note that $C(Kn)$ approaches unity in the continuum limit, and approaches $(\alpha + \beta)Kn$ in the free molecule limit.

Fluid-Drag Force: Assumptions and Practical Considerations. All of the above formulas assume solid, homogeneous, spherical particles, while real particles may be far from spherical in shape (flakes, rods, deformed droplets, etc.) and may be porous or inhomogeneous in composition. These variations from ideality can result in particle rotation, modifications to the drag law, etc. Two methods are commonly used to account for nonsphericity: the use of an equivalent diameter or the correction of the drag law with a dynamic shape factor. Hinds [2] and Fuchs [13] provide discussion of both of these approaches. Techniques are becoming available for more accurately modeling the transport of certain classes of nonspherical particles (including flakes, rods, chains of spheres, etc.), but these greatly increase the complexity of the problem.

In adopting Eq. (4), resistance terms resulting from fluid inertia (e.g., the virtual mass and Basset history integral terms) have been neglected, which has been shown [13,14] to be a reasonable approximation for aerosol particles (where the particle density is much larger than the fluid density) at particle Reynolds numbers not exceeding a few hundred. The uncertainty in Eq. (4) becomes greater at higher Reynolds numbers, but high Reynolds numbers are not expected in semiconductor applications.

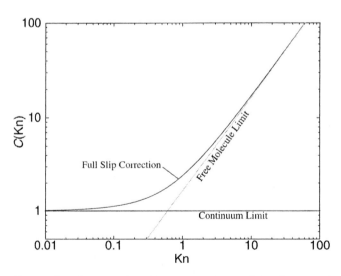

Figure 1 Slip correction factor. Dependence of the slip correction factor on particle Knudsen number (solid line). The free molecule limit for the slip correction factor is also shown (dotted line).

Equation (4) strictly applies to the uniform (nonaccelerating), straight-line motion of a sphere in a quiescent fluid. Often, however, the drag on an accelerating particle in a nonuniform flow field is needed. Fuchs [13, Chap. 3] and Clift et al. [14, Chap. 11] review the issues related to the nonuniform rectilinear motion of aerosol particles. Although it will introduce some inaccuracy, the instantaneous drag acting on an accelerating particle can be estimated with the above constant-velocity drag expression with the fluid and particle velocities taken as their local, instantaneous velocities.

Strictly speaking, the combination of the slip $C(\text{Kn})$ and non-Stokesian $[C_D(\text{Re}_p) \cdot \text{Re}_p/24]$ corrections in Eq. (4) as a product is on flimsy grounds. Henderson [15], for example, presents a correlation that adds the corrections.[2] Unfortunately, there is little or no data available for $\text{Re} \sim 1$ where slip is appreciable, so the proper formulation can't be decided. For typical semiconductor process environments, however, particle Reynolds numbers are likely to be low so that the non-Stokesian correction is near 1 and Eq. (4) is likely to be quite satisfactory.[3] Thus, in the rest of this work, the slip-corrected Stokes drag law [Eq. (4) with $C_D(\text{Re}_p) \cdot \text{Re}_p/24 = 1$] will be used.

Continuum Regime Limit. In the continuum limit (large particles and/or near-atmospheric process pressures) and assuming non-Stokesian effects can be neglected, Eq. (4) for the gas resistance reduces to Stokes law (e.g., see Hinds [2], p. 41):

$$\mathbf{F}_{D,\text{continuum}} = -3\pi\mu d_p(\mathbf{V}_p - \mathbf{U}) \tag{9}$$

The continuum regime drag force is seen to be directly proportional to particle size and to the velocity difference between the particle and the gas; it is independent of process pressure (the dependence of fluid viscosity on pressure is very weak) and depends on temperature only through the temperature dependance of viscosity.

Free Molecular Regime Limit. In the free-molecular limit (small particles and/or low process pressures) and assuming non-Stokesian effects can be neglected, Eq. (4) for the gas resistance reduces to the following free-molecular result (similar to that originally derived by Epstein [16]) in the limit of large Kn:

$$\mathbf{F}_{D,\text{molecular}} = -\frac{3\pi\phi}{2(\alpha + \beta)}\rho\bar{c}d_p^2(\mathbf{V}_p - \mathbf{U}) \tag{10}$$

---

[2] C is removed from Eq. (4) and the expression $1/C$ replaces "1" as the first term on the right-hand side of Eq. (5).

[3] In practice, non-Stokesian effects can be neglected when the particle Reynolds number is less than about 0.3.

As in the continuum limit, free-molecular drag force is directly proportional to the gas-particle velocity difference. Unlike the continuum result, however, free-molecular drag shows a much stronger (squared) dependence on particle diameter. Another difference is that free-molecular drag is directly proportional to process pressure (through the fluid density term). Temperature dependencies arise implicitly through the fluid density and the mean gas velocity.

B. Gravitational Force

The gravitational force acting on a spherical particle is given by Hinds [2, p. 42]:

$$\mathbf{F}_G = \frac{\pi}{6} d_p^3 (\rho_p - \rho) \mathbf{g} \tag{11}$$

where ρ_p and ρ are the particle and gas density, respectively, and $\mathbf{g}$ is the gravitational acceleration vector. The gas density is typically much smaller than the particle density, and can be neglected in Eq. (11). The gravitational force is independent of the gas mean free path.

C. Thermophoretic Force

Because of the thermophoretic force, particles suspended in a gas with a temperature gradient will migrate in the direction opposite to the gradient, i.e., away from hot regions and toward cold regions. This phenomena results in the preferential deposition of particles on a cold wall, and explains the appearance of a particle free zone near a hot wall. A formulation for the thermophoretic force on a spherical particle was developed by Talbot et al. [17] in a review of thermophoresis:

$$\mathbf{F}_T = -3\pi \mu d_p \nu K_T \cdot \frac{\nabla T}{T} \tag{12}$$

where

$$K_T = \frac{2C_s [k_g/k_p + C_t \cdot \mathrm{Kn}]}{[1 + 3C_m \cdot \mathrm{Kn}][1 + 2 \cdot k_g/k_p + 2C_t \cdot \mathrm{Kn}]} \tag{13}$$

and where ∇T is the temperature gradient in the gas, T is the mean gas temperature about the particle, $\nu = \mu/\rho$, k_g and k_p are the gas and particle thermal conductivities,[4] and C_t, C_s, C_m are the thermal creep coefficient, temperature jump coefficient, and velocity jump coefficient, respectively ($C_s = 1.147$, $C_t = 2.20$, and $C_m = 1.146$ are recommended by Batchelor and Shen [19]). Talbot et al. [17] compared their correlation with other experimenters' data over a

---

[4]For polyatomic gases, Talbot et al. [17] recommend the use of the "translational" thermal conductivity, which is given by simple kinetic theory as $k_g = \frac{15}{4} \mu R/M$.

wide range of Knudsen numbers, and found it agrees with available experimental data to within 20%.

Continuum Regime Limit. In the continuum limit of small Kn the thermophoretic force approaches:

$$\mathbf{F}_{T,\text{continuum}} = -6\pi\mu^2 d_p \cdot \frac{C_s k_g}{k_p + 2k_g} \cdot \frac{\nabla T}{\rho T} \tag{14}$$

which depends on temperature, pressure, gas and particle thermal conductivities, and is proportional to particle diameter.

Free Molecule Regime Limit. In this limit Eq. (12) approaches the limit first derived by Waldmann and Schmitt [19]:

$$\mathbf{F}_{T,\text{molecular}} = -\frac{\pi}{2}\phi\mu\bar{c}d_p^2 \cdot \frac{\nabla T}{T} \tag{15}$$

Note that in this limit the thermophoretic force is proportional to diameter squared and is independent of the particle thermal conductivity.[5]

D. Electrostatic Force

A charged particle suspended in a region with an electric field E will experience a force (Hinds [2, p. 286]):

$$\mathbf{F}_E = n_p e E = q E \tag{16}$$

where n_p is the number of elementary charge units e (4.803×10^{-10} esu in cgs) on the particle giving a total charge q. This equation is deceptively simple, in that the determination of either the charge on the particle or the surrounding electric field can be exceedingly challenging (experimentally or theoretically). In a plasma tool, for example, n_p and E will change with time and position and may not be independent.

III. INERTIAL EFFECTS

Using the particle fluid-drag force relationship [Eq. (4) with $C_D(\text{Re}_p) \cdot \text{Re}_p/24 = 1$], the particle force balance of Eq. (3) can be rewritten:

$$\tau\frac{d\mathbf{V}_p}{dt} = \mathbf{U} - \mathbf{V}_p + \frac{C(\text{Kn})}{3\pi\mu d_p}\left(\mathbf{F}_G + \mathbf{F}_T + \mathbf{F}_E + \sum\mathbf{F}_i\right)$$

$$= \mathbf{U} - \mathbf{V}_p + \mathbf{V}_G^t + \mathbf{V}_T^t + \mathbf{V}_E^t + \sum\mathbf{V}_i^t \tag{17}$$

---

[5]The gas conductivity does play a role, but has been eliminated from Eq. (15) by the expression in footnote 4.

where we have introduced a particle response or relaxation time

$$\tau = \frac{\rho_p d_p^2 C(\mathrm{Kn})}{18\mu} \tag{18}$$

and the particle deposition velocity V_i^t which is discussed below in Section IV. Particles characterized by small relaxation times respond rapidly to changes in the flow or in the applied forces, while particles with large relaxation times respond slowly to such changes. For large relaxation times, particle transport is dominated by particle inertia.

A. Nondimensionalization

It is often convenient to nondimensionalize Eqs. (2) and (17). For this purpose, a characteristic length and velocity must be specified. For example, for parallel-plate geometry (see Fig. 5), we could chose the interplate separation S as the characteristic length, and the magnitude of the mean face velocity U_0, as the characteristic velocity, for which Eqs. (2) and (3) become:

$$\frac{d\tilde{\mathbf{x}}_p}{d\tilde{t}} = \tilde{\mathbf{V}}_p \tag{19}$$

$$\mathrm{St}\frac{d\tilde{\mathbf{V}}_p}{d\tilde{t}} = \tilde{\mathbf{U}} - \tilde{\mathbf{V}}_p + (\tilde{\mathbf{V}}_G^t + \tilde{\mathbf{V}}_T^t + \tilde{\mathbf{V}}_E^t + \sum \tilde{\mathbf{V}}_i^t \tag{20}$$

where $\tilde{\mathbf{x}}_p = \mathbf{x}_p/S$, $\tilde{t} = tU_0/S$, $\tilde{\mathbf{V}}_p = \mathbf{V}_p/U_0$, $\tilde{\mathbf{U}} = \mathbf{U}/U_0$, and $\tilde{\mathbf{V}}_i^t = \mathbf{V}_i^t/U_0$ where $\mathbf{V}_i^t$ is the particle deposition velocity described below. We have also introduced the particle Stokes number, St, a dimensionless quantity defined as

$$\mathrm{St} = \frac{\tau U_0}{S} = \frac{\rho_p d_p^2 C(\mathrm{Kn})U_0}{18\mu S} \tag{21}$$

where τ is the particle relaxation time. The Stokes number is a convenient measure of the importance of particle inertia in a specific reactor; for small St, inertial effects can be neglected, while for large St inertial effects must be considered. Physically, the Stokes number can be interpreted as the ratio of the particle stopping distance[6] to the characteristic length of system. The continuum and free molecule limits of St are as follows.

Continuum Regime Limit.

$$\mathrm{St}_{\mathrm{continuum}} = \frac{\rho_p d_p^2 U_0}{18\mu S} \tag{22}$$

---

[6]The particle stopping distance, τU_0, is defined as the distance a particle would travel before stopping if injected into a quiescent fluid at an initial velocity of U_0.

Free Molecule Regime Limit.

$$\text{St}_{\text{molecular}} = \frac{\alpha + \beta}{9\phi} \cdot \frac{\rho_p d_p U_0}{\rho \bar{c} S} = 0.373 \cdot \frac{\rho_p d_p U_0}{P S} \cdot \left(\frac{\pi RT}{8M}\right)^{1/2} \qquad (23)$$

IV. DEPOSITION VELOCITY

The manipulation of Eq. (3) into Eq. (17) resulted in the derivation of the particle deposition velocity, $\mathbf{V}_i^t$, which is the particle velocity at which the ith force (neglecting all others) acting on the particle exactly balances the fluid drag force retarding the motion. At this balance point, the particle acceleration vanishes and the particle would move at a steady (or terminal) velocity—the deposition velocity. The time required for the particle to reach the deposition velocity is short for particles characterized by small particle response times (small Stokes numbers). For small Stokes numbers, particle inertia can be neglected and the acceleration term on the left-hand side of Eq. (17) can be dropped, so that particle velocity can be expressed as

$$\mathbf{V}_p = \mathbf{U} + \mathbf{V}_p^t = \mathbf{U} + \mathbf{V}_G^t + \mathbf{V}_T^t + \mathbf{V}_E^t + \sum \mathbf{V}_i^t \qquad (24)$$

where the net particle deposition velocity is obtained by summing over all external forces, that is,

$$\mathbf{V}_p^t = \mathbf{V}_G^t + \mathbf{V}_T^t + \mathbf{V}_E^t + \sum \mathbf{V}_i^t = \frac{C(\text{Kn})}{3\pi \mu d_p} \left(\mathbf{F}_G + \mathbf{F}_T + \mathbf{F}_E + \sum \mathbf{F}_i\right) \qquad (25)$$

Equations (24) and (25) show that, neglecting inertia, the particle will move with the fluid velocity plus the vector sum of the individual deposition velocities from all of the applied external forces. In the absence of any external forces (and having neglected particle diffusion), a noninertial particle will exactly follow flow streamlines. For isothermal flow between horizontal, parallel plates, all the $\mathbf{V}_i^t$ are constant and are directed normal to the plates. Expressions are given below for the deposition velocities for each of the forces listed in Section II.

A. Gravitational Deposition Velocity

The deposition velocity for a spherical particle settling under gravity can be found from Eqs. (4) and (11) with buoyancy and non-Stokesian effects neglected:

$$\mathbf{V}_G^t = \frac{\rho_p d_p^2 \mathbf{g} C(\text{Kn})}{18\mu} \qquad (26)$$

Although the form of the gravitational force is the same in both the continuum and free molecule regime limits, expressions for the two deposition velocity limits differ due to the differing drag law contribution:

Continuum Regime Limit.

$$\mathbf{V}^t_{G,\text{continuum}} = \frac{\rho_p d_p^2 \mathbf{g}}{18\mu} \tag{27}$$

Free Molecule Regime Limit.

$$\mathbf{V}^t_{G,\text{molecular}} = \frac{\alpha + \beta}{9\phi} \cdot \frac{\rho_p d_p \mathbf{g}}{\rho \bar{c}} = 0.373 \cdot \frac{\rho_p d_p \mathbf{g}}{P} \cdot \left(\frac{\pi RT}{8M}\right)^{1/2} \tag{28}$$

Note that the settling speed of a particle in the free molecule limit varies directly as particle diameter and inversely with pressure, while in the continuum limit the settling speed varies as diameter squared and is independent of pressure.

Example. The gravitational deposition velocity for a unit density spherical particle as a function of particle size is shown in Fig. 2 for six different process pressures in argon at 293 K. For pressures below 100 torr and particle diameters below 1 μm, note that the lines are parallel and straight with a slope

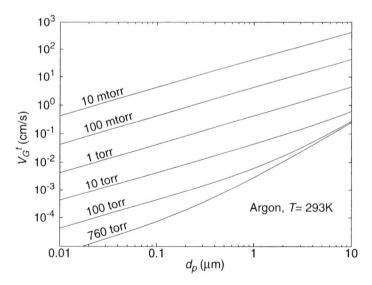

Figure 2 Gravitational deposition velocity. Dependence of the gravitational deposition velocity on particle diameter for six process pressures (argon at 293 K, particle density $\rho_p = 1$ g/cm^3).

of one, as predicted by the free molecule regime limit. Thus, for most of the pressures and particle sizes of interest in semiconductor processing, the free molecule regime limit, Eq. (28), can be used to predict particle settling rates, greatly simplifying calculations.

B. Thermophoretic Deposition Velocity

The thermophoretic deposition velocity resulting from the balance of thermophoretic and drag forces (neglecting non-Stokesian effects) alone can be found by equating Eqs. (4) and (12):

$$\mathbf{V}_T^t = -K_T C(\mathrm{Kn}) \frac{\mu \ \nabla T}{\rho T} = -K_T C(\mathrm{Kn}) \frac{\mu R \ \nabla T}{P M} \tag{29}$$

Interestingly, the deposition velocity given by Eq. (29) depends on particle diameter only implicitly through the slip correction factor and K_T; specifically, in both the large and small particle limits the thermophoretic velocity becomes independent of particle size.

Continuum Regime Limit.

$$\mathbf{V}_{T,\mathrm{continuum}}^t = \frac{2C_s k_g}{k_p + 2k_g} \frac{\mu \ \nabla T}{\rho T} = -\frac{2C_s k_g}{k_p + 2k_g} \frac{\mu R \ \nabla T}{P M} \tag{30}$$

The continuum limit for thermophoretic deposition velocity does not depend on particle size.

Free Molecule Regime Limit. In the free-molecular limit, the thermophoretic deposition velocity reduces to

$$\mathbf{V}_{T,\mathrm{molecular}}^t = -\frac{C_s(\alpha + \beta) \ \mu \ \nabla T}{3C_m \ \ \rho T} = -\frac{C_s(\alpha + \beta) \ \mu R \ \nabla T}{3C_m \ \ P M}$$

$$= -0.549 \frac{\mu R \ \nabla T}{P M} \tag{31}$$

which is the same result as given by Waldmann and Schmitt [19]. The free molecule limit is independent of particle size and gas/particle thermal conductivities. The numerical constant 0.549 in the final equality is very general as the sum $\alpha + \beta$ is nearly constant for most gases and particle surfaces.

Example. The thermophoretic deposition velocity for a spherical particle as a function of particle size and pressure is shown in Fig. 3 for an assumed temperature gradient of $\nabla T = 1$ K/cm. For the calculations, the ratio of gas to particle thermal conductivities was taken as $k_g/k_p = 0.001$, representative of a metal particle suspended in argon (the thermophoretic deposition velocity is not very sensitive to this ratio). Even for this small temperature gradient,

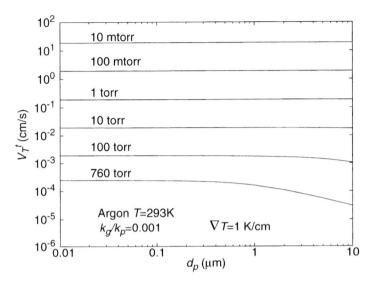

Figure 3 Thermophoretic deposition velocity. Dependence of the thermophoretic deposition velocity on particle diameter for six process pressures (argon at 293 K, $k_g/k_p = 0.001$, $\nabla T = 1$ K/cm).

the deposition velocity can become large at low pressures. For pressures below 100 torr and particle diameters below 1 μm, note that the deposition velocity becomes independent of particle diameter and inversely proportional to process pressure, as predicted for the free molecule regime limit. Thus, for most of the pressures and particle sizes of interest in semiconductor processing, the free molecule regime limit, Eq. (31), can be used to predict particle thermophoretic deposition rates. Because of the simple relationship of deposition velocity to pressure and ∇T, calculating the deposition velocity for other pressures and temperature gradients is easily done in the free molecule regime limit.

C. Electric Deposition Velocity

The deposition velocity for a spherical particle moving under an applied electric field is given by

$$\mathbf{V}_E^t = \frac{C(\mathrm{Kn})}{3\pi\mu d_p} \cdot q E \tag{32}$$

Although the electrical force is independent of process conditions such as temperature and pressure, these quantities enter the expression for deposition velocity through the drag-law contribution.

V. EULERIAN FORMULATION

While the Lagrangian (particle tracking) method predicts particle transport by considering single particle motion, the Eulerian formulation predicts particle transport by viewing the particle concentration field as a continuum. In this case, the solution of the particle transport problem becomes very much like that posed by the flow field; i.e., there is one continuous equation for particle mass (concentration) conservation and one continuous equation for particle momentum (velocity) conservation for each particle size. Particle transport by diffusion (Brownian motion) is naturally included in this formulation. A great simplification is obtained if particle inertia is neglected; i.e., it is assumed that the particle instantaneously reaches the deposition velocity where drag and imposed forces are in balance.[7] In this case, the particle momentum equation is no longer needed, and only the particle continuity equation for particle concentration n (particles/cm^3) remains [13,20]:

$$\frac{\partial n}{\partial t} + \mathbf{U} \cdot \nabla n = \nabla \cdot \mathcal{D} \nabla n - \nabla \cdot (\mathbf{V}_p^t n) + \Lambda \tag{33}$$

where $\mathcal{D}$ (cm^2/s) is the Stokes-Einstein particle diffusion coefficient (discussed below) and Λ (# cm^{-3} s^{-1}) is a particle source/sink term to account for particle generation/consumption. The net deposition velocity, $\mathbf{V}_p^t$, appearing in Eq. (33) is the same one discussed previously. Although we will not make further use of the fact, it is interesting to note that Eq. (33) is applicable to either the Brownian motion of an individual particle or to the diffusion of a particle cloud taken from the continuum point of view [13, p. 191]. For a single particle, n is interpreted as the probability of finding a particle at position (x, y, z) at time t given that the particle was initially located at position (x_0, y_0, z_0) at time t_0. Thus, although semiconductor applications are likely characterized by very low particle concentration levels, the continuum approach can still be applied if we continue to associate the particle concentration with a probability distribution (for example, we may find particle concentrations less than 1 cm^{-3}, which is acceptable from a probabilistic point of view). For boundary conditions, it is assumed that particles which contact the wall stick and are thus instantly removed from the gas, so that the concentration n equals zero at all walls.

Only one-way coupling between the fluid flow and particle concentration fields is used in this work; i.e., the flow field is coupled to particle transport

---

[7]The requirement that the characteristic time for particle diffusion is much longer than the particle relaxation time, $t \gg \tau$, is also essential to the development of the basic equations of particle transport by Brownian motion (see Fuchs [13], Sec. 35).

through the velocity field U which appears in Eq. (33), while the influence of the particle phase upon the flow is neglected. In practice, the flow field is calculated first (in the absence of a particle phase) and the resulting velocity field is supplied to Eq. (33) as a known solution.

A. Particle Diffusion Coefficient

A more complete discussion of the Stokes-Einstein particle diffusion and its derivation is available in any aerosol text (e.g., Hinds [2, chap. 5]). The diffusion coefficient for a spherical particle is

$$\mathcal{D} = \frac{kTC(\text{Kn})}{3\pi \mu d_p} \tag{34}$$

where $C(\text{Kn})$ is defined by Eq. (8). The validity of Eq. (34) rests on several assumptions: 1) The particles move independently of one another and 2) the movements of a particle in consecutive time intervals are independent [2]. The latter assumption is met only if the condition $t \gg \tau$ holds true; in other words, the expression for the diffusion coefficient given in Eq. (34) is only valid for observation times much longer than the particle relaxation time.

Continuum Regime Limit. The continuum regime limit for the diffusion coefficient is

$$\mathcal{D}_{\text{continuum}} = \frac{kT}{3\pi \mu d_p} \tag{35}$$

which is inversely proportional to particle diameter and independent of pressure.

Free Molecule Regime Limit. In the free-molecule limit, the diffusion coefficient reduces to

$$\mathcal{D}_{\text{molecular}} = \frac{\alpha + \beta}{3\phi} \left(\frac{RT}{2\pi M} \right)^{1/2} \frac{kT}{Pd_p^2} \tag{36}$$

which is inversely proportional to particle diameter squared and pressure.

Example. The particle diffusion coefficient for a spherical particle as a function of particle size and pressure is shown in Fig. 4 for a temperature of 293 K. For pressures below 100 torr and particle diameters below 1 μm, note that the lines are parallel and straight with a slope of negative 2, as predicted for the free molecule regime limit. Thus, for most of the pressures and particle sizes of interest in semiconductor processing, the free molecule regime limit, Eq. (36), can be used to calculate the particle diffusion coefficient.

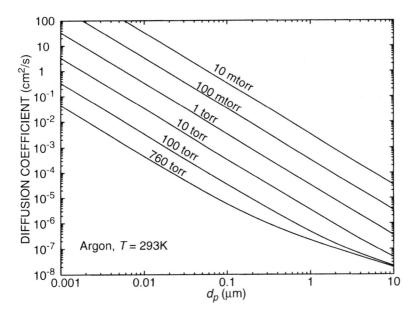

Figure 4 Diffusion coefficient. Dependence of the particle diffusion coefficient on particle diameter for six process pressures in argon at 293 K.

B. Nondimensional Formulation

For generality, Eq. (33) can be nondimensionalized by choosing a characteristic length (taken here as S, the distance between the showerhead and wafer, Fig. 5), velocity (taken here as U_0, the mean inlet velocity of the flow at the showerhead, Fig. 5), and concentration ($n_0 = \Lambda_h/U_0$, the trap source strength divided by inlet velocity. For steady state, and assuming a constant diffusion coefficient, Eq. (33) can be written [20]

$$\tilde{U} \cdot \nabla \tilde{n} = \frac{1}{\mathrm{Pe}} \nabla^2 \tilde{n} - \nabla \cdot (\tilde{V}'_p \tilde{n}) + 1 \tag{37}$$

where the Peclet number (the ratio of convective to diffusive transport) is defined as

$$\mathrm{Pe} = \frac{SU_0}{\mathcal{D}} \tag{38}$$

The *continuum regime limit* for the Peclet number is

$$\mathrm{Pe}_{\mathrm{continuum}} = \frac{3\pi \mu d_p SU_0}{kT} \tag{39}$$

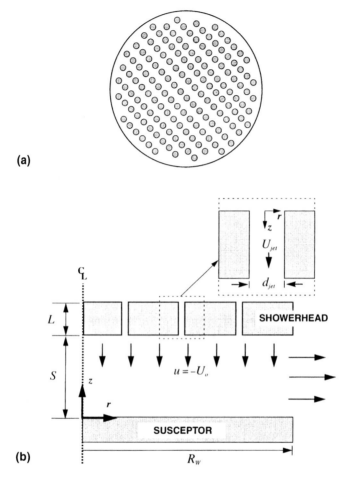

Figure 5 Parallel-plate reactor geometry. Schematic of the reactor geometry assumed in this work: (a) top view of a showerhead and (b) side view of a parallel-plate reactor.

which is proportional to particle diameter. Thus, in the continuum limit, the Peclet number can be used as a dimensionless particle diameter.

In the *free-molecule limit*:

$$\mathrm{Pe}_{\mathrm{molecular}} = \frac{3\phi}{\alpha + \beta} \left(\frac{2\pi M}{RT} \right)^{1/2} \frac{P d_p^2 S U_0}{kT} \tag{40}$$

which is proportional to particle diameter squared and to pressure. In the free-molecular limit, the square root of the Peclet number can be used as a dimensionless particle diameter.

In standard problems of species mass transfer, the Peclet number would be sufficient to completely characterize the problem for a given geometry and flow field. For particles, however, the presence of a deposition velocity term [the second term on the right-hand side of Eq. (37)] means that the Peclet number no longer uniquely specifies the solution and a dimensionless deposition velocity ratio must also be considered:

$$\tilde{\mathbf{V}}_p^t = \frac{\mathbf{V}_p^t}{\mathbf{U}_0} \tag{41}$$

VI. PARTICLE TRANSPORT AND DEPOSITION IN A PARALLEL-PLATE REACTOR

To illustrate an application of the particle transport models, this section analyzes particle transport in an enclosed, parallel-plate reactor geometry characteristic of a wide range of single-wafer process tools. The axisymmetric geometry we consider consists of uniform flow exiting a showerhead separated by a small gap from a parallel susceptor, as shown in Figure 5. The wafer would rest on the susceptor, but for the present analysis the wafer is assumed to be thin enough to be ignored. The showerhead consists of a material (usually a metal or ceramic) through which a large number of holes are drilled (see Fig. 5a). As one major function of the showerhead is to evenly distribute the flow across its face, the holes are usually made very small in diameter and are very numerous (hundreds to thousands for an 8 in. wafer process tool). Ideally, a showerhead would produce a flow characterized by a mean axial (or face) velocity that does not vary in the radial direction; such flow uniformity is needed to accomplish uniform deposition or etching of the wafer surface. In practice, however, commercial showerheads are typically designed empirically to improve process parameters (such as uniformity); the resulting showerhead designs often create nonuniform flow fields which compensate for other system deficiencies—such as radial temperature or reactive species gradients. Various flow fields can be obtained by manipulation of showerhead hole sizes, numbers, and positions.

One common feature of showerhead design is that the area available to the flow is constricted inside the showerhead; consequently, the velocity of the gas inside the holes of the showerhead is much larger than the face velocity in the gap below. Particles originating upstream of the showerhead and suspended in the flow can be dramatically accelerated while passing through the showerhead, so that at the exit of the showerhead particle velocities much larger than the fluid face velocity are possible. Depending on conditions, particle acceleration by the showerhead can lead to inertia-enhanced particle deposition on the wafer below [21]. Thus, a complete description of particle deposition on a wafer in a parallel-plate reactor must include a description of particle transport through the showerhead as well as an analysis of particle transport in the interplate region.

No attempt is made here to analyze particle generation mechanisms; for the present discussion, particles are assumed to originate either 1) upstream of the showerhead with a known concentration or 2) from a specified position between the plates with a fixed number or at a known generation rate.

The determination of particle transport in a reactor must always begin with a determination of the fluid flow and temperature fields. Particle concentrations are assumed to be low enough to allow a dilute approximation, for which the coupling between the fluid and particle phases is one-way. The fluid/thermal transport equations can be solved either analytically or numerically neglecting the particle phase. The resulting velocity and temperature fields are then used as input for the particle transport calculations.[8] In all of the present work isothermal flow is assumed, although small temperature differences are allowed to drive particle thermophoresis. Both analytic and numerical solutions of the flow field are presented.

To provide a single parameter that can be used to compare particle deposition among many cases, a particle collection efficiency is defined as the fraction of particles that deposit on the wafer. Particles are presumed to either enter the reactor through the showerhead (uniformly spread between $r = 0$ and R_W), or to originate in a plane parallel to the wafer. The latter case would correspond to particles being released from a plasma trap upon plasma extinction; in this case the particles are initially assumed to be uniformly spread radially between $r = 0$ and R_W at some distance h from the wafer. Analytic expressions for collection efficiency are presented for the limiting case where external forces control deposition (i.e., neglecting particle diffusion and inertia).

Particle transport is predicted using both a Lagrangian approach (where individual particle trajectories are calculated) and an Eulerian approach (where the particles are modeled as a continuum phase). The strength of the Eulerian formulation is in predicting particle transport resulting from the combination of applied external forces (including the fluid drag force) and the chaotic effect of particle Brownian motion (i.e., particle diffusion), although the current implementation cannot account for particle inertia. In particular, the Eulerian formulation cannot accommodate particle acceleration effects within the showerhead, and is therefore restricted to particle transport in the interplate region. The Eulerian formulation yields an analytic description of particle deposition for the case where the flow field between the plates can be approximated analytically with a creeping-flow assumption and where the particles are assumed to originate from a planar trap located between the plates. The Lagrangian for-

---

[8]The dilute mixture approximation is certainly valid for simulations of commercial semiconductor process tools, as the particle concentrations are typically controlled to very low levels.

mulation can account for inertia-enhanced deposition resulting from particles which originate upstream of the showerhead and which are accelerated while passing through it. The problem is treated in two steps: 1) within a showerhead hole and 2) between the showerhead and susceptor.

A. Fluid Transport Equations

In both of the domains considered (flow within the showerhead and between two parallel plates), the geometry will be axisymmetric. With constant fluid properties, and incompressible, laminar, steady flow, the governing equations for axisymmetric flow are the conservation of mass:

$$\frac{\partial u}{\partial z} + \frac{1}{r}\frac{\partial}{\partial r}(rv) = 0 \qquad (42)$$

and conservation of momentum:

$$\rho\left(\frac{\partial v}{\partial t} + v\frac{\partial v}{\partial r} + u\frac{\partial v}{\partial z}\right) = -\frac{\partial P}{\partial r} + \mu\left(\frac{\partial^2 v}{\partial r^2} + \frac{1}{r}\frac{\partial v}{\partial r} + \frac{\partial^2 v}{\partial z^2} - \frac{v}{r^2}\right)$$

$$\rho\left(\frac{\partial u}{\partial t} + v\frac{\partial u}{\partial r} + u\frac{\partial u}{\partial z}\right) = -\frac{\partial P}{\partial z} + \mu\left(\frac{\partial^2 u}{\partial r^2} + \frac{1}{r}\frac{\partial u}{\partial r} + \frac{\partial^2 u}{\partial z^2}\right)$$

$$(43)$$

where u and v are the axial and radial components of the fluid velocity, P is the pressure, ρ is the fluid density, and μ is the fluid viscosity [22, p. 85].

Boundary conditions are needed to complete the problem specification. In all of the following, no-slip (zero velocity) conditions are taken at all solid walls (i.e., on the showerhead, susceptor, and walls of the showerhead holes), zero radial velocity is assumed along the centerline, and zero traction is assumed at all outflows. For the two flow domains, specific boundary conditions and methods for solving the governing equations are discussed in greater detail below. Note that for convenience, a different coordinate system is used for describing the flow through a showerhead hole than in the region between the parallel plates (see Fig. 5b).

The generality of the present results are improved if the fluid equations are solved in nondimensional form. Because there are two domains of interest, there are two choices for a characteristic length and velocity. For flow in the showerhead holes, the hole diameter d_{jet} and the magnitude of the mean velocity $\overline{U}_{jet}$ are used as the characteristic length and velocity, for which the tube Reynolds number is defined as $\text{Re}_{jet} = \rho\overline{U}_{jet}d_{jet}/\mu$. For the flow between two plates, the interplate separation S and the magnitude of the mean face velocity U_0 are the appropriate choices, and the interplate flow is then characterized by a separate Reynolds number: $\text{Re} = \rho U_0 S/\mu$.

1. Flow Field in the Showerhead Holes

The idealized geometry and the coordinate system for fluid and particle transport in the showerhead holes is shown in the insert in Figure 5b. As seen, z is taken as increasing in the direction of flow (toward the wafer), so that all fluid and particle axial velocities considered in this part of the solution are positive.

The flow in the showerhead holes is assumed laminar with parallel streamlines, thereby neglecting any axial variations in velocity. For laminar flow in a tube with a uniform inlet velocity, however, it is well known that a fully developed parabolic velocity profile develops over an entrance length given approximately by $0.04\,d_{jet}\,Re_{jet}$ [5]. For many showerheads this entrance length is much less than the hole length (showerhead thickness) and so may be safely neglected; for thin showerheads, however, this may not be the case. In the present analysis, we consider two limiting velocity profiles that meet the above assumptions: 1) plug flow (constant velocity profile) and 2) fully developed laminar flow (parabolic velocity profile). For laminar flow, the velocity profile anywhere along the hole will fall somewhere between these two limiting cases. In case 1, the velocity is constant throughout the tube and, for incompressible flow, is equal to the mean velocity in the hole, $\overline{U}_{jet} = 4Q/(N_{jet}\pi d_{jet}^2)$, where Q is the total gas volumetric flow rate through the showerhead and N_{jet} is the number of individual holes in the showerhead. By mass conservation it can be shown that the ratio of the mean axial velocity in a hole and the face velocity is the ratio of the showerhead area to the total hole area:

$$\frac{\overline{U}_{jet}}{U_0} = \frac{A_{showerhead}}{\Sigma A_{jet}} = \frac{D_W^2}{N_{jet}d_{jet}^2} \tag{44}$$

For case 2, fully developed laminar flow in a tube is given by

$$U_{jet}(r) = 2\overline{U}_{jet}\left[1 - \left(\frac{r}{a_{jet}}\right)^2\right] \tag{45}$$

where r is the radial distance from the tube centerline and a_{jet} is the radius of the hole. The maximum velocity for parabolic flow is twice the mean velocity and occurs on the centerline.

2. Fluid Transport Between Parallel Plates

Both analytic and numerical techniques have been used to calculate the fluid flow between the showerhead and susceptor. For both methods we assume axisymmetric, incompressible, constant property, laminar, steady flow between two parallel plates (Fig. 5b). The flow enters through the showerhead ($z = S$) and is assumed to spread immediately, so that the inlet boundary condition is assumed to be a uniform axial velocity, $-U_0$, with zero radial velocity. Both the radial and axial components of velocity vanish at the lower plate ($z = 0$). The analytic approach assumes that the plates are infinite in the radial direction and that the Reynolds number is small; the numerical technique is used for finite plates and is valid for higher Reynolds numbers. The numerical technique is relied on here to define the conditions over which the simpler analytic solution is valid.

The assumption of constant-property flow requires further comment. Although the numerical methods used here can solve for coupled fluid-thermal transport, we have not used this capability in the present analysis. For cases where temperature differences are large or where more accurate solutions are needed for a specific application, the reader is advised to solve the coupled fluid-temperature problem. A modest temperature gradient is allowed to drive particle thermophoresis (the magnitude of all particle deposition velocities will be calculated using fluid properties evaluated at the susceptor temperature).

Under the above assumptions, the flow between the plates is entirely determined by the geometry and the Reynolds number Re $= \rho U_0 S/\mu$. For the range of conditions encountered in semiconductor reactor processes, the associated Reynolds numbers are typically less than 1, and seldom greater than 10. For small Reynolds numbers, viscous effects dominate fluid inertial effects and the "creeping flow" or Stokes flow regime is encountered; it is this regime that allows an analytic solution. The numerical method used for calculating the fluid velocity field was the commercial fluid dynamics analysis code FIDAP (Version 7, Fluid Dynamics International, Evanston, Illinois). The numerical technique was used to calculate flow fields for Reynolds numbers up to 8.

With fixed values for plate separation ($S = 1$), mean inlet velocity ($U_0 = 1$), and fluid viscosity ($\mu = 1$), the fluid density ρ was varied to obtain flow field solutions for Reynolds numbers between 1 and 8. The FIDAP option of solving the Stokes flow equations (Re $= 0$) was also used. The results of these calculations are given in Fig. 6, which shows axial and radial velocity profiles at $r = 1$ for Re $= 0, 1, 2, 4,$ and 8 as a function of the dimensionless axial coordinate z/S. Note that all velocities have been normalized by the magnitude of the inlet velocity U_0, and that the radial velocity is also normalized by radius. For Re $= 0$, the radial velocity profile is found to be parabolic and symmetric around $z/S = 0.5$. As the Reynolds number increases, the symmetry vanishes and the maximum in the radial velocity moves closer to the plate ($z/S = 0$). Variations in the axial and radial velocity profiles are seen to be quite small for Reynolds numbers less than 2. As in previous work in a similar geometry [23], it was found that the flow was quasi-1-D: the axial velocity is independent of radius, while the radial velocity is found to scale with radius such that v/r is independent of radius.

An analytic simplification is gained if the flow between the plates can be approximated as a quasi-1-D stagnation point flow. Terrill and Cornish [24] give an asymptotic solution to the problem of axisymmetric, laminar, incom-

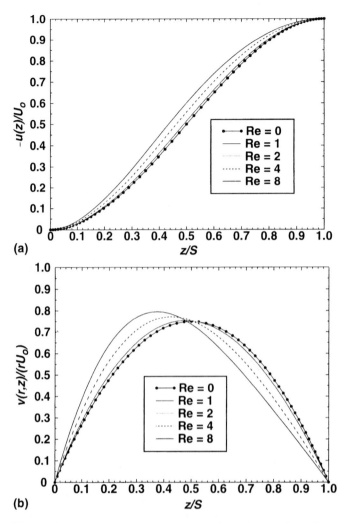

Figure 6 Flow field results for various Reynolds numbers, Re. Axial (a) and radial (b) velocity profiles for Re = 0, 1, 2, 4, and 8 calculated on a refined (30 elements) mesh. ($r/S = 1$ for all curves.)

pressible, constant property and steady flow between two coaxial infinite parallel disks with constant injection across the disks (a uniform gas inlet velocity across the showerhead). Under these assumptions, a similarity solution reduces the 3-D Navier-Stokes equations to a system of ordinary differential equations; for low Reynolds numbers, these equations can be solved with a power series in Reynolds number [24]. The first two terms of their asymptotic expansion (translated into the present problem definition) are

$$\tilde{u}(\tilde{z}) = \frac{u(\tilde{z})}{U_0} = 2\tilde{z}^3 - 3\tilde{z}^2 + \frac{\text{Re}}{70}(2\tilde{z}^7 - 7\tilde{z}^6 + 18\tilde{z}^3 - 13\tilde{z}^2)$$

$$\tilde{v}(\tilde{r}, \tilde{z}) = \frac{v(\tilde{r}, \tilde{z})}{U_0} = \tilde{r}\left[3\tilde{z} - 3\tilde{z}^2 - \frac{\text{Re}}{70}(7\tilde{z}^6 - 21\tilde{z}^5 + 27\tilde{z}^2 - 13\tilde{z})\right]$$

(46)

where $\tilde{z} = z/S$ and $\tilde{r} = r/S$. These two equations exactly satisfy all boundary conditions. The quasi-1-D nature of the result is clearly seen as the axial velocity is independent of radius, while the radial velocity scales linearly with radius.

In the limit of vanishingly small Reynolds number, Eq. (46) reduces to a symmetric, parabolic profile for radial velocity, in excellent agreement with the Stokes flow solution (Re = 0) obtained by FIDAP (this limit has also been previously reported [25].

Equation (46) does a very good job of approximating the axial velocity profile—agreeing with FIDAP solutions to better than 1% for Reynolds numbers less than 4, and to better than about 4% for Reynolds numbers up to 8. The success of Eq. (46) in predicting radial velocity is not nearly so good. Although the error is better than 1% for Re < 1, the maximum observed error quickly grows, reaching 15% for Re = 4 and 70% at Re = 8. As can be seen, the largest errors are found near the showerhead ($z/S = 1$), where the magnitude of the radial velocity is quite small. Thus, although the relative error is quite large, the absolute error is small. In any case, our treatment of the region near the showerhead is only approximate because we have neglected the effect of the discrete jets issuing from the showerhead. Strictly from a fluid velocity point of view, Eq. (46) provides a very good approximation of the flow for Reynolds numbers less than 2, and a reasonable approximation up to a Reynolds number of 4.

3. Summary: Fluid Flow Analysis for the Parallel-Plate Geometry

This section has defined the parallel-plate geometry which will be used to approximate the flow inside a showerhead-type etch or CVD reactor. The acceleration of the gas flow as it passes through the showerhead will later be found to play a key role in enhancing particle deposition by particle inertia; for this reason, solutions for flow within the showerhead holes have been presented. The two limiting cases which were considered, plug and fully developed parabolic flow, should bracket the range of flows likely to be encountered in semiconductor applications. Laminar, incompressible, constant-property flow between two infinite, parallel plates was used to approximate the interplate flow in real reactors which are certainly more complicated. Reynolds number and edge effects were discussed, and an analytic solution was found that should provide a fairly accurate description of the flow for Reynolds numbers less than about 4 when the plate separation is much smaller than the overall system radial dimension. Variable temperature effects have not been considered.

B. Particle Collection Efficiency

Particle collection efficiency is defined as the fraction of particles present in the interplate region that deposit on the wafer. The collection efficiency is introduced to provide a single parameter that can be used to compare particle transport and deposition results among many cases. Note that the use of collection efficiency side-steps the important issue of the particle source term. Thus, while the present transport analysis addresses the question of the *fraction* of gas-borne particles that deposit on the wafer, a prediction of the *number* of particles that deposit on the wafer additionally requires a clear understanding of the controlling particle generation mechanisms. In practical terms, the present analysis helps identify strategies for reducing the probability that particles are transported to and deposit on a wafer; a complete strategy for reduction of total particle-on-wafer counts also requires that particle source terms be understood and controlled.

Three particle-source scenarios are considered: 1) a continuous source of particles entering the interplate region through the showerhead with known concentration (such as for contaminated process gases), 2) a discrete number of particles that are originally trapped between the plates (such as by a plasma) but are subsequently released (such as at plasma extinction), and 3) a continuous source of particles which are created between the plates at a known generation rate (such as by particle nucleation). General collection efficiency expressions for these cases are defined below. In addition, analytic expressions are provided for the limiting case where external forces control particle deposition—i.e., both particle inertia and Brownian motion are neglected. In the absence of particle inertia and Brownian motion, Robinson [26] has shown that particle concentration is constant along particle trajectories if 1) the flow is incompressible and 2) the external forces acting on the particle are all divergence free. For the infinite parallel-plate geometry with constant-property flow, the flow is clearly incompressible and the second condition is met for the gravitational and Coulombic electric particle forces (which are each constant between the plates).

1. Particles Entering Through the Showerhead

For this case, collection efficiency is defined as the fraction of particles entering the interplate region through the showerhead (between $r = 0$ and R_W) that deposit on the wafer. These particles are presumed to originate upstream of the showerhead and are assumed to be evenly distributed across the showerhead. Particle acceleration through the showerhead will be considered in Section VI.C.1. The present Lagrangian formulation accounts for the coupling between particle inertia and external forces in determining particle transport in the interplate region. The calculation of a collection efficiency with a Lagrangian technique requires the determination of the critical radius, R_{crit}, which is the starting radial position (at the showerhead) of a particle that follows a trajectory that leads it to deposit at the edge of the wafer, R_W (see Fig. 7). All particles starting closer to the centerline will deposit on the wafer, while those starting farther out will exit the reactor. For a uniform concentration across the showerhead, the collection efficiency, η, can be written as

$$\eta = \left(\frac{R_{crit}}{R_W} \right)^2 = \left(\frac{R_i}{R_f} \right)^2 \tag{47}$$

In general, the critical trajectory must be found by a trial-and-error method. The second equality of Eq. (47) is a simplification that only applies under our quasi-1-D approximation. In this case, all the factors that influence particle deposition (e.g., axial fluid velocity profile, the particle initial velocity, and particle axial deposition velocity) are independent of radial position; thus, the question of whether a particle will hit the wafer must not depend on its initial radial position, R_i (although the *radial* position at which the particle hits the wafer, R_f, will depend on R_i). It can be shown for our quasi-1-D case that the ratio R_i/R_f is independent of initial radial position. Thus, efficiency in the

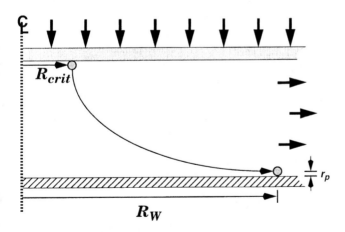

Figure 7 Critical trajectory. Diagram of a critical trajectory for a particle which starts at the showerhead at radial position R_{crit} and deposits at the wafer edge R_W.

Lagrangian framework is calculated by starting the particle at a particular radial position ($R_i = 1$) and calculating its trajectory to determine the radial position of contact with the wafer[9]; the efficiency is then $(R_i/R_f)^2 = (1/R_f)^2$ as given in Eq. (47). The total number of particles depositing on the wafer is the product of efficiency times the total flux of particles entering through the showerhead.

External Force Limit. For the case where external forces control particle deposition (neglecting inertia, interception, and diffusion), Rader et al. [27,28] used a Lagrangian analysis to obtain the following expression for deposition efficiency in isothermal, quasi-1-D parallel-plate flows:

$$\eta = \left(\frac{R_{crit}}{R_W}\right)^2 = \left(\frac{R_i}{R_f}\right)^2 = \frac{|V_p^t|}{|V_p^t| + U_0} \qquad V_p^t \leq 0 \qquad (48)$$

where V_p^t is the z component of the net particle deposition velocity (the resultant of all external forces in the axial direction).[10] For net deposition velocities greater than zero (net external force pushing particles away from the wafer), no particle deposition on the wafer is predicted (although it will be shown later that particle inertia or diffusion can cause deposition even in this case). Equation (48) provides a lower bound for particle deposition, as inertial and diffusional effects can only increase deposition from what is predicted. Interestingly, as the particle net deposition velocity is typically much smaller than the fluid entrance velocity, Eq. (48) predicts that (in the absence of inertia and diffusion) the particles which land on the wafer originate from near the reactor centerline.

That Eq. (48) is independent of the flow field (consider that the flow Reynolds number does not appear) can be more easily understood by applying Robinson's [26] result (as discussed above). Neglecting diffusion and inertia, the concentration over the lower plate must equal the inlet concentration, n_0. The rate at which particles deposit on the lower plate becomes $|V_p^t|n_0\pi R_W^2$, while the rate at which particles enter through the showerhead is $(|V_p^t| + U_0)n_0\pi R_W^2$. Taking the ratio of these expressions gives the same efficiency as Eq. (48) (see also Ref. [29]). Similar results, including Eq. (48), were found by Ramarao and Tien [30] for plane-stagnation flow.

Interception effects were neglected in Eq. (48), which implies that a particle is collected only when its center of mass reaches the wafer surface. A better assumption is that particle collection occurs when the particle comes within one

---

[9]In the event that the particle axial velocity becomes zero, or begins to move away from the wafer, then the trajectory calculation is terminated and the efficiency is set to zero.

[10]When inertia is neglected, the particles enter the reactor with the mean gas velocity so that $V_{p0} = -U_0$. Also, note that U_0 is the magnitude of the face velocity, and so is positive.

particle radius ($r_p = d_p/2$) of the wafer surface.[11] The derivation of Eq. (48) can be easily modified to include interception, with the following result:

$$\eta = \left(\frac{R_{\text{crit}}}{R_W}\right)^2 = \left(\frac{R_i}{R_f}\right)^2 = \frac{|V_p^t| + |u(z = r_p)|}{|V_p^t| + U_0} \qquad V_p^t \le |u(z = r_p)|$$

(49)

Note that particle collection is now expected in the absence of external forces (or even for weak repulsive forces); the physical interpretation is that collection occurs when the flow brings the particle within one particle radius of the wall. The inclusion of interception also has the effect that Eq. (49) (unlike the previous equation) depends on the flow field through the term $u(z = r_p)$. Because the gas velocity one particle radius away from the wafer is typically vanishingly small, interception effects are generally neglected in the following discussion, and Eq. (48) is used.

2. Particle Traps/In Situ Nucleation

Another source of wafer contamination is from particles that start somewhere between the plates, and are subsequently transported to the wafer. For example: 1) particles generated in situ by nucleation and 2) particles that are originally trapped between the plates during a plasma process, which are subsequently released at plasma extinction. While the plasma is on, contaminant particles generally accumulate in specific regions of the radio frequency (RF) discharge. Roth et al. [31] first used laser light scattering to observe that particles accumulate near the bulk plasma-sheath boundary in these discharges. Sommerer et al. [32] and Barnes et al. [33] first proposed that particle transport in the discharge is dominated by two forces: electrostatic and viscous ion drag. The electrostatic force accelerates negatively charged particles toward the center of electropositive plasmas, while viscous ion drag accelerates particles in the direction of net ion flux (generally toward plasma boundaries). Particle "traps" occur in regions where the sum of forces acting on the particle vanishes. In many cases these traps are approximately planar and parallel to the plates [34]; a schematic of a planar trap is shown in Fig. 8, where the particles are uniformly distributed at a distance h above the lower plate. Only planar traps are considered in this work, although a variety of other trap structures (rings, domes, etc.) are well known in the literature. For any trap structure more complicated than an infinite plane, the problem becomes inherently 2-D, which is beyond the scope of the present analysis.

---

[11]The inclusion of particle interception effects is somewhat overkill, as we have neglected wafer surface roughness/structure which is likely characterized by dimensions similar to particle sizes.

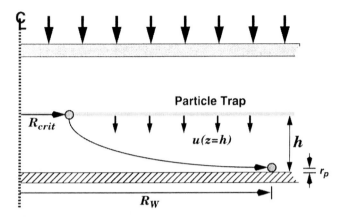

Figure 8 Trap schematic. Diagram of particles in a planar trap located a distance h from the lower plate; a critical trajectory is also shown for a particle which starts at radial position R_{crit} and deposits at the wafer edge R_W.

At the end of the process step, the discharge is extinguished and the plasma-induced forces responsible for particle trapping are assumed to dissipate rapidly (compared to particle transport times) in the afterglow. In this work, we assume that the charged particles are rapidly neutralized after the plasma extinction and can therefore be treated as neutral particles as experimentally observed by Jellum et al. [35], Shiratani et al. [36] and Yeon et al. [37].[12] Under the assumption of rapid neutralization, the particles are released from the traps and can deposit on the wafer as a result of external forces, inertia, or Brownian motion (diffusion). To analyze the extent of deposition, both the Lagrangian and Eulerian formulations have been used. Although the physical interpretation of efficiency (fraction of particles starting in the trap that end up on the wafer) is the same for both approaches, the methods of calculating the efficiency are quite different.

Efficiency for the Lagrangian Formulation. In the Lagrangian formulation, Brownian motion is neglected and calculation of particle trajectories is determined from the coupling between particle inertia and external forces. Consequently, the determination of a collection efficiency reduces to the determination of a critical trajectory just as defined in Eq. (47), except that the particle starting position is now at axial position h and the particle initial velocity is assumed to be zero. As before, it can be shown for our quasi-1-D case that

---

[12]However, a recent study by Collins et al. [38] suggests that some particles might retain a few residual charges (positive or negative) in the afterglow.

the ratio R_i/R_f is independent of initial radial position within the trap. Thus, efficiency in the Lagrangian framework is calculated by starting the particle at a particular radial position ($R_i = 1$) in the trap ($z = h$) and calculating its trajectory to determine the radial position of contact with the wafer; the efficiency is then calculated by Eq. (47).

Efficiency for the Eulerian Formulation. For small particles and/or at low pressure, the effects of Brownian motion on particle transport must be considered. Brownian motion results from random variations in the force exerted on the particle by background-gas molecular bombardment, and gives rise to particle diffusion along concentration gradients. Also, Brownian motion implies that particle trajectories are no longer deterministic; i.e., identical particles started at the same initial location with the same initial conditions will not follow the same path through the reactor. In this case, an Eulerian formulation of particle transport is used, in which the particles are treated as a continuum or cloud and the particle concentration field is calculated (inertia is neglected). Particle deposition is determined in terms of a particle flux at the wafer's surface, J_0 (# cm^{-2}s^{-1}), which is calculated from the surface concentration gradient (where particle interception is neglected):

$$J_0 = \mathcal{D}\frac{dn}{dz}\bigg|_{z=0} \tag{50}$$

where n is the particle concentration and $\mathcal{D}$ is the particle diffusion coefficient. Note that a general expression for particle flux would include both a diffusional term, given by Eq. (50), and a deposition-velocity term, given by $n\tilde{V}_p^t$. In Eq. (50) only the diffusional term is shown because, under our assumption that particle concentration vanishes at surfaces, the deposition-velocity contribution must also vanish at the susceptor. Thus, even when external forces are controlling deposition, a thin boundary layer must exist near the susceptor wherein the concentration drops from the free-stream value to zero at the susceptor's surface. The particle collection efficiency is then calculated as the ratio of particle flux to the wafer divided by the particle source term (number of particles being released from the trap).

Equation (50) can be extended to account for particle interception by evaluating the concentration derivative at r_p (instead of zero):

$$J_0 = \mathcal{D}\frac{dn}{dz}\bigg|_{z=r_p} \tag{51}$$

External Force Limit. As in the previous section, an analytic result can be derived for deposition efficiency in the limiting case where external forces control particle deposition (particle inertia, interception, and diffusion are all neglected):

$$\eta = \left(\frac{R_{\text{crit}}}{R_W}\right)^2 = \left(\frac{R_i}{R_f}\right)^2 = \frac{|V_p^t|}{|V_p^t| + |u(z = h)|} \qquad V_p^t \leq 0 \qquad (52)$$

which is the same as Eq. (48) except that the gas axial velocity at the trap location replaces the mean gas velocity in the denominator. For net deposition velocities greater than zero (net external force pushing particles away from the wafer), no particle deposition on the wafer is predicted. Equation (52) provides a lower bound for particle deposition, as inertial and diffusional effects can only increase deposition from what is predicted. As expected, the collection efficiency tends toward unity as the particle trap moves closer to the lower plate ($h \to 0$) because the axial gas velocity must approach zero at the plate surface. For particles which ultimately deposit on the wafer, Eq. (52) also can be used to determine the radial position on the wafer at which particles are collected— R_f—based on starting position $r = R_i$ and $z = h$. As discussed in the previous section, particles which deposit on the wafer are those which start nearest to the reactor centerline. It should be noted that both the Eulerian and Lagrangian collection efficiencies defined above must tend to Eq. (52) in the limit when particle diffusion, inertia, and interception effects are all negligible.

Equation (52) can be extended to include particle interception as in the previous section:

$$\eta = \left(\frac{R_{\text{crit}}}{R_W}\right)^2 = \left(\frac{R_i}{R_f}\right)^2 = \frac{|V_p^t| + |u(z = r_p)|}{|V_p^t| + |u(z = h)|} \qquad V_p^t \leq 0 \qquad (53)$$

As before, particle collection is now predicted in the absence of (or for weak) external forces, and is seen to depend on the flow field through flow velocity terms in both the numerator and denominator. Both the Eulerian and Lagrangian collection efficiencies defined above must tend to Eq. (53) in the limit where particle diffusion and inertial effects are negligible.

3. Diffusion-Enhanced Deposition from Traps or In Situ Nucleation

One difference between these two particle-source scenarios is that the source term resulting from nucleation is continuous, while the source term for a plasma-trap release is a transient event characterized by the number of particles in the trap at the time of release. In either case, the particles of interest are likely to be quite small and chamber pressures may be low, so that the effect of particle Brownian motion must be considered. Although these very small particles are not currently considered to reduce yield, the trend toward smaller feature sizes on integrated circuits is continually reducing the size of a killer defect. Thus, the industry will inevitably be faced with the need to understand the role of diffusion in particle transport and deposition.

The analysis of this section closely follows the previous work of Peters et al. [20] who investigated the diffusive deposition of particles onto disks in an infinite stagnation point flow (such as for a wafer exposed to the downward flow in a cleanroom). In their work, Peters and coworkers assume axisymmetric viscous stagnation point flow, while in this work an analytic asymptotic result for flow between two axisymmetric infinite parallel plates is used. Peters et al. also used different particle concentration boundary conditions than in this work: particle concentration was assumed to be zero at the disk and to approach a constant infinitely far away from it. Here the two plates are considered perfectly absorbing (vanishing particle concentration), and a planar particle source is assumed to be located somewhere between them.

Problem Definition. We assume the geometry shown in Figure 9: axisymmetric flow between two infinite, parallel plates (a showerhead and a susceptor) separated by a distance S. The effect of jetting out of the showerhead holes is neglected, so that the flow is assumed to be uniformly distributed across the bottom of the showerhead with velocity $-U_0$. The flow is assumed to be isothermal (constant gas properties), steady, laminar, incompressible, and viscous such that the quasi-1-D analytic result, Eq. (46), can be used.

As discussed, Eq. (46) is reasonably accurate for flow Reynolds numbers less than about 4 based on comparison to more accurate numerical finite-element simulations.

The particles are assumed to enter the domain at a steady volumetric rate Λ (# cm^{-3} s^{-1}) from a planar source located a distance h from the suscep-

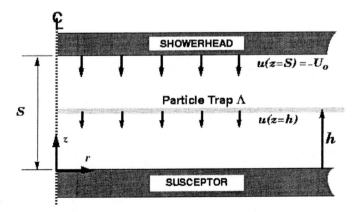

Figure 9   Geometry. Diagram of parallel, infinite-plate geometry with particles in a planar trap located a distance h from the lower plate.

tor. Although this description adequately applies to a continuous source such as particle nucleation, it is not immediately obvious that a steady state analysis is applicable to the transient case where a finite number of particles are simultaneously released from a trap at time $t = 0$. Analysis of the governing equations reveals that the particle collection efficiencies from the steady state and transient problems are in fact the same under the following conditions: 1) steady flow field, 2) infinite parallel-plate (1-D) domain, 3) radially uniform distribution of initial particle positions for the transient problem, and 4) radially uniform particle source for the steady state problem. To confirm this contention, particle transport calculations using the present steady state Eulerian approach have been compared with the Brownian dynamics simulations (BDS) of Choi et al. [39]. Choi and coworkers solved the Langevin equation directly using a massively parallel numerical Lagrangian particle tracking model which included a fluctuating Brownian force; transport calculations were presented for particles that were initially distributed in planar traps in a parallel-plate geometry similar to that assumed here. The BDS method is inherently transient in nature, in that a large number of particles were initially distributed uniformly throughout the trap, and their trajectories followed in time until the particles either deposited on a plate or left the calculation domain. For comparison with the present approach, Brownian dynamics simulations were performed with the analytic velocity field given by Eq. (46). As expected, BDS results for particle collection efficiency were in excellent agreement with the steady state Eulerian formulation presented here. Thus, the analytic result for particle collection efficiency given below applies equally well for a steady state particle planar source as for the case of a cloud of particles released from a planar trap.

Solution of the Eulerian Particle Transport Equation. Neglecting particle inertia, the Eulerian expression for particle concentration, n (#/cm^3), is Eq. (33) which is reproduced here:

$$\frac{\partial n}{\partial t} + \mathbf{U} \cdot \nabla n = \nabla \cdot \mathcal{D} \nabla n - \nabla \cdot (\mathbf{V}_p^t n) + \Lambda \tag{54}$$

where $\mathcal{D}$ (cm^2/s) is the Stokes-Einstein particle diffusion coefficient, Λ (# cm^{-3} s^{-1}) is the particle source term, and $\mathbf{V}_p^t$ is the net deposition velocity vector. Consistent with our flow assumptions, the concentration field is assumed to be steady and one-dimensional (depending only on axial position). Also, for isothermal flow, the diffusion coefficient and particle deposition velocity are constant. With these assumptions and simplifications, Eq. (33) may be rewritten as

$$u\frac{dn}{dz} = \mathcal{D}\frac{d^2n}{dz^2} - V_p^t\frac{dn}{dz} + \Lambda \tag{55}$$

In Eq. (55), V_p^t is the z component of the net deposition velocity vector and u is the axial velocity field given by Eq. (46). For boundary conditions, the assumption of perfectly absorbing walls is made which implies that the particle concentration is zero at both the upper and lower plates, i.e., $n(z = 0) = n(z = S) = 0$.[13] Note that by assuming an absorbing surface at the showerhead we are neglecting the showerhead holes, but this is reasonable as the holes typically account for only a few percent of the total showerhead surface.

For this analysis, the particle source is assumed to be infinitely thin so that

$$\Lambda = \Lambda_h \delta(z - h) \tag{56}$$

where Λ_h (# $cm^{-2}s^{-1}$) is a constant area source term and δ (cm^{-1}) is the Dirac delta function. Although the following derivation also could be followed for a finite-thickness source term, the resulting analytic expression for particle collection efficiency would be much more complicated than that given below.

To nondimensionalize Eq. (55), the appropriate characteristic length and velocity scales are S and U_0, respectively. A characteristic particle concentration, n_0, can be defined based on the particle source strength and gas inlet velocity:

$$n_0 = \frac{\Lambda_h}{U_0} \tag{57}$$

Using these definitions, Eq. (55) becomes

$$\tilde{u}\frac{d\tilde{n}}{d\tilde{z}} = \frac{1}{\text{Pe}}\frac{d^2\tilde{n}}{d\tilde{z}^2} - \tilde{V}_p^t\frac{d\tilde{n}}{d\tilde{z}} + \delta(\tilde{z} - \tilde{h}) \tag{58}$$

where $\tilde{n} = n/n_0$, $\tilde{u} = u/U_0$, $\tilde{z} = z/S$, $\tilde{V}_p^t = V_p^t/U_0$, $\text{Pe} = SU_0/\mathcal{D}$, and $\tilde{h} = h/S$. As discussed above, the solution for the dimensionless concentration is completely determined by the dimensionless groups Pe, $\tilde{h}$, $\tilde{V}_p^t$, and Re (which enters implicitly as $\tilde{u}$ depends on Reynolds number).

After defining a dimensionless concentration gradient

$$\tilde{G} = \frac{d\tilde{n}}{d\tilde{z}} \tag{59}$$

Eq. (58) can be rewritten as

$$\frac{d\tilde{G}}{d\tilde{z}} - \text{Pe}(\tilde{u} + \tilde{V}_p^t)\tilde{G} = -\text{Pe}\delta(\tilde{z} - \tilde{h}) \tag{60}$$

The solution to Eq. (60) is

$$\tilde{G} = \tilde{G}_0 \exp(A) - \exp(A) \int_0^{\tilde{z}} \text{Pe}\ \delta(\tilde{z} - \tilde{h}) \exp(-A)\ d\tilde{z} \tag{61}$$

---

[13]Deposition by interception, due to the finite size of the particle, is neglected. To account for interception requires that the boundary conditions be given as $n(z = r_p) = n(z = S - r_p) = 0$ where r_p is the particle radius.

where

$$A(\tilde{z}) = \text{Pe} \left[\frac{1}{2}\tilde{z}^4 - \tilde{z}^3 + \frac{\text{Re}}{70}\left(\frac{1}{4}\tilde{z}^8 - \tilde{z}^7 + \frac{9}{2}\tilde{z}^4 - \frac{13}{3}\tilde{z}^3\right) + \tilde{V}_p'\tilde{z} \right] \quad (62)$$

and $\tilde{G}_0$ is the dimensionless concentration gradient at the lower plate, $\tilde{z} = 0$. Note that $\tilde{G}_0$ is frequently referred to as the Sherwood number, Sh (see, for example, Ref. 20). To determine $\tilde{G}_0$, apply the boundary conditions $\tilde{n}(\tilde{z} = 0) = \tilde{n}(\tilde{z} = 1) = 0$ after integrating Eq. (59):

$$\int_0^1 \frac{d\tilde{n}}{d\tilde{z}}d\tilde{z} = \int_0^1 d\tilde{n} = 0 = \int_0^1 \tilde{G} \, d\tilde{z} \quad (63)$$

Solving Eq. (63) for $\tilde{G}_0$ gives

$$\tilde{G}_0 = \left(\text{Pe} \cdot \int_{\tilde{h}}^1 \exp\left(\text{Pe}\left\{ \frac{1}{2}(t^4 - \tilde{h}^4) - (t^3 - \tilde{h}^3) + \frac{\text{Re}}{70} \right.\right.\right.$$

$$\times \left[\frac{1}{4}(t^8 - \tilde{h}^8) - (t^7 - \tilde{h}^7) + \frac{9}{2}(t^4 - \tilde{h}^4) - \frac{13}{3}(t^3 - \tilde{h}^3) \right]$$

$$\left.\left.\left. + \tilde{V}_p'(t - \tilde{h}) \right\} \right) dt \right) \Bigg/$$

$$\left(\int_0^1 \exp\left\{ \text{Pe}\left[\frac{1}{2}t^4 - t^3 + \frac{\text{Re}}{70}\left(\frac{1}{4}t^8 - t^7 + \frac{9}{2}t^4 - \frac{13}{3}t^3\right) + \tilde{V}_p't \right] \right\} dt \right)$$

$$(64)$$

Particle Collection Efficiency.   The particle collection efficiency is found as the ratio of particle flux to the lower plate divided by the total number of particles entering the reactor:

$$\eta = \frac{A_w \mathcal{D} \dfrac{dn}{dz}\Big|_{z=0}}{A_w \displaystyle\int_0^S \Lambda_h \delta(z - h) \, dz} = \frac{\dfrac{\mathcal{D}\Lambda_h}{SU_0}\dfrac{d\tilde{n}}{d\tilde{z}}\Big|_{\tilde{z}=0}}{\Lambda_h} = \frac{\tilde{G}_0}{\text{Pe}} \quad (65)$$

where A_w is area in the $r\theta$ plane. Thus, an analytic result for the particle collection efficiency is given by Eqs. (64) and (65)—although the solution requires numerical quadrature. The dependence of the collection efficiency on the four dimensionless groups, Re, $\tilde{h}$, $\tilde{V}_p'$, and Pe, is clearly shown in Eqs. (64) and (65). For comparison to previous work in the literature, the particle collection efficiency defined by Eq. (65) can also be expressed as the ratio of the Sherwood to Peclet numbers, $\eta = \text{Sh/Pe}$. The use of collection efficiency as the dimensionless number characterizing the deposition process is preferred to the Sherwood number for this application for two reasons: 1) Determination of the collection efficiency is also straightforward for Lagrangian formulations to be applied to showerhead-enhanced inertial deposition in which critical trajectories can be calculated, and 2) efficiency is a commonly accepted concept within the semiconductor industry (e.g., yield). In practical terms, an efficiency of unity indicates that all particles in the chamber are depositing on the wafer, while an efficiency of zero indicates that no particles are depositing on the wafer.

Particle Flux. The particle deposition rate on the wafer is found as the product of collection efficiency times the number of particles released from the trap (or generated by nucleation):

$$J_0 = \mathcal{D}\frac{dn}{dz} = \frac{\mathcal{D}\Lambda_h}{SU_0}\frac{d\tilde{n}}{d\tilde{z}}\bigg|_{\tilde{z}=0} = \Lambda_h\frac{\tilde{G}_0}{\text{Pe}} = \Lambda_h\eta \tag{66}$$

When the nature of the source term Λ_h is not known, the best strategy for reducing the number of defects on a wafer is to choose conditions that inhibit particle transport to the wafer, i.e., minimize the collection efficiency. One potential weakness of this strategy is if process conditions selected to reduce the collection efficiency result in a corresponding increase in the particle generation rate; this possibility is certainly of concern for particle nucleation.

4. Nondimensional Results

This section presents calculations of particle collection efficiency using numerical quadrature of Eqs. (64) and (65) based on a fourth-order Runge-Kutta technique with automatic error control. Both local and global error control parameters can be set, and convergence tests showed that the resulting integrals were unchanged in the fifth place when these parameters were set to 10^{-10}. Efficiency is found to be a function of four dimensionless parameters: Re, $\tilde{h}$, $\tilde{V}_p^t$, and Pe (a fifth—the interception parameter d_p/S—is neglected in this work). Calculations of particle collection efficiency versus Peclet number are shown in Fig. 10 for creeping flow (Re $= 0$) for three attractive forces (characterized by $\tilde{V}_p^t = -0.1$, -0.5, and -1.0), for no external force ($\tilde{V}_p^t = 0$), and for three repulsive forces ($\tilde{V}_p^t = 0.1$, 0.5, and 1.0). Plots are shown for three different trap heights, where the particles are trapped: 1) near the wafer ($h/S = 0.1$), 2) midway between the wafer and showerhead ($h/S = 0.5$), and 3) near the showerhead ($h/S = 0.9$).

Efficiency Intermediate Peclet Numbers. Although the small- and large-Pe asymptotic limits for the collection efficiency are well described by analytic expressions, the shape of the collection efficiency curves for intermediate Peclet numbers can be quite complex and requires the full numerical integration of Eqs. (64) and (65). For example, while the efficiency-curve transition between

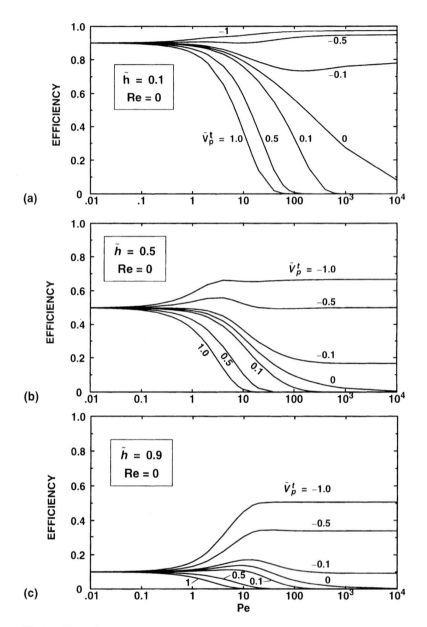

Figure 10 Efficiency versus Pe for various deposition velocities. Figures show calculated efficiencies for particles starting in traps at (a) $h/S = 0.1$, (b) $h/S = 0.5$, and (c) $h/S = 0.9$.

the small- and large-Pe asymptotic limits is generally monotonic (e.g., Fig. 10a for $\tilde{V}_p^t > 0$ or Fig. 10c for $\tilde{V}_p^t < -0.5$), in some cases there may be a local minimum (e.g., Fig. 10a for $\tilde{V}_p^t = -0.1$) or maximum (e.g., Fig. 10c for $\tilde{V}_p^t = -0.1$). The exact shape of the efficiency curve depends on the magnitudes of the three parameters, $\tilde{h}$, $\tilde{V}_p^t$, and Pe, and although the interaction among them can be complex, a few simple observations can be made. First, moving the particle trap away from the wafer (i.e., increasing $\tilde{h}$) always tends to lower the collection efficiency. Although this effect is most notable for low or intermediate Pe values where diffusional effects are strong, it is also true for large Pe values where deposition is controlled by external forces. The latter claim is supported by noting that gas velocity at the trap location, $\tilde{u}(\tilde{h})$, increases with increasing distance from the wafer so that collection efficiency decreases. Trap manipulation can be accomplished in practice under some conditions. For example, in plasma processing, the trap location is determined by process parameters such as pressure, rf power, and flow rate; while these parameters may be fixed during etch by process requirements, they could be adjusted to manipulate the particle trap location just prior to plasma extinction. Similarly, trap position when particle nucleation is present could be controlled by pressure, wall temperature, flow rates, or chemistry selections.

Second, reducing the dimensionless attractive external force or increasing the dimensionless repulsive force always lowers the collection efficiency. This trend is clearly evident in Figure 10 for intermediate and large Peclet numbers, although the benefit becomes less significant at low Pe where diffusion dominates deposition. It is interesting to note that for attractive forces such as gravity (wafer facing up) or thermophoresis (wafer cooler than the surrounding gas), the dimensionless deposition velocity increases as pressure is decreased (for constant U_0). In this case, the tendency toward processing at lower pressures ultimately must lead to an increase in the fraction of particles which end up on the wafer. If a process recipe is selected that maintains the wafer warmer than its surroundings, however, the velocity resulting from this repulsive external force will increase with decreasing operating pressure (for constant U_0) and thereby reduce the fraction of particles depositing on the wafer.

To explore the effect of Reynolds number on collection efficiency, calculations were made for Reynolds numbers of 0 and 8 using the analytic approximation for the flow field given in Eq. (46)[14]; the results are shown in Figure 11. As shown, even this relatively large variation in Re (spanning the Re range of the majority of low-pressure commercial tools) produces only modest variations in the collection efficiency. Reynolds number effects are most apparent in the

---

[14]Note that Re $= 8$ is beyond the range over which the analytic approximation was found to be accurate, but the analytic result is used here beyond its range to qualitatively investigate any Reynolds number dependencies.

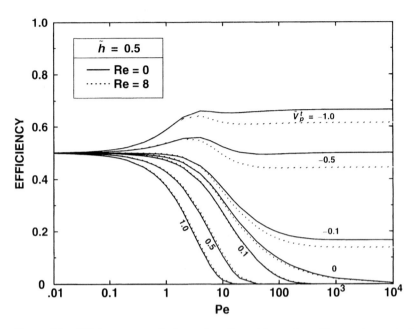

Figure 11 Efficiency versus Pe for various Reynolds numbers. Figures show calculated efficiencies for Re = 0 and 2 ($h/S = 0.5$).

large-Pe limit for attractive forces. This effect can be quantified by noting that 1) the large-Pe efficiency limit of Eq. (52) depends on gas velocity at the trap location, $\tilde{u}(\tilde{h})$, and 2) the value of $\tilde{u}(\tilde{h})$ changes from -0.5 at Re = 0 to a value of -0.625 at Re = 8. This 25% increase in $\tilde{u}(\tilde{h})$ associated with Re being increased from 0 to 8 leads to an approximately 25% decrease in collection efficiency for a weak attractive force, i.e., $|\tilde{V}_p^t| \ll |\tilde{u}(\tilde{h})|$. As $\tilde{V}_p^t$ is increased to a magnitude comparable to $\tilde{u}(\tilde{h})$, this effect diminishes; for example, for $\tilde{V}_p^t = -1$ the efficiency for Re = 8 is only 8% less than for Re = 0.

For small Peclet numbers, the effect of Reynolds number vanishes for all values of the external force. In this case, particle transport by diffusion dominates convective transport, so that the collection efficiency is decoupled from the details of the flow field. Figure 11 also shows that Reynolds number effects are negligible in the presence of repulsive external forces; although this conclusion is true in an absolute sense, Reynolds number effects do become important in a relative sense for larger Peclet numbers where the collection efficiency becomes small. For example, for $\tilde{V}_p^t = 1.0$ and Pe = 15, the collection efficiency for Re = 8 ($\eta = 3.18 \cdot 10^{-3}$) is approximately 50% larger than that for Re = 0 ($\eta = 2.11 \cdot 10^{-3}$). While such differences may not be detectable at low particle concentrations, the difference in the number of particles depositing on the wafer may become quite large when particle concentrations are high—such as is typical of systems in which particle nucleation is occurring.

5. Dimensional Results

This section presents several example calculations of collection efficiency in dimensional terms when gravity and diffusion act simultaneously. The solution scheme described in the previous section is again used here, except that the dimensional inputs are first converted into the required nondimensional groups before the numerical integrations are performed. The particle diameter replaces the Peclet number as the independent variable in all of the following. All of the examples assume a 200 mm diameter susceptor with a showerhead-to-wafer gap of 2.54 cm. For these calculations, a baseline process is taken as argon flowing at a mass flow rate of 1000 sccm (standard cubic centimeters per minute) at a pressure of 1 torr and temperature of 300 K. Constant gas properties are assumed (isothermal flow) along with a particle density of 1 g/cm^3. The Reynolds number for these conditions is 0.984, which indicates viscous dominated flow and is well inside the range for which the analytic flow field expression can be used. In the following examples, trap height, pressure, flow rate, and pressure are individually varied about the baseline value. Note that these parameters may not be independent; for example, trap height may depend on both pressure and gas flow rate. Finally, the section concludes with a demonstration of the reduction in deposition that can be obtained by introducing a force that opposes deposition, such as heating the wafer relative to the showerhead to take advantage of thermophoretic protection.

Trap Height Effects. Plots of calculated particle collection efficiency as a function of particle size are shown in Figure 12 for dimensionless trap heights of 0.1, 0.3, 0.5, 0.7, and 0.9. All of the curves exhibit a minimum near 0.1 μm, with increasing efficiency for both smaller and larger sizes. This shape has commonly been reported in previous deposition studies: for example, see Figure 3 of Ref. 40 which shows the net stagnation-point deposition velocity based on additivity of convective-diffusion, electrostatic, and gravitational velocities. In Figure 12, the increase in efficiency below 0.1 μm is associated with increasing diffusional deposition, while the increase in efficiency above 0.1 μm is associated with increasing gravitational deposition. Note that in the present geometry the diffusional branch does not increase without bound, but instead asymptotically approaches the highly diffusive limit $\eta \rightarrow 1 - \tilde{h}$. As seen in Figure 12, however, this limit is not quite achieved even for particles as small as 0.001 μm. As expected based on the previous discussion in nondimensional terms, the trapping height plays a key role in net particle deposition. Clearly, it is always advantageous to manipulate the particle trap to a location as far from the wafer as possible. It is clear from Figure 12 that as the size of IC-killing

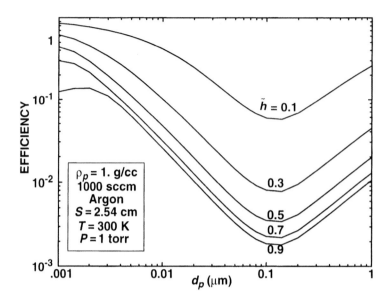

Figure 12 Efficiency vs. particle diameter and trap position (isothermal case). Collection efficiencies for particle transport over a 200 mm wafer including gravitational settling and diffusion for dimensionless trap locations of 0.1, 0.3, 0.5, 0.7, and 0.9 ($S = 2.54$ cm, $Q = 1000$ sccm of argon, $P = 1$ torr, $T = 300$ K, Re = 0.98, and particle density of 1 g/cm^3).

particles shrinks below 0.1 μ that particle collection efficiencies will climb as a result of increased particle diffusion.

Pressure Effects. Plots of calculated particle collection efficiency as a function of particle size are shown in Figure 13 for reactor pressures of 1, 10, 100, and 760 torr. For these calculations the mass flow rate is held constant at the baseline value of 1000 sccm, and the trap height is assumed to be 1.27 cm which is exactly half-way between the wafer and showerhead. As before, all of the curves show the characteristic "U" shape resulting from the combination of deposition from convective-diffusion and gravitational settling. It is evident, however, that the diffusional branch of the efficiency curves are independent of pressure for this example. This result is explained by considering that the Peclet number (which along with geometry completely specifies the convective-diffusive problem) depends on the ratio of inlet gas velocity to the particle diffusion coefficient, and that both of these quantities are inversely proportional to pressure at a constant mass flow rate. The pressure dependence of the diffusion coefficient is evident in the free molecule limit given in Eq. (36), which applies at low pressures and for small particle size. The velocity used in scaling

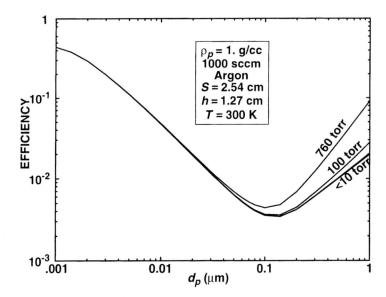

Figure 13 Efficiency vs. particle diameter and pressure (isothermal case). Collection efficiencies for particle transport over a 200 mm wafer including gravitational settling and diffusion for reactor pressures of 1, 10, 100, and 760 torr ($S = 2.54$ cm, $Q = 1000$ sccm of argon, $h = 1.27$ cm, $T = 300$ K, Re $= 0.98$, and particle density of 1 g/cm^3).

the problem, U_0, is the linear gas velocity at the reactor pressure, and for a presumed constant mass flow rate this also must scale inversely proportional to pressure. Thus, the collection efficiency resulting from particle diffusion is nearly independent of pressure for a fixed gas mass flow rate.

The branch of the efficiency curve resulting from gravitational settling is also seen to be independent of pressure below 10 torr. This result is explained by considering the limiting expression for external-force dominated deposition, which depends only on $\tilde{u}(\tilde{h})$ and $\tilde{V}_p^t$. The term $\tilde{u}(\tilde{h})$ is independent of pressure as it depends only on trap position and Reynolds number [see Eq. (46)], and, as at constant mass flow rate the Reynolds number is independent of pressure. In the free molecule limit given by Eq. (28), the gravitational deposition velocity, $\tilde{V}_G^t$, varies inversely with pressure so that its ratio to the inlet velocity, $\tilde{V}_p^t$, must also be independent of pressure. Thus, all of the terms describing the collection efficiency are independent of pressure in the particle free molecule limit. As pressure increases above 10 torr and for particles larger than 1 μm, however, the free molecule limit for the settling velocity no longer applies and the full expression of Eq. (26) must be used. As the particle mean free path decreases, the inverse pressure dependence of the dimensional settling velocity diminishes, until the continuum regime limit equation (27) is reached which is independent of pressure. As V_p^t becomes independent of pressure, $\tilde{V}_p^t$ becomes proportional to pressure, resulting in the marked increase in collection efficiency with pressure as seen in Figure 13 for larger particles and at higher pressures.

Mass Flow Rate Effects.   Plots of calculated particle collection efficiency as a function of particle size are shown in Figure 14 for argon mass flow rates of 10, 100, and 1000 sccm. For these calculations the reactor pressure is held constant at the baseline value of 1 torr, and the trap height is assumed to be 1.27 cm (half-way between the wafer and showerhead). As before, all of the curves show the characteristic "U" shape resulting from the combination of deposition from convective-diffusion and gravitational settling. These results show that the collection efficiency is a strong function of gas mass flow rate except for very small particle sizes, for which diffusion dominates and all curves must tend toward the same limit. Note that the point at which the diffusion-dominated limit is achieved varies with mass flow rate: for the lowest flow rate of 10 sccm the limit ($\eta = 0.5$) is reached for particles less than about 0.01 μm, while at the highest flow rate of 1000 sccm the collection efficiency is still

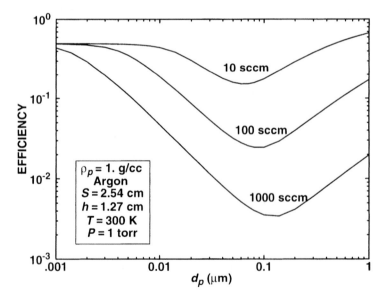

Figure 14   Efficiency vs. particle diameter and flow rate (isothermal case). Collection efficiencies for particle transport over a 200 mm wafer including gravitational settling and diffusion for gas mass flow rates of 10, 100, and 1000 sccm (argon, $S = 2.54$ cm, $h = 1.27$ cm, $T = 300$ K, $P = 1$ torr, $0.01 < \text{Re} < 0.98$, and particle density of 1 g/cm^3).

below the limit for 0.001 μm. Thus, a significant reduction in particle collection efficiency can be achieved by increasing the mass flow rate at constant pressure. The effect of mass flow rate on the actual *number* of particles depositing on the wafer is less obvious. For example, consider the case in which the mass flow rate is increased from 100 to 1000 sccm. Although the collection efficiency drops by approximately a factor of 10, the flow increases by a factor of 10, so that if the number of particles entering the domain scales with flow rate, then the total number of particles depositing on the wafer should be about the same for the two flow rates. If, however, the number of particles present is independent of the flow rate, then increased flow rates should reduce the number of particles on the wafer.

For calculations where pressure is fixed, it should be noted that the Reynolds number increases proportionately with mass flow. However, even at the highest flow rate considered, 1000 sccm, the Reynolds number is less than 1 and the analytic flow approximation is excellent.

Effect of Thermophoresis. This final section explores the role of thermophoresis in determining particle collection efficiency. For these calculations the showerhead temperature has been held constant at 300 K, and the wafer temperature varied to produce a temperature gradient that drives thermophoretic deposition. To accommodate our assumption of constant properties, only small temperature differences are considered. Plots of calculated particle collection efficiency as a function of particle size are shown in Figure 15 for wafer temperatures of 280, 290, 300, 310, and 320 K. The baseline conditions described above are used for all of these calculations: reactor pressure 100 mtorr, argon mass flow rate of 1000 sccm, wafer-to-showerhead gap of 2.54 cm, and the trap height is assumed to be 1.27 cm. A particle of density of 1 g/cm^3 was assumed, and the ratio of the gas to particle thermal conductivity was taken as 0.02 to approximate a fused silica particle. As before, all of the curves show the characteristic "U" shape resulting from the combination of deposition from convective-diffusion and external forces, where the net external forces contains contributions from both gravitational and thermophoretic forces. The minimum collection efficiency for the case of thermophoretic protection (wafer hotter than showerhead) becomes vanishingly small and so is not shown; note, however, that thermophoretic protection the collection efficiency never reaches absolute zero, even with thermophoretic protection, as there always is some contribution from diffusion/convection.

A key result of these calculations is that even modest temperature differences can lead to dramatic changes in collection efficiency. For example, deposition is nearly eliminated in the 0.1–1.0 μm range when the wafer is kept only 10–20 degrees warmer than the showerhead. On the other hand, the collection efficiency increases by an order of magnitude when the wafer temperature is decreased 10 degrees as compared to the isothermal case. In addition, when the

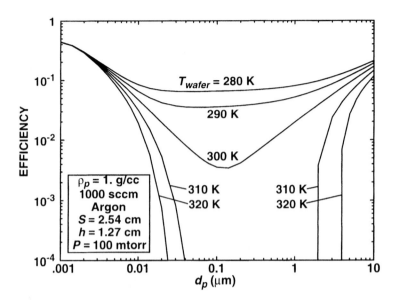

Figure 15   Efficiency versus particle diameter and wafer temperature. Collection efficiencies including gravity, thermophoresis, and diffusion for wafer temperatures of 280, 290, 300, 310, and 320 K (200 mm wafer, 1000 sccm argon, $S = 2.54$ cm, $h = 1.27$ cm, $T_{showerhead} = 300$ K, $P = 100$ mtorr, and particle density of 1 g/cm^3).

wafer temperature is kept below the showerhead temperature, the depth of the collection efficiency minimum at about 0.1 μm becomes shallower; this results from the fact that the thermophoretic deposition velocity is nearly independent of particle diameter as indicated by Eqs. (30) and (31). These calculations clearly demonstrate the importance of keeping the wafer warmer than its surroundings at all times.

6.   Summary: Diffusion-Enhanced Deposition

This section explored particle deposition resulting from external forces and Brownian motion in a parallel-plate geometry characteristic of a wide range of semiconductor process tools. The need to properly account for diffusion-enhanced particle deposition becomes increasingly important as the semiconductor industry moves toward smaller feature sizes and becomes concerned with smaller-sized particles. Particle transport was modeled using the Eulerian approach of Section V, so that the continuum convective-diffusion equation was solved for particle flux. One strength of the Eulerian formulation is in predicting particle transport resulting from the combination of applied external forces (including the fluid drag force) and the chaotic effect of particle Brownian motion (i.e., particle diffusion), although the current implementation neglected particle inertia. Furthermore, particles were assumed to originate in a planar trap located between the plates, and only transport in the interplate region was considered (showerhead acceleration was neglected). Flow between infinite parallel plates was assumed as described by the quasi-1-D creeping flow approximation, where the showerhead was treated as a porous plate.

The key result of the analysis of this section was the derivation of expressions for the particle collection efficiency—which is the fraction of trapped particles which end up on the wafer. An analytic, integral expression was derived that gives the particle collection efficiency as a function of four dimensionless parameters: Re, $\tilde{h}$, $\tilde{V}_p^t$, and Pe (a fifth—the interception parameter d_p/S—was neglected). The first parameter, the Reynolds number, completely specifies the flow field under the present assumptions. The second parameter, the dimensionless trap height, $\tilde{h} = h/S$, specifies the position of the particle source term. The influence of external forces enters through the third parameter, the dimensionless particle deposition velocity, $\tilde{V}_p^t = V_p^t/U_0$, which is defined as the z component of the net deposition velocity. The fourth parameter is the particle Peclet number, Pe $= SU_0/\mathcal{D}$, which is a measure of the relative importance of particle Brownian motion. In the free-molecular limit the Peclet number is proportional to diameter squared, so that $\text{Pe}^{1/2}$ can be thought of as a dimensionless particle size.

Numerical quadrature of Eqs. (64) and (65) using a fourth-order Runge-Kutta technique was used to calculate particle collection efficiency in terms of the controlling dimensionless parameters. Initial calculations showed the numerical results to be in good agreement with the various analytic limits, providing confidence in the current implementation. In general, the highly diffusive limit was approached for Peclet numbers less than about 0.1, while the nondiffusing limit was essentially reached for Peclet numbers larger than $\sim 10^2$ or $\sim 10^3$ depending on the strength of the external force and the initial particle trapping position. For intermediate Peclet numbers, the shapes of the collection efficiency curves were often found to be complex; for example, some conditions gave efficiency curves which showed local minima or maxima. Despite this complexity, a few simple observations were made: moving the particle trap away from the wafer (i.e., increasing $\tilde{h}$) always lowered the collection efficiency as did reducing the attractive external force or increasing the repulsive force. Finally, calculations made for Reynolds numbers of 0 and 8 showed only modest variations in particle collection efficiency, suggesting that this parameter plays only a minor role over the range likely to be encountered in realistic processing environments.

Example calculations of collection efficiency were presented in dimensional terms for one representative set of process conditions (200 mm wafer, showerhead-to-wafer gap of 2.54 cm, mass flow rate of 1000 sccm argon, 1 torr, 300 K). In all cases, the efficiency curves exhibited a minimum near 0.1 μm, with increasing efficiency for both smaller and larger sizes. Trapping height was found to play a key role in deposition, and in all cases it was advantageous to manipulate the particle trap to a location as far from the wafer as possible. Trap manipulation can be accomplished in practice under some conditions. For example, in plasma processing, the trap location is determined by process parameters such as pressure, rf power, and flow rate; while these parameters may be fixed during etch by process requirements, they could be adjusted to manipulate the particle trap location just prior to plasma extinction. Similarly, trap position when particle nucleation is present could be controlled by pressure, wall temperature, flow rates, or chemistry selections.

At constant mass flow rate, collection efficiency was found to be independent of pressure for the low pressures and small particle sizes of interest. At constant pressure, collection efficiency was found to decrease significantly with increasing mass flow rate. Thus, from a particle transport point of view, reduction of particle-on-wafer counts could be obtained by increasing the mass flow rate in a process, assuming particle concentration varies inversely with mass flow rate rather than remaining independent.

Another key result was that even modest temperature differences can lead to dramatic changes in the collection efficiency of 0.05–1.0 μm particles due to the thermophoretic force. For example, deposition is nearly eliminated when the wafer is kept only 10–20 degrees warmer than the showerhead. On the other hand, the collection efficiency increases by an order of magnitude when the wafer temperature is decreased 10 degrees below the showerhead temperature.

Caution is suggested in implementing any of these strategy since the effect on the particle source term is not known.

C. Inertia-Enhanced Deposition

The use of a showerhead restricts the area available to the flow inside the showerhead; consequently, the velocity of the gas inside the holes is much larger than the face velocity in the gap below. Particles originating upstream of the showerhead and suspended in the flow can be dramatically accelerated while passing through the showerhead, so that, at the exit of the showerhead, particle velocities much larger than the fluid face velocity are possible. Depending on conditions, particle acceleration by the showerhead can lead to inertia-enhanced particle deposition on the wafer below. Thus, a complete description of particle deposition on a wafer in a parallel-plate reactor must include a description of particle transport through the showerhead as well as an analysis of particle transport in the interplate region.

This section explores the role of inertia-enhanced deposition using the Lagrangian particle transport formulation given in Section II. The strength of the Lagrangian formulation is in predicting particle transport resulting from the combination of applied external forces and particle inertia; the current implementation does not account for particle diffusion. The problem is separated into two domains in which particle and fluid transport are determined: 1) within a showerhead hole and 2) between the showerhead and susceptor.

1. Particle Transport in the Showerhead Holes

Particles are assumed to be evenly distributed across the showerhead hole inlet and to enter with zero radial velocity and with an initial axial velocity equal to the face velocity U_0. The particle will immediately see a fluid velocity $U_{jet}(r)$, and will either be accelerated or deaccelerated by fluid drag depending on the magnitude of U_{jet}. Because of inertia, however, the particle will require a finite time to respond. In particular, the particle will accelerate to U_{jet} only if the showerhead hole is sufficiently long or the particle sufficiently small. In practice, the particle velocity at the exit of the showerhead, V_{p0}, will fall somewhere between U_0 and U_{jet}, depending on the relative magnitudes of these two velocities, the showerhead thickness, and the particle relaxation time τ.

Assuming fully developed flow at the inlet and neglecting lift forces, the particle will remain at its initial radial position while in the tube; consequently, the axial fluid velocity driving the particle through the tube will also remain constant during the traverse. For plug flow, all particles will be accelerated by the same fluid velocity, $\overline{U}_{jet}$, independent of radial starting position. For fully developed parabolic flow, the local fluid velocity for each particle will depend on its starting position; particles near the wall will be slowed by drag, while particles near the centerline will be significantly accelerated. For parabolic flow, the assumption that the particles are evenly distributed (i.e., that the particle flux is constant) across the tube inlet is based on the assumption that the fluid entrance length is so short that particles do not have time to migrate radially as the flow is developing.[15]

The problem of a particle of given initial velocity experiencing a step-function change in the local fluid velocity is a classic problem. Although the problem could be solved in dimensional form, it is convenient for matching with the second part of this analysis if we solve the nondimensionalized particle equations of motion, Eqs. (19) and (20). As stated previously, we choose the mean chamber axial velocity U_0 as the characteristic velocity and the interplate spacing S as the characteristic length scale. External forces are assumed negligible compared to the large drag forces encountered in the showerhead holes. Given a particle of Stokes number St entering the hole with initial velocity U_0,

---

[15] An alternative assumption is that the particles follow streamlines, in which case the particle concentration is constant across the tube inlet so that the local flux of particles through the tube depends on radial position.

and an axially constant fluid velocity $U_{\text{jet}}(r)$, Eqs. (19) and (20) can be solved analytically for the particle velocity at the showerhead exit, $z = L$:

$$\frac{L}{S} = \frac{U_{\text{jet}}}{U_0}\tilde{t} + \text{St}\left(\frac{U_{\text{jet}}}{U_0} - 1\right)\left(e^{-\tilde{t}/\text{St}} - 1\right) \tag{67}$$

$$\frac{V_{p0}}{U_0} = \frac{U_{\text{jet}}}{U_0} - \left(\frac{U_{\text{jet}}}{U_0} - 1\right)e^{-\tilde{t}/\text{St}} \tag{68}$$

The solution procedure is as follows: 1) Calculate U_{jet}/U_0, L/S, and St from process and particle parameters, 2) solve Eq. (67) for the dimensionless time $\tilde{t}$, and 3) use $\tilde{t}$ in Eq. (68) to solve for the dimensionless particle velocity at the showerhead exit. Equation (67) must be solved iteratively; for this purpose we use a Newton root-finding technique.

Some complexity has been added to Eqs. (67) and (68) by our use of one characteristic length and one characteristic velocity for both the showerhead and parallel-plate domains. An interesting result is found if L and U_{jet} (natural choices for characterizing transport through the showerhead) are substituted for S and U_0 as the characteristic length and velocity. With the appropriate redefinitions of the nondimensional terms, a set of equations similar to Eqs. (19) and (20) can be derived that depend on the jet Stokes number, which is related to our earlier definition by

$$\text{St}_{\text{jet}} = \frac{\tau U_{\text{jet}}}{L} = \text{St}\frac{U_{\text{jet}}}{U_0}\frac{S}{L} \tag{69}$$

Using the same assumptions and initial conditions as above, we can derive a set of equations analogous to Eqs. (67) and (68) that are independent of L/S, that is, $V_{p0}/U_{\text{jet}} = f(U_{\text{jet}}/U_0, \text{St}_{\text{jet}})$. In addition, the functional dependence on the velocity ratio U_{jet}/U_0 is very weak (entering only as a result of the assumption that the initial particle velocity is U_0), and vanishes for the limiting case where $U_{\text{jet}}/U_0 \gg 1$ (which includes the case where the particle initial velocity is zero). In this limiting case, the ratio of the showerhead-exit velocity of the particle to U_{jet} depends only on the jet Stokes number. This result suggests plotting the results of Eqs. (67) and (68) as V_{p0}/U_{jet} against St_{jet}, such as shown in Fig. 16. As can be seen, the dimensionless particle velocity at the exit of the showerhead, V_{p0}/U_{jet}, is reasonably insensitive to the velocity ratio when $U_{\text{jet}}/U_0 > 10$.

Thus, the analysis of particle transport in the showerhead domain is complete: given the three inputs $\overline{U}_{\text{jet}}/U_0$, L/S, and St (and particle initial radial position to determine $U_{\text{jet}}(r)$ for parabolic flow), we can calculate the dimensionless velocity V_{p0}/U_0 at which the particle exits the showerhead. Interestingly, both the mean velocity ratio and the length ratio *depend only on reactor geometry*; process conditions such as temperature, pressure, or flow rate enter only through the Stokes number. Thus, for a given Stokes number, the extent of particle acceleration in the showerhead is entirely determined by hardware and

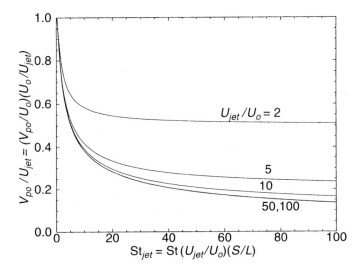

Figure 16 Acceleration of particles through showerhead. Dimensionless velocity of particles exiting showerhead tubes, V_{p0}/U_{jet}, as a function of jet Stokes number, St_{jet}, for a range of velocity ratios ($U_{jet}/U_0 = 2$, 5, 10, 50, and 100).

is thus a characteristic of a specific tool design. Finally, in moving to the calculation of particle trajectories between the plates, the sign of the particle velocity V_{p0}/U_{jet} must be switched to account for the different coordinate systems used in the two domains (see Fig. 5b and inset).

2. Particle Transport Between Parallel Plates

In this section, a Lagrangian formulation is used to calculate particle trajectories in the interplate region using both numerical FIDAP and analytic solutions of the flow field. Under the present assumptions, four dimensionless parameters uniquely determine particle transport in the inter-plate region: Re, St, V_p^t/U_0, and V_{p0}/U_0 (a fifth, the interception parameter d_p/S, is neglected in this section). The first parameter, the Reynolds number, completely specifies the flow field for the infinite parallel-plate geometry—as demonstrated in the analytic low-Re approximation to the flow field given in Eq. (46). For a finite-plate geometry, the aspect ratio is also needed to specify the flow. The second parameter, the particle Stokes number, is used in this work as a dimensionless particle diameter—as suggested by the free molecule limit Eq. (23). The influence of external forces enters through the third parameter, the dimensionless particle deposition velocity, which parameterizes the forces via the z component of the net deposition velocity, V_p^t. The fourth parameter, the dimensionless particle velocity at the showerhead exit, is determined by the strength of the showerhead acceleration effect as described in Section I above. Initially, however, V_{p0}/U_0 will be taken as an independent variable; later we discuss the coupling of the showerhead and parallel-plates domains.

The effect of the these dimensionless parameters is shown in Figure 17, where particle collection efficiency is plotted against Stokes number for Re = 8. For Figure 17a, an initial dimensionless particle velocity $V_{p0}/U_0 = -1$ is assumed (no particle showerhead acceleration), while in Figure 17b the initial velocity is taken as -100 (substantial showerhead acceleration characteristic of commercial reactors). In each plot the influence of external forces is explored by varying the deposition velocity: curves for $V_p^t/U_0 = -0.5, -0.1, -0.01, 0, 0.1,$ and 1.0 are shown. Negative values of the deposition velocity correspond to an external force directed toward the wafer (attractive, enhancing deposition), while positive values correspond to an external force directed away from the wafer (repulsive, inhibiting deposition). Several important features of these plots will now be explored.

First, inertial effects lead to particle deposition even in the absence of external forces as shown by the curves for $V_p^t/U_0 = 0$ in Figure 17. In this case there is a critical Stokes number, St_{crit}, below which no deposition occurs. At St_{crit} there is a sharp jump in efficiency, which then increases toward unity with increasing St. The jump is steeper, higher (approaching unit collection efficiency), and occurs for a much smaller St_{crit} in the case with substantial showerhead acceleration than when the particles enter with the fluid face velocity. This effect is discussed in greater detail below. As seen in Figure 17, particle inertia can also lead to deposition even when an external force is pushing particles away ($V_p^t/U_0 > 0$). The extent of external force "protection" is significantly reduced for large initial velocities: compare the $V_{p0}/U_0 = 0$ and 1 curves in Figure 17a and b.

Second, when an external force is directed toward the wafer ($V_p^t/U_0 < 0$), particle deposition occurs at all values of the Stokes number. In the small-St limit (negligible particle inertia), the collection efficiency should tend to Eq. (48) for external forces that can be described by a potential; this trend is clearly evident in Figure 17. For example, for $V_p^t/U_0 = -0.5$, Eq. (48) predicts an efficiency of 1/3 which is the observed asymptote of the appropriate curves in Figure 17a and b. The presence of an attractive external force smooths the shape of the efficiency curves for large attractive forces; however, when the magnitude of the attractive force is small (e.g., $V_p^t/U_0 = -0.01$), the efficiency still exhibits a sharp increase in the neighborhood of St_{crit} (from the no-force case). The rise in efficiency near St_{crit} is much steeper for high initial particle velocities.

Finally, in the large-St limit, the collection efficiency must approach unity for all cases. That is, for large enough Stokes numbers, particle inertia leads to straight trajectories and complete deposition independent of the details of

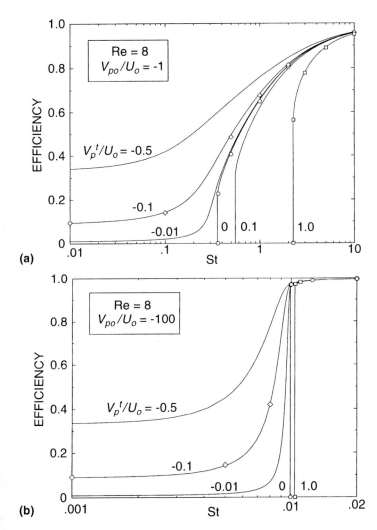

Figure 17 Efficiency versus Stokes number for various deposition velocities for Re = 8. Solid lines are calculated using an analytic flow field and a Runge-Kutta integrator, while the symbols are calculated using numerical flow solutions and the FIDAP particle tracking post-processing routines. (a) Particle dimensionless initial velocity of −1; (b) particle dimensionless initial velocity of −100.

the flow field, the external forces, or particle initial velocity (assuming it is not zero). This limit is approached in all of the calculations shown in Figure 17.

Asymptotic Limit of Critical Stokes Number. An interesting result is suggested by Figure 17b for the case of no external force: as the particle initial velocity becomes large, the collection efficiency tends toward a step function which jumps from zero to unity at a critical Stokes value equal to the inverse of the dimensionless initial particle velocity. To confirm this result, a series of calculations were made to explore the dependence of the critical Stokes number on initial particle velocity in the absence of an external force. These results are shown in Figure 18, where St_{crit} is plotted as a function of V_{p0}/U_0 for fluid Reynolds numbers of 0, 4, and 8. For a given value of V_{p0}/U_0, particles with $St < St_{crit}$ (below the line) will exit the reactor, while particles with $St > St_{crit}$ (above the line) will impact. The effect of Reynolds number is negligible for large values of V_{p0}/U_0; in fact, for $V_{p0}/U_0 > 10$ the three Re curves approach the same asymptotic limit. For large values of V_{p0}/U_0 particle inertia dominates deposition and the details of the flow field become unimportant. Inspection of the large initial-velocity asymptotic limit reveals the following relationship:

$$St_{crit} \rightarrow \frac{U_0}{V_{p0}} \tag{70}$$

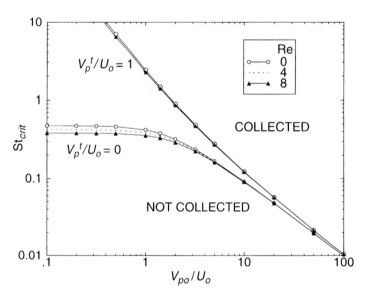

Figure 18 Critical Stokes number versus particle dimensionless inlet velocity. Values of the critical Stokes number were calculated using the analytic approximation to the flow field for Reynolds numbers of 0, 4, and 8. One set of curves applies for no external force ($V_p^t/U_0 = 0$), the other set applies for a strong force resisting deposition ($V_p^t/U_0 = 1$).

A simple explanation of this limit is readily illustrated by rearranging Eq. (70) to give $St_{crit} V_{p0}/U_0 = \tau V_{p0}/S = 1$, which states that impaction occurs when the particle stopping distance based on its *initial* velocity V_{p0} equals the showerhead-to-wafer gap.[16]

Figure 18 also shows the variation of the critical Stokes number when a large external force opposing deposition is applied ($V_p^t/U_0 = 1$). Although the value of St_{crit} is greatly increased for small V_{p0}/U_0 (compared to the case with no external force acting), all curves approach the same asymptote, Eq. (70), in the limit of very large initial particle velocity. The St_{crit} value for a large external force was found to be $\sim4\%$ higher than without an external force for $V_{p0}/U_0 = 100$. As noted above, the influence of Reynolds number on St_{crit} is greatly reduced when a repulsive force is acting.

Thus, for no, or repulsive, external forces, a great simplification results for large values of the initial particle velocity (say for $V_{p0}/U_0 > 100$): the collection efficiency can be closely approximated by a step function (from zero to unity) at a critical Stokes number calculated by Eq. (70). When an attractive force is present, the concept of a critical Stokes number breaks down, as there is some deposition at all Stokes numbers. Even in this case, however, a sharp increase in efficiency near St_{crit} is stil seen (such as shown in Fig. 17b).

3. Coupled Transport: Nondimensional Results

In this section, showerhead-enhanced inertial deposition is explored by coupling the transport of particles through the showerhead and in the interplate region. The procedure is as follows: 1) For given values of U_{jet}/U_0, L/S, and St, calculate the dimensionless velocity of the particle exiting the showerhead, V_{p0}/U_0, using Eqs. (67) and (68); and then 2) using V_{p0}/U_0 as the initial particle velocity, and the parameters Re, St, and V_p^t/U_0, integrate the particle trajectory between the plates to determine the particle collection efficiency. Thus, the coupled particle transport problem (for an infinite parallel-plate geometry and under the present assumptions) is completely specified by five independent dimensionless parameters (note that V_{p0}/U_0 is dependent). Efficiency results from these coupled calculations should look qualitatively like those shown in Figure 17, although some variations are expected as the initial particle velocity is no longer fixed but depends on the degree of particle acceleration through the showerhead.

It is valuable at this point to clarify the use of the jet velocity to face velocity parameter U_{jet}/U_0, which is the local fluid velocity that a particle experiences while passing through a showerhead hole. For the assumption of plug flow through the showerhead, $U_{jet}/U_0 = \overline{U}_{jet}/U_0 = A_{showerhead}/\Sigma A_{jet}$, where $\overline{U}_{jet}$ is the mean velocity in the tube. In the plug-flow case, U_{jet}/U_0 must

---

[16]A further implication of Eq. (70) is that for large values of V_{p0}/U_0 a more appropriate choice for the characteristic velocity in defining particle Stokes number would have been V_{p0}.

always be larger than unity and is constant across each showerhead tube cross section. In commercial reactors, values of $\overline{U}_{jet}/U_0$ are seldom less than 20, and can range up to several hundred. The other limit considered in this work is parabolic flow through the showerhead holes. In this case, U_{jet}/U_0 is function of both the area ratio and the radial starting position of the particle in the showerhead tube. For example, a particle starting on the tube centerline would experience a fluid velocity twice the mean, so that $U_{jet}(r = 0)/U_0 = 2\overline{U}_{jet}/U_0$. Because the fluid velocity must vanish at the tube wall, jet velocity to face velocity ratios less than 1 are possible for the parabolic case for particles starting near the wall. In the following, results are parameterized with the most general form U_{jet}/U_0.

Critical Stokes Numbers. One set of coupled efficiency calculations is shown in Figure 19, which plots efficiency vs. Stokes number for jet-to-face velocity ratios of 0.1, 1, 10, 100, and 1000. For these calculations, the showerhead thickness was assumed equal to the plate gap ($L/S = 1$) for the case of Re = 0 and no external forces acting ($V_p^t/U_0 = 0$). The critical Stokes number (the smallest St for which collection occurs) is found to decrease with increasing values of U_{jet}/U_0. This result is not surprising: as U_{jet}/U_0 increases the particle

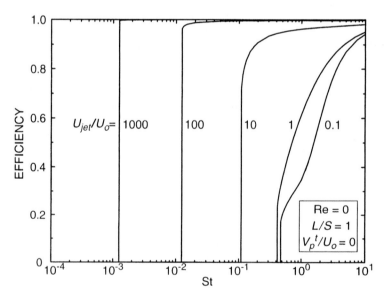

Figure 19   Effect of jet-to-face velocity ratio. Efficiency versus Stokes number for $U_{jet}/U_0 = 1$, 10, 100, and 1000 including coupling between showerhead and interplate transport (for this calculation $L/S = 1$, Re = 0, $V_p^t/U_0 = 0$, and $r_p/S = 1 \times 10^{-7}$).

velocity at the showerhead exit (V_{p0}/U_0) must also increase, and we have shown in Section 2 above that increasing values of V_{p0}/U_0 lead to smaller values for St_{crit} (see Fig. 18). In particular, we have shown in the limit of large V_{p0}/U_0 that $St_{crit} \rightarrow U_0/V_{p0}$. It is interesting to note that, for coupled transport, the large U_{jet}/U_0 limit of St_{crit} is *not* U_0/U_{jet}, but a slightly higher value (e.g., for $U_{jet}/U_0 = 100$, $St_{crit} = 0.01237$). This difference is explained by the fact that, because of inertia, the particle can't accelerate to the jet velocity before exiting the showerhead (i.e., $V_{p0}/U_0 \le U_{jet}/U_0$); consequently, a larger Stokes number is required to initiate deposition. In the slow-jet limit, say $U_{jet}/U_0 < 1$, the critical Stokes number becomes less sensitive to the particle inlet velocity, although the shapes of the efficiency curves can be quite different (e.g., compare the curves for $U_{jet}/U_0 = 0.1$ and 1). Here, the details of the efficiency curve result from a complicated interplay among the parameters. Overall, reducing the value of U_{jet}/U_0 (more and/or larger showerhead holes) increases St_{crit}—with the favorable result of increasing the minimum size for which inertial effects lead to particle deposition on the wafer.

The showerhead thickness also plays a role in determining the magnitude of the critical Stokes number. The effect of showerhead thickness is presented in Figure 20, in which collection efficiency is plotted against St for $L/S = 0.1, 0.5,$

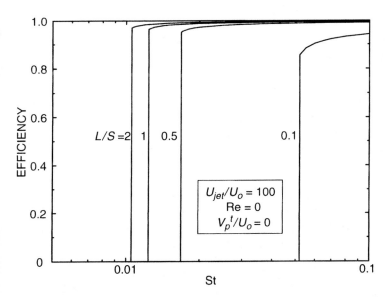

Figure 20 Effect of showerhead thickness. Efficiency versus Stokes number for L/S = 0.1, 0.5, 1, and 2 including coupling between showerhead and interplate transport (for this calculation $U_{jet}/U_0 = 100$, Re = 0, $V_p^t/U_0 = 0$, and $r_p/S = 1 \times 10^{-7}$).

1, and 2 (for $U_{jet}/U_0 = 100$, Re $= 0$, and $V_p^t/U_0 = 0$). For large L/S values, there is sufficient time in the showerhead for the particle to accelerate to the jet velocity ($V_{p0}/U_0 \to U_{jet}/U_0$), which by Eq. (70) gives the asymptotic limit $St_{crit} \to U_0/U_{jet}$. As seen in Figure 20, this limit is approached for $L/S > 2$. For very thin showerheads, the particles spend only a short time in the showerhead, and will exit the showerhead with a velocity much less than the jet velocity. In this case, larger Stokes numbers are needed to initiate deposition—as seen in Figure 20 for the curves with $L/S = 0.1$ and 0.5. Based on these results, St_{crit} can be increased (and inertial deposition reduced) by reducing the dimensionless showerhead thickness L/S.

Grand Design Curves. If the effect of a repulsive force is neglected, the value of St_{crit} (for a given flowfield) is determined solely by the values of U_{jet}/U_0 and L/S. Interestingly, both of these parameters are geometrical in nature. The geometric interpretation of L/S is obvious (the ratio of showerhead thickness to interplate gap), while that for U_{jet}/U_0 requires some explanation. Under the assumption of plug flow within the showerhead holes, it has already been shown that $U_{jet}/U_0 = \overline{U}_{jet}/U_0 = A_{showerhead}/\Sigma A_{jet}$ where $\overline{U}_{jet}$ is the mean velocity in the tube. Thus, in the plug-flow limit the velocity ratio is completely specified by the number and size of the showerhead holes and by the diameter of the showerhead—purely geometric properties of the showerhead. Under the assumption of parabolic flow, the radial starting position of the particle within the showerhead hole must also be considered, but this is another geometrical parameter. Thus, for a specific flowfield and neglecting external forces, the critical Stokes number is uniquely specified by chamber and showerhead geometry (and possibly an assumed particle starting position), and is independent of process parameters (e.g., gas temperature, pressure, or flow rate).

This simplification leads to the idea of the grand design curves, which give critical Stokes number as a function of the velocity ratio U_{jet}/U_0 for various dimensionless showerhead thicknesses L/S. An example of a grand design curve over a wide range of these two parameters is shown in Figure 21 for Re $= 0$ and $V_p^t/U_0 = 0$. The qualitative trends are consistent with earlier discussion: the critical Stokes number decreases with increasing values of U_{jet}/U_0 and L/S. The curves for all values of L/S intersect at $U_{jet}/U_0 = 1$, which corresponds to the case for which the particle enters (and exits) the showerhead at the same velocity. Variations in St_{crit} with L/S become small for dimensionless showerhead thicknesses larger than about 2; in this case, the particle has had sufficient time to accelerate to near the gas velocity, so that making the showerhead longer has little effect. For velocity ratios less than 1, the critical Stokes number is essentially independent of velocity ratio and the dimensionless showerhead thickness. For large velocity ratios, the curves in Figure 21 for different L/S values become parallel with a slope of unity—suggesting that in this limit St_{crit}

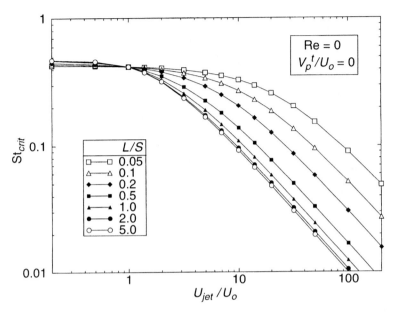

Figure 21 Grand design curve. Grand design curve for estimating the critical Stokes number based on showerhead parameters (for this calculation $Re = 0$, $V_p^t/U_0 = 0$, and $r_p/S = 1 \times 10^{-7}$).

becomes proportional to U_0/U_{jet}. As shown, the value of U_{jet}/U_0 at which this asymptotic limit is reached depends on L/S: for thick showerheads the linear limit is achieved at much smaller velocity ratios than for thin showerheads.

For design applications, the grand design curves are used with the parameters U_{jet}/U_0 and L/S to find the critical Stokes number for the proposed reactor geometry. To minimize particle deposition on the wafer, it is desirable to choose parameters that give as large a critical Stokes number as possible, as increasing St_{crit} increases the minimum size at which inertial deposition begins. Based on Figure 21, larger values of St_{crit} are obtained by decreasing U_{jet}/U_0 (use more and/or larger showerhead holes) or by decreasing L/S (use a thin showerhead or a large interplate gap). Introducing a force that opposes deposition (such as by heating the wafer relative to the showerhead) will always help, but as discussed above, the effect is fairly small under realistic conditions. Once St_{crit} has been determined, the corresponding particle critical size, $d_{p,\text{crit}}$, can be found from Eq. (21). Note that although the critical Stokes number depends only on geometric parameters, the critical particle diameter depends on geometric, process, and particle parameters. Thus, particle density and gas pressure, temperature, type, and flow rate all play a role in determining $d_{p,\text{crit}}$. The effects of process

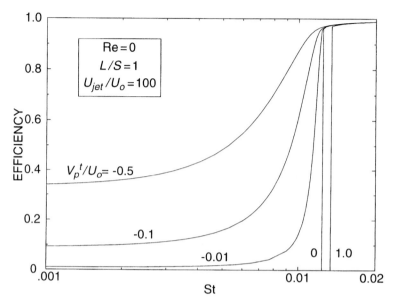

Figure 22   Efficiency versus Stokes number for various deposition velocities. Collection efficiencies for fully coupled particle transport for V_p^t/U_0 values of -0.5, -0.1, -0.01, 0, and 1 (for this calculation Re $= 0$, $L/S = 1$, and $U_{jet}/U_0 = 100$, and $r_p/S = 1 \times 10^{-7}$).

parameters on showerhead-enhanced inertial deposition are discussed in a later section.

External Forces.   Although particle deposition only occurs for St > St$_{crit}$ for repulsive or zero external forces, inertial deposition will always take place when the net external forces are attractive. To demonstrate this behavior, particle collection efficiencies calculated using fully coupled particle transport (i.e., including showerhead acceleration) are shown for various values of the external force in Figure 22 (for Re $= 0$, $U_{jet}/U_0 = 100$, $L/S = 1$). The results are very similar to those in Figure 17b, which gives efficiency for a fixed particle inlet velocity ($V_{p0}/U_0 = 100$) instead of the present case where the showerhead exit velocity is calculated based on showerhead parameters.[17] The large and small Stokes limits are the same: for small St inertial effects vanish and the efficiency must tend to Eq. (48), while for large St inertia dominates and efficiency must tend to unity. At intermediate values of St there are some differences. For example, it is seen that the critical Stokes number for the coupled analysis is slightly larger than for the case where V_{p0}/U_0 is held constant; as

---

[17]Note that although Re $= 8$ in Figure 17b and Re $= 0$ in Figure 22, Reynolds number effects are small.

discussed above, for a finite-length showerhead the particle velocity at the showerhead exit must be slightly less that U_{jet}/U_0, so that a larger Stokes number
is needed to initiate inertial deposition. Note that the case of constant V_p^t/U_0
(i.e., V_p^t/U_0 is independent of particle size or Stokes number) corresponds to
the physically meaningful situation in which a thermophoretic force is acting
(as the thermophoretic deposition velocity is independent of particle diameter).
For other external forces—such as gravity—the net deposition velocity will not
be a constant but will vary with particle diameter (and hence with St).

Parabolic Profile. For fully developed parabolic flow in the showerhead
hole, the gas velocity varies with radial position within the hole as given by
Eq. (45); in this case the local velocity experienced by a particle in a showerhead hole depends on its radial starting position and the mean velocity $\overline{U}_{jet}$ [as
given by Eq. (44)]. For example, a particle starting on the hole centerline will
experience the highest local gas velocity ($2\overline{U}_{jet}$), while particles starting near the
hole wall will experience much lower velocities. As the amount of acceleration
the particle experiences within the showerhead depends on the local gas velocity,
the particle collection efficiency must vary with particle radial position within
the showerhead hole. Thus, the calculation of the net collection efficiency for the
parabolic flow case requires integrating the local efficiency radially across the
showerhead hole. Assuming that the particles are uniformly distributed across
the showerhead hole (i.e., that the flux of particles across the tube cross-section
is constant), the net efficiency is

$$\eta_{net}(St) = \frac{2}{R_{jet}^2} \int_0^{R_{jet}} \eta[U_{jet}(r), St] r \, dr \tag{71}$$

The numerical integration of Eq. (71) is computationally expensive, as each
evaluation of the integrand $\eta[U_{jet}(r), St]$ requires a coupled calculation of the
particle acceleration through the showerhead [with local gas velocity $U_{jet}/(r)$]
along with the corresponding numerical integration of the particle trajectory
between the two plates. The integration of Eq. (71) is further complicated by the
fact that, in some cases, efficiency can change significantly with small changes
in St or $U_{jet}(r)$. In the present work an adaptive Gauss integration scheme with
automatic error control has been used to evaluate Eq. (71).

An example of a net efficiency curve for a parabolic velocity profile is
shown in Figure 23 for the case where there is no external force (Re = 0,
$\overline{U}_{jet}/U_0 = 100$, $L/S = 1$, and $V_p^t/U_0 = 0$). For comparison, efficiency curves
are also shown for plug flow (where all the particles experience the mean velocity) and for the hypothetical case where all of the particles experience the
centerline velocity (essentially a plug flow moving at twice the mean velocity).
These three curves show three interesting effects. First, the critical Stokes number for the centerline case is approximately one half that for the plug flow case;

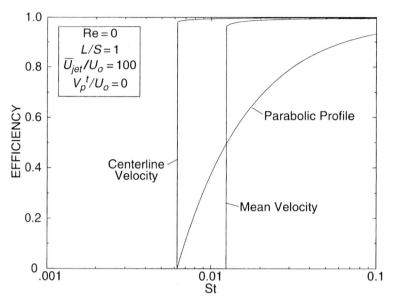

Figure 23 Efficiency versus Stokes number for parabolic showerhead profile (no external force). Collection efficiencies for fully coupled particle transport for particles experiencing the showerhead hole centerline and mean velocities (plug flow assumption), and integrated over the parabolic velocity profile in the showerhead holes in the absence of external forces (Re $= 0$, $L/S = 1$, $V_p^t/U_0 = 0$, and $U_{jet}/U_0 = 100$, and $r_p/S = 1 \times 10^{-7}$).

this result is explained by the fact that the critical Stokes number is inversely proportional to jet velocity in the limit of large values of U_{jet}/U_0 (i.e., doubling the local showerhead gas velocity halves St_{crit}).[18] Second, the efficiency curve for the parabolic case approaches unity much slower than either of the plug flow curves. This is expected, as particles located near the tube wall experience very low local velocities; for some region very close to the wall, particles actually deaccelerate while passing through the showerhead. Note that although this slow approach to complete collection is favorable from a defect reduction point of view, it is not of practical significance because in this large Stokes regime the majority of particles entrained in the flow would still be deposited.

Third, the critical Stokes number for the parabolic case is the same as for the centerline case. The smallest particles deposited are those experiencing the highest velocity in the showerhead hole; thus, for parabolic flow inertia-enhanced deposition begins with those particles moving along the showerhead-hole cen-

---

[18]The critical Stokes number becomes less dependent on the velocity ratio for small U_{jet}/U_0.

terline. The slope of the parabolic case is not as steep as for the centerline case because few particles are contained in the small-area region near the centerline, whereas in the centerline case we have assumed all of the particles are moving at the centerline velocity. The centerline case therefore serves as a limit of the smallest Stokes value for which inertial enhancement to deposition becomes important in parabolic flow. In fact, the centerline case serves as a lower limit for all laminar flow conditions in the tubes, since even for developing flow the maximum velocity in the tube will always be less than $2\overline{U}_{\text{jet}}$. Thus, the most conservative practice for predicting the effects of inertia-enhanced deposition in real reactors is to use $U_{\text{jet}}/U_0 = 2\overline{U}_{\text{jet}}/U_0$ as the characteristic velocity ratio in the grand design curves.

The present results show that, for parabolic flow in the showerhead, the best practice for determining net efficiency is to perform the full integration of Eq. (71). However, since this calculation can be computationally expensive, the following approximations are suggested: for attractive external forces (say, $V_p^t/U_0 < -0.01$), use the mean velocity approximation, otherwise (say, $V_p^t/U_0 > 0$) use the more conservative centerline approximation.

4. Coupled Transport: Dimensional Results

The two approaches to reduce inertia-enhanced particle deposition are[19]: 1) design equipment with as large a value of St_{crit} as possible, and 2) select process conditions that give as high a value of $d_{p,\text{crit}}$ as possible. Based on our previous analysis, the only three ways to increase St_{crit} is to design for minimum $\overline{U}_{\text{jet}}/U_0$ and/or L/S, and to apply an external force that opposes deposition. To minimize $\overline{U}_{\text{jet}}/U_0$, very porous showerhead designs are needed to reduce the constriction of the flow [decrease the area ratio given by Eq. (44) by either increasing the number or the size of holes]. The ratio L/S can be reduced by reducing the showerhead thickness or by increasing the showerhead-to-wafer gap. Intuitively, a short showerhead thickness reduces the time available to accelerate the particle, and a large showerhead-to-wafer gap provides the particle more opportunity to slow down. Finally, an opposing external force could be used to inhibit inertial deposition such as by keeping the wafer warmer than the showerhead to take advantage of thermophoresis but remember that the opposing force typically had a fairly weak effect on reducing inertia-enhanced deposition.

Once a hardware design is fixed, St_{crit} is fixed, but it is still possible to minimize inertial deposition by selecting process conditions that give as high a value of $d_{p,\text{crit}}$ as possible. Based on the free molecule definition of Stokes number in Eq. (23), we can see that low face velocities U_0 and large showerhead-to-wafer gaps S are preferred. Mean gas velocities can be reduced by reducing

---

[19]It is impossible to eliminate inertial effects, as one can always imagine a particle large enough so that it will impact.

mass flow rates (at constant pressure) or by operating at higher pressures (for a fixed mass flow rate). As seen in Eq. (23), operating at higher pressures also directly increases $d_{p,\text{crit}}$. Thus, for a constant mass flow rate, a twofold increase in pressure produces a fourfold increase in $d_{p,\text{crit}}$ (one factor of 2 directly from the pressure reduction, and an additional factor of 2 from lowering the face velocity).[20] Recall that Eq. (23) strictly applies to particles in the free molecular limit (small sizes and/or low pressures) which is a reasonable assumption for the pressure and particle size regimes in typical semiconductor manufacturing reactors.

An example of particle collection efficiency as a function of particle size is shown in Figure 24 for a hypothetical 200 mm (diameter) reactor characterized by an argon flow rate of 1000 sccm (standard cubic centimeters per minute) through a showerhead 2.54 cm thick with 1000 holes of diameter 0.0635 cm with a showerhead-to-wafer gap of 2.54 cm. The flow is assumed isothermal and a particle density of 1 g/cm^3 used. These reasonable physical parameters give the dimensionless quantities $L/S = 1$, $\overline{U}_{\text{jet}}/U_0 = 99.2$, and Re $= 0.984$; the dimensionless deposition velocity varies with size according to the expression for gravitational settling velocity, Eq. (26). Efficiency curves are shown for reactor pressures (i.e., pressure between the two plates) of 0.5, 1.0, 2.0, and 10 torr. For pressures less than 2 torr, inertial enhancement to deposition is clearly evident by the abrupt jump in efficiency from nearly zero to unity in the vicinity of a critical size. As seen, this critical size is a strong function of pressure, with an approximately fourfold decrease in the critical size for a twofold decrease in chamber pressure. For chamber pressures above about 10 torr inertial effects no longer contribute to deposition; the increase in efficiency for increasing particle size seen in Figure 24 for the $P = 10$ torr case results because the gravitational deposition velocity increases with size.

The same reactor geometry and process conditions have been used to calculate the effect of thermophoresis on collection efficiency as shown in Fig. 25. The showerhead temperature has been held constant at 300 K while the wafer (lower plate) temperature is made colder (280 K), isothermal (300 K), or hotter (320 K). Fairly modest temperature differences have been used in accordance with our isothermal (constant gas properties) assumption for flow calculations, but small temperature differences are allowed to drive particle thermophoresis. The isothermal case shows that, for small particle sizes, deposition decreases with decreasing size because the gravitational deposition velocity is decreasing.

---

[20]For higher pressures and/or larger particle diameters the quadratic relationship between pressure and critical diameter fails as the free molecule expression for Stokes number given by Eq. (23) becomes inaccurate. In the continuum limit the particle relaxation time becomes independent of pressure and proportional to diameter squared, so that a fourfold pressure increase would result in a twofold increase in $d_{p,\text{crit}}$. Thus, the influence of pressure on $d_{p,\text{crit}}$ is most pronounced in the free molecule regime.

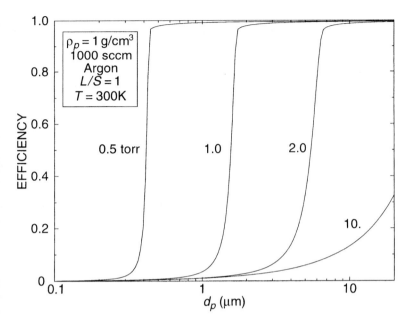

Figure 24 Efficiency versus particle diameter and pressure (isothermal case). Collection efficiencies for fully coupled particle transport assuming plug flow through the showerhead hole for reactor pressures of 0.5, 1.0, 2.0, and 10. torr (Re = 0.98, $L = S = 2.54$ cm, $\overline{U}_{jet}/U_0 = 99.2$, $Q = 1000$ sccm argon, $T = 300$ K, and particle density of 1 g/cm^3).

When the wafer temperature is less than the showerhead temperature, thermophoresis acts to increase deposition compared to the isothermal case. Although the differences become small for sizes larger than $d_{p,crit}$ (about 1.5 μm), the thermophoretic contribution is clear in the small-particle limit where the efficiency approaches a constant.[21] Heating the wafer relative to the showerhead eliminates all small-particle deposition as the thermophoretic resistive force overwhelms the attractive gravitational force. The critical diameter at which inertial effects dominate is shifted to a slightly larger size than for the isothermal case, although the shift is fairly small since the effect of an external force on St_{crit} is fairly weak. By using thermophoretic protection and by designing a reactor with a sufficiently large $d_{p,crit}$, particle deposition can be mitigated over the particle size range of interest.

---

[21] Since the thermophoretic deposition velocity is independent of particle diameter [Eq. (31)] while the settling velocity is proportional to diameter, for small particles thermophoretic deposition dominates and the efficiency becomes constant.

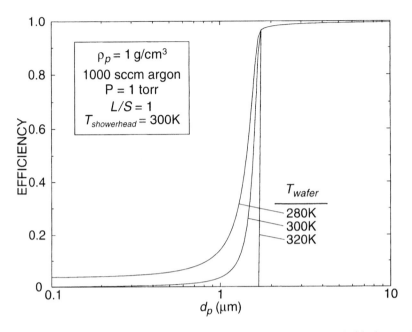

Figure 25 Efficiency vs. particle diameter and wafer temperature (with thermophoresis). Collection efficiencies for fully coupled particle transport assuming plug flow through the showerhead hole for wafer temperatures of 280, 300, and 320 K (Re = 0.98, $L = S = 2.54$ cm, $\overline{U}_{jet}/U_0 = 99.2$, $Q = 1000$ sccm argon, $P = 1$ torr, $T_{showerhead} = 300$ K, and particle density of 1 g/cm^3).

The two previous examples highlight two important points: 1) Inertia-enhanced deposition becomes dramatically more important at low pressure, and 2) thermophoretic protection coupled with careful control of inertial effects can significantly reduce particle deposition. Dimensional calculations could also be used to demonstrate all of the effects reported in Section VI.C.3, however, because of the wide variability in reactor geometry and particle and process parameters, only these two examples are presented.

VII. CHAPTER SUMMARY AND PRACTICAL GUIDELINES

This chapter reviewed the basic phenomena controlling particle transport and the underlying general equations with an emphasis on conditions encountered in semiconductor process tools (i.e., subatmospheric pressures and submicron particles). The discussion included expressions for the following particle forces: fluid drag, gravity, thermophoresis, and electrophoresis. The concepts of particle deposition velocity and stopping distance were introduced, and issues of continuum vs. free-molecular particle transport were outlined. Particle concentrations were assumed to be low enough to allow a dilute approximation, for which the coupling between the fluid and particle phases is one way. In this case, the fluid/thermal transport equations can be solved either analytically or numerically neglecting the particle phase; the resulting velocity and temperature fields were then used as input for the particle transport calculations. Isothermal flow was assumed, although small temperature differences were allowed to drive particle thermophoresis; both analytic and numerical solutions of the flow field representative of a parallel plate geometry were presented. Particle collection efficiency was defined as the fraction of particles present in the interplate region of the reactor that deposit on the wafer. Particles were presumed to either enter the reactor through the showerhead (uniformly spread between $r = 0$ and R_W), or to originate in a plane parallel to the wafer. The latter case corresponds to particles being released from a plasma trap upon plasma extinction or being formed in a nucleation process; in this case, the particles are initially assumed to be uniformly spread radially between $r = 0$ and R_W at some distance h from the wafer. Analytic expressions for collection efficiency were presented for the limiting case where external forces control deposition (i.e., neglecting particle diffusion and inertia).

Particle transport in the parallel-plate geometry was predicted using both the Lagrangian approach (where individual particle trajectories are calculated) and the Eulerian approach (where the particles are modeled as a cloud). The Eulerian formulation yielded an analytic, integral description of particle deposition for the case where the flow field between the plates can be approximated analytically. The strength of the Eulerian formulation is in predicting particle transport resulting from the combination of applied external forces (including the fluid drag force) and the chaotic effect of particle Brownian motion (i.e., particle diffusion), although the current implementation cannot account for particle inertia. In particular, the Eulerian formulation cannot accommodate particle acceleration effects within the showerhead, and is therefore restricted to particle transport in the interplate region.

The need to properly account for diffusion-enhanced particle deposition becomes increasingly important as the semiconductor industry moves toward smaller feature sizes and becomes concerned with smaller-sized particles. Based on the Eulerian analysis, the following guidelines are intended to help tool operators and designers reduce particle deposition when diffusional effects are important:

- Keep traps as far from the wafer as possible.
- Take advantage of repulsive forces, such as by thermophoretic protection gained by keeping the wafer warmer than the showerhead.

- Reduce attractive forces.
- For a specific pressure, use as high a mass flow rate as possible.

The strength of the Lagrangian formulation is in predicting particle transport resulting from the combination of applied external forces and particle inertia, although the current implementation cannot account for particle diffusion. It is the Lagrangian formulation that can properly account for inertia-enhanced deposition resulting from particle acceleration in the showerhead. The problem was treated in two steps, in which both particle and fluid transport were determined: 1) within a showerhead hole, and 2) between the showerhead and susceptor. For fluid and particle transport in the showerhead, approximate analytic expressions were derived based on a few assumptions. The output of this first step was the particle velocity at the exit of the showerhead, as a function of showerhead geometry, flow rate, and gas and particle properties. The particle showerhead exit velocity was next used as an initial condition required for particle transport between the plates. The output of the second step was a prediction of particle collection efficiency by the susceptor (wafer), as a function of showerhead-exit particle velocity, the plate separation, flow rate, and gas and particle properties. Based on the Lagrangian analysis, two approaches were identified to help tool operators and designers reduce particle deposition when inertial effects are important: 1) design equipment with as large a value of St_{crit} as possible, and 2) select process conditions that give as high a value of $d_{p,crit}$ as possible. Based on this analysis, St_{crit} is determined primarily by reactor and showerhead geometry. Only three methods were identified for increasing St_{crit}.

- Decrease the showerhead velocity ratio $\overline{U}_{jet}/U_0$ by increasing the number of and/or enlarging the size of the showerhead holes.
- Decrease the showerhead thickness ratio L/S by making the showerhead very thin or the interplate gap large.
- Apply an external force that opposes particle deposition (such as keeping the wafer warmer than the adjacent gas).

Given a specific hardware design (and corresponding St_{crit}), inertial deposition can be reduced by selecting process conditions that give as high a value of $d_{p,crit}$ as possible. Based on the free molecule limit of Stokes number, the following general guidelines are offered to increase the critical diameter (for a given St_{crit}):

- Increase the gap between the showerhead and wafer.
- Use low mass flow rates.
- Raise chamber pressure.
- Use a high molecular weight gas.

The previous recommendations specifically pertain to reducing particle deposition given an assumed dominant deposition mechanism; note that one set of guidelines (e.g., for inertia) may conflict with those intended to reduce deposition by other mechanisms (e.g., gravity or diffusion). In order to reduce particle deposition in real tools, it is up to equipment designers/operators to first identify the dominant deposition mechanism so that an effective improvement strategy can be identified. Note that the guidelines given above are not intended to replace detailed calculations (using the proper analysis with the actual process conditions), but to provide the user with a general feel for inherently clean practices. In addition, equipment designers should be aware that while these recommendations should improve particle performance, the effect of any changes on process performance must also be investigated.

REFERENCES

1. Allen, M. D., and Raabe, O. G. (1982). Re-evaluation of Millikan's oil drop data for the motion of small particles in air. J Aerosol Sci 6:537–547.
2. Hinds, W. C. (1982). Aerosol Technology. Wiley, New York.
3. Donovan, R. P., Yamamoto, T., Periasamy, R., and Clayton, A. C. (1993). Mechanisms of particle transport in process equipment. J Electrochem Soc 140(10): 2917–2922.
4. Rader, D. J., and Geller, A. S. (1994). Particle transport modeling in semiconductor process environments. Plasma Sources Sci Technol 3:426–432.
5. Friedlander, S. K. (1977). Smoke, Dust and Haze. John Wiley and Sons, New York.
6. Turton, R., and Levenspiel, O. (1986). A short note on the drag correlation for spheres. Powder Technol 47:83–86.
7. Cunningham, E. (1910). On the velocity of steady fall of spherical particles through fluid medium. Proc R Soc A-83, 357–365.
8. Knudsen, M., and Weber, S. (1911). Ann Phys 36:981–994.
9. Ishida, Y. (1923). Determination of viscosities and of the Stokes-Millikan Law constant by the oil drop method. Phys Rev 21:550–563.
10. Rader, D. J. (1990). Momentum slip correction factor for small particles in nine common gases. J Aerosol Sci 21(2):161–168.
11. Millikan, R. A. (1923). The general law of fall of a small spherical body through a gas, and its bearing upon the nature of molecular reflection from surfaces. Phys Rev 22:1.
12. Eglin, J. M. (1923). The coefficients of viscosity and slip of carbon dioxide by the oil drop method and the law of motion of an oil drop in carbon dioxide, oxygen, and helium, at low pressures. Phys Rev 22:161–170.
13. Fuchs, N. A. (1964). The Mechanics of Aerosols. Dover Publications, New York.
14. Clift, R., Grace, J. R., and Weber, M. E. (1978). Bubbles, Drops, and Particles. Academic Press, New York.
15. Henderson, C. B. (1976). Drag coefficients of spheres in continuum and rarefied flows. AIAA J 14(6):707–708.
16. Epstein, P. S. (1924). On the resistance experienced by spheres in their motion through gases. Phys Rev 23:710–733.

17. Talbot, L., Cheng, R. K., Schefer, R. W., and Willis, D. R. (1980). Thermophoresis of particles in a heated boundary layer. J Fluid Mech 101(4):737–758.

18. Batchelor, G. K., and Shen, C. (1985). Thermophoretic deposition of particles in gas flowing over cold surfaces. J Colloid Interface Sci 107(1):21–37.

19. Waldmann, L., and Schmitt, K. H. (1966). Chapter VI. Thermophoresis and Diffusiophoresis of Aerosols in Aerosol Science (C. N. Davies, ed.), Academic Press, New York.

20. Peters, M. H., Cooper, D. W., and Miller, R. J. (1989). The effects of electrostatic and inertial forces on the diffusive deposition of small particles onto large disks: viscous axisymmetric stagnation point flow approximations. J Aerosol Sci 20(1):123–136.

21. Rader, D. J., and Geller, A. S. (1998). Showerhead-enhanced inertial particle deposition in parallel-plate reactors. Aerosol Sci Technol 28:105–132.

22. Bird, R. B., Stewart, W. E., and Lightfoot, E. N. (1960). Transport Phenomena. Wiley and Sons, New York.

23. Rader, D. J., and Geller, A. S. (1994). Particle transport modeling in semiconductor process environments. Plasma Sources Sci Technol 3:426–432.

24. Terrill, R. M., and Cornish, J. P. (1973). Radial flow of a viscous, incompressible fluid between two stationary, uniformly porous discs. J Appl Math Phys (ZAMP) 24:676–688.

25. Houtman, C., Graves, D. B., and Jensen, K. F. (1986). CVD in stagnation point flow: an evaluation of the classical 1D treatment. J Electrochem Soc 133(5):961–970.

26. Robinson, A. (1956). On the motion of small particles in a potential field of flow. Communications on Pure and Applied Mathematics 9:69–48.

27. Rader, D. J., Geller, A. S., Choi, S. J., and Kushner, M. J. (1994a). Application of numerical models to predict particle contamination in semiconductor process environments. 1994 Proceedings Institute of Environmental Sciences 40th Technical Meeting: 308–315.

28. Rader, D. J., Geller, A. S., Choi, S. J., and Kushner, M. J. (1994b). Particle transport in plasma reactors. Microcontamination 94 Conference Proceedings, Oct. 4–6, San Jose, CA: 39–48.

29. Cooper, D. W., Miller, R. J., Wu, J. J., and Peters, M. H. (1990). Deposition of submicron aerosol particles during integrated circuit manufacturing: theory. Part Sci. Technol 8:209–224.

30. Ramarao, B. V., and Tien, C. (1989). Aerosol deposition in two-dimensional laminar stagnation flow. J Aerosol Sci 20(7):775–785.

31. Roth, R. M., Spears, K. G., Stein, G. D., and Wong, G. (1985). Spatial dependence of particle light scattering in an rf silane discharge. Appl Phys Lett 46:253–255.

32. Sommerer, T. J., Barnes, M. S., Keller, J. H., McCaughey, M. J., and Kushner, M. J. (1991). Monte Carlo-fluid hybrid model of the accmulation of dust particles at sheath edges in radio-frequency discharges. Appl Phys Lett 59(6):638–640.

33. Barnes, M. S., Keller, J. H., Forster, J. C., O'Neill, J. A., and Coultas, D. K. (1992). Phys Rev Lett 68:313.

34. Choi, S. J., Ventzek, P. L. G., Hoekstra, R. J., and Kushner, M. J. (1994). Plasma Sources Sci Technol 3:418.

35. Jellum, G. M., Daugherty, J. E., and Graves, D. B. (1991). Particle thermophoresis in low pressure glow discharges. J Appl Phys 69:6923–6934.

36. Shiratani, M., Matsuo, S., and Watanabe, Y. (1991). In Situ Observation of Particle Behavior in rf Silane Plasmas. Jpn J Appl Phys 30:1887.

37. Yeon, C.-K., Kim, J.-h., and Whang, K.-W. (1995). Dynamics of particulates in the afterglow of a radio frequency excited plasma. J Vac Sci Technol A. 13:927–930.

38. Collins, S., O'Hanlon, J. F., Carlile, R., and Kang, J. (1996). Presented at Dusty Plasma-95 (Wickenburg, AZ, Oct. 1–7, 1995) (to appear in J Vacuum Sci Technol).

39. Choi, S. J., Rader, D. J., and Geller, A. S. (1996). Massively parallel simulations of Brownian dynamics particle transport in low pressure parallel-plate reactors. J Vac Sci Technol A 14(2):660–665.

40. Cooper, D. W., Peters, M. H., and Miller, R. J. (1989). Predicted deposition of submicrometer particles due to diffusion and electrostatics in viscous axisymmetric stagnation-point flow. Aerosol Sci Technol 11:133–143.

7

Particulate Deposition in Liquid Systems

Deborah J. Riley
Advanced Micro Devices, Inc., Austin, Texas

Liquid-based processing (i.e., cleaning, etching, developing) is prominent in modern microelectronics fabrication facilities. These liquid processes can often prove to be significant sources of wafer contamination. Wet cleaning is especially problematic due to the large number of wet cleaning steps that are utilized to fabricate a completed device; while film deposition operations contribute more particles per process step than wet cleaning, the use of multiple cleaning steps during fabrication makes exposure of wafers to liquid cleaning solutions the top defect contributor in the cleanroom [1].

Once particles deposit onto a wafer surface, subsequent removal of the particulate matter can prove to be extremely difficult (see Chap. 9); for that reason, avoidance of contamination is an important goal to any facility seeking maximized yield. To prevent or minimize contamination as much as possible, it is crucial that a wet process engineer have a strong understanding of liquid-phase particle deposition levels and mechanisms in cleanroom solutions. Armed with a strong understanding of particle deposition behavior, an engineer can make intelligent decisions regarding process behaviors (which chemistries should be suspect when particle counts jump?), process monitoring decisions (where should particle monitoring efforts be focused?), and process change implications (what particle impact would be expected if chemical dilution led to increased pH and decreased ionic strength?).

I. AN OVERVIEW OF PARTICLE DEPOSITION

When deposition occurs from ambient air (or a process gas), the primary deposition mechanisms encountered are sedimentation, diffusion, impaction, and electrostatic interaction. Sedimentation, which is brought about by gravitational attraction, is of primary importance for particles larger than 1 μm in diameter. For particles smaller than 0.1 μm, diffusion is the dominant mechanism [2]. A detailed treatment of aerosol deposition mechanisms is available in Chapter 6 of this text.

In liquids, all of the transport mechanisms which are active for aerosol deposition can theoretically play a role. The significance of some of these transport mechanisms, however, is greatly reduced. Sedimentation rates, for example, depend upon the difference between particle density and fluid density; in liquids, this difference tends to be small and the gravitational settling velocity is low. For the submicron particles of interest to the microelectronics industry, Brownian diffusion tends to be the dominant mechanism by which suspended particles are brought into the vicinity of wafer surfaces.

To model and understand liquid-based particle deposition, however, consideration of transport mechanisms alone is not sufficient. Deposition rates that are much smaller than those predicted by diffusion rate calculations are often observed. The interaction forces which exist between the particle and the wafer surface can play a critical role in the deposition process. These interactions can create a deposition barrier under certain conditions; it cannot be assumed that all particles which could reach the vicinity of a wafer surface (based upon mass transfer considerations) will actually deposit. Both particle transport to the wafer surface and surface interaction behavior must be considered when predicting deposition rate; either consideration can prove to be rate determining.

It is critical that those wishing to understand particle deposition from liquid systems have a strong understanding of the source, behavior, and influence of wafer/particle interaction energy. Discussion of this topic will comprise a large portion of this chapter. Readers seeking additional insight beyond that available in this text are referred to some of the numerous colloid science texts available [3–8]. While colloid scientists tend to work in conditions which significantly deviate from those found in a cleanroom environment [9–13], the implications of their findings can be readily applied to problems of cleanroom-liquid particle deposition.

Deposition which occurs while a wafer surface remains submerged in a liquid environment will serve as the focus for this chapter. It should be realized, however, that there are other sources of deposition which are related to liquid-based processing. Anytime that a wafer must pass between a gaseous environment and a liquid environment, a gas-liquid interface is encountered. Surface forces such as those due to surface tension can contribute to particle deposition during passage through these interfaces; gas-liquid interfaces are also encountered when wafers are submerged into highly effervescent solutions. Difficulties in cleanly removing the last remnants of liquid from a wafer surface are another source of contamination. Residue deposition from tiny water droplets has been proposed as a contamination source on hydrophobic surfaces [14], and as a source of water spot formation on patterned wafers [15]. Deposition from the thin layer of rinse water covering a hydrophilic surface as it enters the drying process has also been shown to be a potential source of particle addition [16].

II. SURFACE CHARGING IN LIQUID ENVIRONMENTS

When solid surfaces are placed into aqueous environments, they develop a surface charge. This surface charge is acquired by either the dissociation of surface groups or by the adsorption of potential-determining ions. An example of how surface charge is acquired can be given by considering silicon dioxide, a surface of particular interest for microelectronics applications.

The mechanism by which surface charge is established on an oxide surface can be viewed as a two-step process: surface hydration followed by dissociation of surface hydroxyl groups. When the oxide is placed into an aqueous environment, exposed cations (Si) attempt to complete their coordination shell of nearest neighbors by attracting OH^- ions or water molecules; exposed oxygen ions pull protons from the aqueous phase [17]. The end result is that the oxide surface becomes covered with a layer of hydroxyl groups not found in the bulk oxide. The surface density of these hydroxyl groups is typically reported as being between 3 and 6 OH^- per nm^2 for amorphous silica [18–20].

Surface charge develops through the amphoteric reaction of these hydroxyl groups. Dissociation of hydroxyl groups and/or the adsorption of H^+ and OH^- leaves behind a surface with both negatively charged sites and positively charged sites. H^+ and OH^- are **potential-determining ions** for oxides and the pH of the solution will be critical in defining the overall net charge of the surface.

The actual value of the surface charge σ_0 will also be influenced by the concentration and nature of any specifically adsorbing or indifferent ions in solution (see Ref. 21 for an extended discussion). Specifically adsorbing ions are those which have a strong nonelectrostatic affinity for the surface while indifferent ions are those which experience only a coulombic interaction. Figure 1 illustrates the influence of electrolyte concentration on σ_0 (pH) curves for systems without specific adsorption. A general trend is that the magnitude of charge at a given pH increases as the concentration of indifferent electrolyte increases.

Experimental determination of surface charge at an oxide/aqueous interface can be made using a potentiometric titration procedure. A concentrated suspension of colloidal oxide is titrated with acid or base while the pH of the

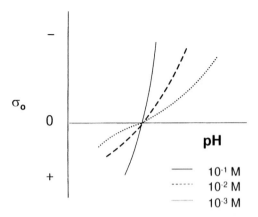

Figure 1   General appearance of the surface charge on oxides as a function of pH when no specific adsorption occurs in the system. Note that at the point of zero charge, electrolyte concentration is inconsequential.

suspension is monitored using a glass electrode. Assuming that the surface area of the oxide is known and the oxide has a stochiometric surface, σ_0 is readily found from the relationship $\sigma_0 = F(\Gamma_{H^+} - \Gamma_{OH^-})$. In this expression, Γ_{H^+} and Γ_{OH^-} are the surface adsorption densities of H^+ and OH^-, while F is Faraday's constant.

The pH at which $\Gamma_{H^+} = \Gamma_{OH^-}$ and $\sigma_0 = 0$ is known as the **point of zero charge** (PZC). At pHs above the PZC, negative sites should dominate and a negative surface charge should exist; at pHs below the PZC, a net positive charge should result. SiO_2 is somewhat unique among the oxides in that it does not show a significant tendency to develop a positive surface charge in the absence of specific adsorption. The PZC of SiO_2 is generally accepted as being pH ≈ 2; a review of the literature, however, shows that values between 1.5 and 3 are routinely reported. As the pH of the solution decreases, the magnitude of any negative surface charge on SiO_2 drops. Even at pHs below 2, however, silicon dioxide surfaces don't show a strong experimental tendency to acquire a positive charge [22]. It is speculated that the basic dissociation of surface hydroxyl groups does not readily occur on silicon dioxide, and thus no excess positive surface charge results [23].

Experimental techniques for determining the PZC of a solid are described by Lyklema [20] and Hunter [3, p. 380; 4, p. 225]. These techniques generally involve the measurement of σ_0 (pH) curves at different concentrations of an indifferent electrolyte. The only pH at which indifferent electrolyte concentration will have no influence on the data is at the PZC where the surface is uncharged. In Figure 1, the PZC is shown to be the only pH at which electrolyte concentration has no consequence. Be aware that if an electrolyte contains an ion which specifically adsorbs, the common intersection points of the curves will not be at the PZC.

Charging mechanisms in nonaqueous liquids are not as well understood as charging mechanisms in aqueous solutions. Surface charge in nonaqueous liquids is believed to come from either preferential adsorption of dissociated cations and anions or from acid-base reactions between particles and solvent [24,25]. While the remainder of this chapter focuses upon the behavior of aqueous systems, many of the theoretical principles can be extended to nonaqueous media. In general, charges tend to be lower in nonaqueous systems, yet the lack of ions in solution prevents screening of the charge that exists; electrical repulsions tend to be longer range interactions in nonaqueous media [21].

III. ELECTROSTATIC DOUBLE LAYERS

While particles suspended in an aqueous solution exhibit a surface charge, the system as a whole must maintain electrical neutrality. An excess of counterions in the solution adjacent to the solid surface balances the surface charge so that, from a distance, the charged particle appears to be electrically neutral. The surface charge and countercharge together form the **electrostatic double layer**. The separation of charge that exists within the double-layer environment leads to the existence of an electrostatic surface potential. The electrostatic potential decays from a finite value at the surface of the solid to a value of zero beyond the outer edge of the double layer (Fig. 2).

Describing the distribution of charge and potential in the double layer is the goal of various double-layer theories. A classic model which tends to perform well in many situations is the Gouy-Chapman-Stern-Grahame (GCSG) model. In certain circumstances, however, GCSG theory tends to be problematic. The double-layer structures of oxides, in particular, appear to be better described by alternative theories. A thorough review of alternative double-layer theories is beyond the scope of this chapter. Emphasis in this review will be placed on the traditional double-layer treatment of the GCSG model; readers wishing for more detailed information are referred to a few of the numerous texts and reviews on the subject [3, Chap. 6; 20,26–29].

A. Gouy-Chapman-Stern-Grahame Theory

In the classical GCSG model, the electrical double layer is regarded as consisting of two regions: the outer diffuse region in which ions undergo thermal motion and the inner Stern layer where adsorbed ions are immobile. Figure 2 illustrates these different regions.

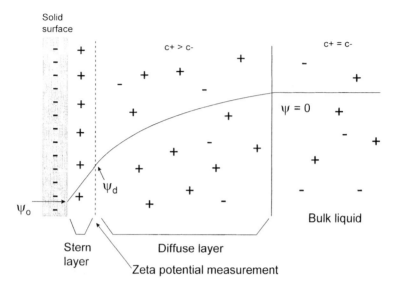

Figure 2   The electrostatic double layer around a solid surface.

1. The Diffuse Layer

In the diffuse region of the double layer, counterions and co-ions are distributed such that a balance exists between their thermal motion and the forces of electrical attraction. The potential distribution in the diffuse layer is described by the Poisson-Boltzmann equation. This equation combines classical expressions from the fields of electrostatics (Poisson equation) and statistical thermodynamics (Boltzmann distribution). The Poisson-Boltzmann equation is given as

$$\nabla^2 \psi = -\frac{e}{\varepsilon} \sum_i z_i n_{io} \exp\left(\frac{-z_i e\psi}{kT}\right) \tag{1}$$

where the mean potential is ψ, ε is the dielectric constant, n_{io} is the ion concentration in the bulk solution, z_i is the ion valency, e is the electronic charge, k is Boltzmann's constant, and T is the absolute temperature.

An exact analysis solution to the Poisson-Boltzmann equation which is generally valid does not exist. System geometry and electrolyte type strongly influence the solution derivation. The most common approximations used to simplify the Poisson-Boltzmann equation are assumptions of small surface potentials and/or symmetrical electrolytes. A symmetrical electrolyte is one in which the valence of the negative ion is equivalent to the valence of the positive ion. For the case of a flat surface, the assumption of a symmetrical electrolyte permits an analytical solution. If the Debye-Hückel approximation is then made (potentials are low so that $|z_i e \psi| < kT$), the solution to Eq. (1) reduces to

$$\psi = \psi_d \exp(-\kappa x) \tag{2}$$

where

$$\kappa = \left(\frac{2z^2 e^2 n_0}{\varepsilon kT} \right)^{1/2} = 3.29\sqrt{I}(nm^{-1}) \qquad \text{at 298 K}$$

In Eq. (2), I is the ionic strength (moles/liter) and ψ_d is the potential at the inner edge of the diffuse layer (see Fig. 2). Note that ε equals the product of ε_0 (the permittivity of a vacuum $= 8.85 \times 10^{-12}$ J m^{-1} V^{-2}) and ε_r (which for water at 25°C is 78.54). The exponential nature of potential decay with distance from the solid surface is obvious from Eq. (2). The parameter κ is known as the reciprocal of the **Debye length**; $1/\kappa$ is often considered a representation of the double-layer thickness. The potential drops to ~2% of the surface value at a distance of ~$3/\kappa$; thus the Debye length serves as a good relative gauge for estimating the range of electrostatic interaction [4, p. 26]. Note that the Debye length depends only on the temperature and the electrolyte concentration in the solution; properties of the surface do not influence this important parameter.

Equation (2) serves as a solution to Eq. (1) only when the assumptions of a low potential, a symmetrical electrolyte, and a flat interface are valid. Solutions under several alternative assumption sets are readily available [3,5,26,30].

2. The Stern Layer

The diffuse layer model above was originally developed by Gouy [31] and, independently, Chapman [32]. In their original work, ions were treated as point charges. If the diffuse layer model is assumed to represent the entire double layer, neglect of finite ionic size leads to poor correlation between experimentally observed double-layer capacitance and theoretical prediction. In 1924, Stern proposed a refinement to the original Gouy-Chapman model whereby ionic dimension is considered in the region closest to the surface [33].

Through Stern's refinement, the double layer is divided into two distinct regions: the inner Stern layer and the outer diffuse layer. The Gouy-Chapman theory is assumed in the diffuse layer region with the Poisson-Boltzmann equation being used to describe the ionic distribution. In the Stern layer, however, the charge and potential distribution is controlled by the finite size of the ions and the short-range interactions that exist between the ion, the solid surface, and the adjoining ions and dipoles. The Stern layer thickness is typically considered to equal about one hydrated ion radius.

Readers wishing for detail on analysis of the Stern layer are referred to a review by Wiese et al. [26]. In that review, equations for charge density in the Stern layer, charge density in the diffuse layer, and the relationship be-tween ψ_0 and ψ_d are available. It's notable that Stern's model permits the first layer of adsorbed ions to specifically interact with the solid surface. The equa-tions describing Stern's double-layer theory include specific chemical adsorption potentials for anions and cations. This specific interaction offers a theoretical explanation for measurements in which ψ_0 and ψ_d are found to be of opposite sign. If specific adsorption of counter-ions is strong enough, it's possible for the potential to change signs within the Stern layer. Alternatively, if the specific adsorption of co-ions is strong, it's possible that ψ_d could be of the same sign as ψ_0, but of larger magnitude (Fig. 3).

Stern's model for the nondiffuse part of the double layer was refined by Grahame in 1947, leading to the current GCSG model [3, p. 381]. Grahame introduced a distinction between the distance of closest approach of fully hy-drated ions in solution (the outer Helmholtz plane = the Stern plane) and the distance of closest approach of specifically adsorbed ions (the inner Helmholtz plane). In terms of understanding particle deposition behavior, the potential at the outer Helmholtz plane (= the Stern plane) is of particular significance as the outer Helmholtz plane marks the beginning of the diffuse layer. The value of

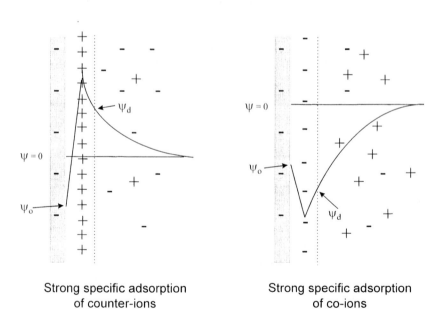

Strong specific adsorption   Strong specific adsorption
of counter-ions   of co-ions

Figure 3   The electrostatic double layer around a solid surface when specific adsorption is significant. If adsorption of counter-ions is strong, the potential at the inner edge of the diffuse layer (ψ_d) can be the opposite sign of the potential at the solid surface (ψ_0). If adsorption of co-ions is strong, ψ_d can be of larger magnitude than ψ_0.

ψ_d (and the associated **zeta potential**, ζ) will be shown to be prominent in calculations of the interaction energy between two solids suspended in an aqueous solution.

B. Oxide Double-Layer Theories

While GCSG theory has been found to work well for nonoxide surfaces, its application to oxide interfaces has been somewhat problematic. A number of research groups [34,35] have had some success in using GCSG theory to fit experimental data for oxide surfaces; if a modified Nernst equation for oxides is used to correlate between the concentration of potential determining ions in solution and the surface potential, the agreement is found to be better than when a classical Nernst equation is considered [34]. There are difficulties which arise in applying classical GCSG theory to oxide/aqueous interfaces, however, which persist even with the modified Nernst equation.

With the classic GCSG vision of the double-layer environment, it would not be possible for the surface charge to exceed that which would develop from the ionization of all hydroxyl groups on the oxide surface. Experimentally, however, measurements of surface charge that significantly exceed this upper limit are obtained. In contrast to the high surface charges measured for oxides, diffuse double-layer potentials ψ_d are found to be relatively low. Application of GCSG theory to oxide/aqueous interfaces would imply that a significant portion of the surface charge must be balanced by counterions residing within the inner Helmholtz layer.

Theories that have been postulated to describe oxide/aqueous interfaces include the porous surface model [20,29] and surface complexation models [27,28,33,36]. In the porous surface model it is postulated that the solid surface is permeable to both potential-determining ions and counter-ions. Because the charge is not restricted to the outer surface plane, experimental surface charges can exceed the charge that corresponds to surface hydroxyl groups; because charge compensation can begin within the porous surface layer, relatively low values for ψ_d are explained. While this model has been quite successful at analyzing SiO_2 data, it does not do as well with other oxides.

In surface complexation theory, it is assumed that counter-ions in the solution are able to form weak, ion-pair–like complexes with charged surface sites. As a result, part of the negative surface charge due to SiO^- groups will be neutralized by complexation with cations from the solution. Similarly, positively charged sites can be neutralized with solution anions. Charges in the diffuse layer of the solution must be large enough to balance only those surface charges that were not neutralized by complexation. Surface complexation theory has generated considerable interest as an alternative theory for oxide interfaces [3,33].

IV. ZETA POTENTIALS

In order to study electrical double layers, it would be ideal to directly measure the potential or charge distribution which exists within the double layer. In practice, this isn't possible. What can be measured is a quantity known as the **zeta potential** (ζ). The zeta potential is measured at a small distance from the solid surface using various techniques based upon electrokinetic phenomena.

A. Particle Zeta Potentials

Particle zeta potentials are typically measured using the technique of electrophoresis. In electrophoresis, an applied field is used to induce movement of a particle through a liquid solution. When a potential gradient is introduced to a liquid suspension, charged particles begin to move with a fixed velocity toward the oppositely charged electrode. As the particle moves, the portion of the double layer closest to the particle travels with the particle.

The zeta potential of a particle (also referred to as its **electrokinetic potential**) is defined as the potential at the shear or slipping plane between the particle and the bulk liquid solution. Countercharge that is located between the particle surface and the shear plane moves with the particle in solution; beyond the slipping plane, charge remains fully mobile. As an approximation, liquid within the Stern layer is usually considered to be rigidly attached to a solid while the diffuse layer is not. As such, $\zeta \cong \psi_d$ is usually considered to be a valid relationship (see Fig. 2). The approximation that $\zeta \cong \psi_d$ is best at low potentials; at higher potentials and high-electrolyte concentrations, the shear plane shifts so that $|\zeta|$ will be less than $|\psi_d|$ [26].

It is not possible to directly "measure" a particle's zeta potential. While tools are commercially available for determining zeta potentials of colloidal particles, these tools actually measure the colloid's **electrophoretic mobility** (i.e., how fast the particle moves in an applied field). Once the electrophoretic mobility is known, calculations can be performed to deduce the apparent particle zeta potential.

A common expression used to relate electrophoretic mobility to zeta potential is Henry's equation for spherical particles. The full form of Henry's equation can be found elsewhere [4, p. 71]. For this review, it is sufficient to realize that Henry's equation approximates the form

$$U = C \frac{\varepsilon \zeta}{\eta} \tag{3}$$

where C is a function of κa. In Eq. (3), U is the measured electrophoretic mobility, ζ is the zeta potential, and η is the liquid viscosity. Recall that the parameter κ is the inverse of the solution Debye length [see Eq. (2)] while a is the particle radius. When double layers are very thin and/or when particles are large ($\kappa a > 100$), C approaches 1. This limit is referred to as the Smoluchowski approximation. When double layers are thick and/or when particles are very small ($\kappa a < 0.1$), C approaches 2/3; this is the Hückel approximation. Because κa values of less than 0.1 are relatively uncommon, the Smoluchowski equation is often a better general assumption that the Hückel equation. The Hückel approximation may have applicability if nonaqueous media are being considered.

A variety of textbooks are available which delve more deeply into a discussion of the relationship between electrophoretic mobility and zeta potential, and plots of C versus κa are readily available in those sources [4,5]. It should be realized that Henry's equation is in itself an approximation. General limitations of Henry's equation are that the electrostatic potential is assumed to be low, and the ionic atmosphere around the particles is assumed to remain undistorted by the applied external field. For a full discussion of these limitations and their impact, the reader is referred to discussions in Hunter [4] or Hiemenz [5].

B. Wafer Zeta Potentials

The technique of electrophoresis is not suitable for measuring zeta potentials on large flat surfaces such as semiconductor wafers. The alternative technique of streaming potential measurement, however, has been successfully applied to the measurement of wafer zeta potential [25,37].

In streaming potential methodology, liquid is forced at a controlled pressure through a narrow gap between two flat plates (wafers surfaces in this case). Excess mobile charge near the walls of the plates (wafers) is carried along by the liquid and accumulates downstream; electrodes are placed at either end of the gap and the steady state potential difference which develops across the system is measured. This recorded potential difference is the streaming potential which can then be related to the zeta potential of the wafer surfaces making up the walls of the cell.

Researchers at the University of Arizona have used streaming potential methodology extensively to monitor zeta potentials on a variety of different Si, SiO_2, and Si_3N_4 wafer surfaces. Figures 4 and 5 illustrate wafer zeta potential measurements made using their custom streaming potential apparatus (illustrated in Fig. 6). In Fig. 4, the zeta potential/pH profiles for Si and SiO_2 wafers are displayed [25]; in Fig. 5, the impact of surface cleaning technique on apparent zeta potential is illustrated for LPCVD nitride wafers [37].

C. Suspension Media Impact on Zeta Potential

Figures 4 and 5 help to illustrate the impact of the surrounding media on zeta potential. As is readily apparent, zeta potentials are generally more negative in basic media and more positive in acidic media. This trend would be anticipated

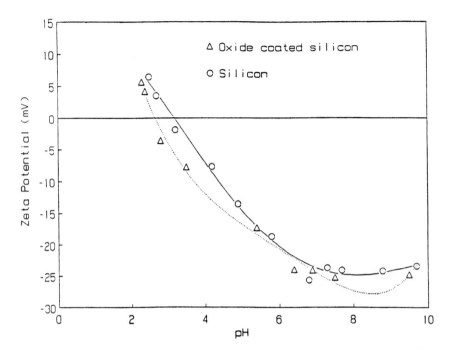

Figure 4   Zeta potential/pH profiles for Si and SiO$_2$ wafers as measured through streaming potential methodology. (From Ref. 25.)

from the impact of pH on surface charge discussed previously. While the zeta potential and the underlying surface charge are typically of the same sign, realize that it is possible for their signs to differ if specific adsorption occurs within the Stern layer.

The pH at which the zeta potential equals zero is known as the **isoelectric point** (IEP). In the absence of specific adsorption, the isoelectric point (IEP) and the point of zero charge (PZC) will be equivalent. Table 1 summarizes the isoelectric points for various metal oxides as reported by Kitahara and Furusawa [38]. Below these IEP values, the oxides will exhibit a positive zeta potential, while a negative zeta potential exists at higher pH levels. For a larger compilation of IEP values, see the review by Parks [39].

While the IEP and the PZC will be equivalent in the absence of specifically adsorbing ions, the presence of such species will cause the two values to shift in opposite directions. Specific adsorption of a cation, for instance, will cause the PZC to shift toward lower pH values while the IEP shifts toward higher pH values. The presence of the cation encourages OH$^-$ adsorption to the solid surface while discouraging H$^+$ adsorption; as a result, it is necessary to increase

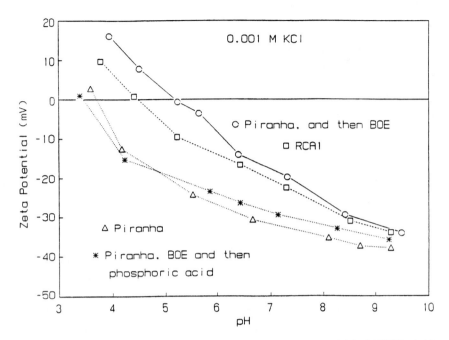

Figure 5   The impact of surface cleaning technique on zeta potential for LPCVD nitride wafers. Measurements are made using steaming potential methodology. (From Ref. 37.)

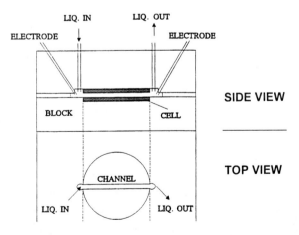

Figure 6   A schematic illustration of the streaming potential apparatus employed by Jan and Raghavan. (From Ref. 37.)

Table 1 IEPs of Common
Metal Oxides

Metal oxide	Isoelectric point
SiO_2	$1.5 \sim 3.7$
Fe_2O_3	$5.7 \sim 6.9$
TiO_2	$6.0 \sim 6.7$
Al_2O_3	$7.4 \sim 9.5$
ZnO	$9.3 \sim 10.3$

Source: Ref. 38.

the solution H^+ concentration in order to return to a condition where $\Gamma_{H^+} = \Gamma_{OH^-}$ (the PZC). Where the zeta potential is measured, however, the presence of the cation makes the effective charge more positive than in the case of no specific adsorption. In order to reach the point where the "effective charge" at the slipping plane equals zero (IEP), it is necessary to compensate for the cation by making the underlying surface charge more negative (less positive); this is done by increasing OH^- concentration in solution. If an IEP is determined in the absence of specific adsorption, a positive zeta potential will be measured at that pH in the presence of specifically adsorbed cation. A zeta potential value of zero will be recorded at a higher pH in the presence of adsorbed cation than when specifically adsorbing ions are not present [21].

In the presence of indifferent electrolyte, the IEP and the PZC do not shift. The measured zeta potential value at other pHs, however, will be influenced by the presence of the electrolyte. The common observation made is that the magnitude of the zeta potential decreases with increasing concentration of indifferent electrolyte. The reader is referred to Hunter [4] for additional discussion on the influence of solution media on zeta potential.

V. INTERACTION POTENTIAL ENERGIES

Discussion thus far has focused upon the environment around a single solid surface. To move toward understanding particle deposition behavior, it's necessary to turn attention toward the interaction of multiple surfaces. Interaction forces between solid surfaces in liquid suspensions have been studied extensively by colloid scientists, and it is well-known that the total potential energy of interaction between suspended particles has a dramatic influence on colloid stability. These same concepts can be readily applied to the problems of cleanroom liquid particulate deposition.

The total potential energy of interaction between two solids can be represented with the well-established DLVO theory. The theory derives its name from the two research teams which independently developed it: Derjaquin/Landau and Verwey/Overbeek.

In DLVO theory, the total potential energy of interaction between two solids is represented by the sum of attractive and repulsive potential energy terms. Primary components of the interaction energy are the van der Waals interaction energy V_A and the electrostatic double-layer interaction V_E. The total potential energy of interaction V_T is represented by the sum of these two terms:

$$V_T = V_A + V_E \tag{4}$$

A. Van der Waals Interaction

The van der Waals force is an intermolecular force which acts between all atoms and molecules. This force can be understood by considering a nonpolar atom such as helium: while the time averaged dipole moment of the atom is zero, at any given instant it possesses a finite dipole moment which can be described by the instantaneous positions of the atom's electrons. This instantaneous dipole moment will generate an electric field which subsequently polarizes a neighboring atom, inducing a dipole moment in it. The interaction that results between these two dipoles gives rise to an instantaneous attractive force; when these dispersive interactions are "summed" over all of the atoms in interacting bodies, a finite attractive force results which is know as the van der Waals–London force [40].

One method of estimating van der Waals forces involves the assumption that individual atomic interactions are linearly additive. This method involves integrating over all pairs of atoms and molecules and leads to van der Waals forces being defined relative to a term known as the **Hamaker constant**. A more rigorous approach for defining van der Waals interactions is based upon Lifshitz theory; this approach depends upon the optical properties of interacting macroscopic bodies and is found to be more accurate. Despite its shortcomings, the older Hamaker approach tends to be more widely used as it is more convenient and gives results that are often not significantly in error. Readers wishing for more information on the derivation and theory of van der Waals forces are referred elsewhere [40–42]. Visser [43] offers a full discussion on the comparison of Hamaker constants with Lifshitz–van der Waals constants.

For interaction between a sphere and a flat plate (i.e., a particle and a wafer surface), the Hamaker expression for the van der Waals interaction term is given as

$$V_A = -\frac{A}{6} \left(\frac{a}{x} + \frac{a}{x + 2a} + \ln \frac{x}{x + 2a} \right) \tag{5}$$

In Eq. (5), A is the Hamaker constant, a is the particle radius, and x is the distance between the particle and the wafer surface. This relationship applies only to the interaction of a sphere with a flat plate; expressions for other geometries are readily available [3]. Equation (5) is commonly found in the literature and has been adopted by some researchers exploring particle deposition onto semiconductor wafers [44]. An expression which accounts for retardation effects, however, can offer greater precision when studying particle deposition theory.

Retardation effects occur due to the fact that van der Waals interactions are electromagnetic in nature and there is a finite time of propagation between interacting solid surfaces. This retardation leads to a reduced interaction for solids that are separated by more than a few nanometers. As the separation between solids increases, the significance of retardation grows. While retardation effects are of little consequence when considering van der Waals forces between solids which are already in contact, the effects can impact particle deposition theory. Gregory [45] gives an approximate expression for calculating retarded van der Waals interactions as

$$V_A = -\frac{Aa}{6x}\left(\frac{1}{1 + 14x/\lambda}\right) \tag{6}$$

This expression works well for x values that are less than ~20% of the particle radius a. While Eq. (6) starts to lose accuracy when the separation between the two surfaces becomes large, the magnitude of V_A becomes so small at these separations that V_A effectively ceases to influence the overall interaction energy. In Eq. (6), λ is a characteristic wavelength for the interaction (often assumed to be ~90–100 nm).

Perhaps the greatest difficulty in accurately estimating van der Waals forces is the selection of an appropriate value for the Hamaker constant A. The value A will be strongly dependent upon the composition of materials involved in the interaction as well as the fluid media in which the interaction occurs. The van der Waals interaction energy will be notably smaller in a liquid than it will be in a gas. Hamaker constants for many materials are available in Visser [43] where it becomes apparent that there is a good deal of scatter in Hamaker values reported in the literature; as a result, there is some degree of uncertainty that exists in the choice of an appropriate A value. For SiO_2 particles interacting with an SiO_2-covered wafer in water, suggested values of A fall between 0.3 × 10^{-20} J and 1.7 × 10^{-20} J. Mixing rules that can be used to calculate effective Hamaker constants when dissimilar materials are involved in the interaction are discussed in Israelachvili [40] and Visser [43].

Finally, it will be mentioned here that under some conditions, negative values can result for the Hamaker constant. This can occur when the Hamaker constant for the suspension media has a value that is intermediate between the Hamaker constants for the two interacting solids. The result is that the solids interact more strongly with the media in which they are suspended than with one another [3]. This is a relatively rare occurrence, but if conditions are created in which the necessary criteria are met, the van der Waals force will actually be repulsive in nature. In situations of general interest to the semiconductor industry, the van der Waals force can be considered an attractive interaction.

B. Electrostatic Double-Layer Interaction

As two solid surfaces approach one another, their diffuse double-layer regions begin to overlap. As a result, the individual double layers of the involved solids become restricted; the limited space available between the approaching solids prevents potential decay from proceeding as it does in bulk solution. The overlap of the diffuse double layers results in a contribution to the total interaction energy between approaching solids. When solid surfaces are of like sign, this double-layer interaction is repulsive in nature; when solid surfaces are oppositely charged, double-layer attraction can occur.

Calculation of the electrostatic double-layer interaction energy V_E is complex. To develop an approximation, numerous assumptions are necessary. A general expression for electrostatic double layer interaction energy that has been applied for microelectronics applications [44,46] is the expression developed by Hogg et al. [47]. This formula, valid for symmetric electrolytes, is given in SI units by:

$$V_E = \pi \varepsilon a (\psi_1^2 + \psi_2^2)$$
$$\times \left\{ \frac{2\psi_1 \psi_2}{\psi_1^2 + \psi_2^2} \ln \left[\frac{1 + \exp(-\kappa x)}{1 - \exp(-\kappa x)} \right] + \ln[1 - \exp(-2\kappa x)] \right\} \quad (7)$$

To reconcile that this is the same expression employed by Saito et al. [44], slight rearrangement is necessary and it must be noted that Saito et al. did not chose to work with SI units; see appendix A of Hunter [4] for a discussion of SI versus CGS electrical units.

In Eq. (7), ψ_1 and ψ_2 represent effective surface potentials of the wafer and the particle (volts). Because it is the diffuse layers which interact as solids approach one another, it is usually considered reasonable to use the diffuse layer potentials (ψ_d) in calculations of interaction potential energy. Recall that ψ_d is, in most cases, represented well by the measured zeta potential, ζ. Substitution of zeta potential values for ψ_1 and ψ_2 in Eq. (7) is common practice.

It should be realized that derivation of Eq. (7) was based upon numerous assumptions and that there are limitations to its applicability. In deriving Eq. (7), it was assumed that solid surfaces approaching one another (i.e., a wafer and a particle) maintain a constant surface potential during the approach. Alternative expressions exist if the assumption is made that the solids maintain a constant surface charge during the interaction [48]; these alternative derivations would have applicability when the surface charge results from ionization of surface groups. When surface charge results from the adsorption of potential determining ions (as with oxide surfaces), it is usually best to assume that the potential remains constant and that the surface charge density adjusts to the double-layer overlap. In reality, it is likely that neither constant surface potential nor constant surface charge models hold exactly.

It was also assumed in the derivation of Eq. (7) that the potentials of the surfaces involved do not exceed a potential of 25 mV and that κa is much greater than 1. Evidence suggests that the expression holds up well in practice as long as surface potential magnitudes don't greatly exceed 50 or 60 mV and as long as κa is kept at a value of 5 or greater [47]. For particle diameters of interest to the microelectronics industry, Eq. (7) becomes somewhat suspect in very low ionic strength solutions (e.g., pure deionized water).

C. Total Interaction Energy

Using appropriate expressions for V_A and V_E, Eq. (4) can be solved to determine the total interaction energy between a particle and a wafer surface. The interaction energy will be dependent upon the distance between the wafer and the particle x, with interaction energies vanishing as separation distances become large.

If total interaction energy becomes positive as a particle approaches a wafer surface, a potential energy barrier to deposition exists. Figure 7 illustrates this situation. To create this figure, Eqs. (6) and (7) were solved for a particle diameter of 0.6 μm, a Hamaker constant of 1.11×10^{-20} J, a particle potential of -20 mV, a wafer potential of -10 mV, an ionic strength of 0.001 M, and a solution temperature of 298 K. Note that Fig. 7 is presented with all interactions shown relative to the thermal energy kT of the system, where k is Boltzmann's constant and T is the solution temperature.

For most conditions of interest, the van der Waals interaction term V_A will be negative; negative interaction energies are attractive in nature. Van der Waals forces tend to be short range and usually approach zero within a few hundred angstroms of the wafer surface. If the electrostatic interaction V_E is positive (as in Fig. 7), the total interaction energy V_T can suggest either a net attraction or a net repulsion depending upon the magnitude of V_E relative to V_A. Depending upon specific system parameters, electrostatic interactions can subside within a few hundred angstroms of the wafer surface or they can extend as much as 1 or 2 μm into the surrounding solution. If a positive (repulsive) V_E dominates V_A, a potential energy peak results; any particle which approaches the wafer surface must surmount this positive V_T peak for deposition to occur.

From Eqs. (6) and (7) it is readily apparent that a variety of different system parameters influence the total interaction energy of a system. Alteration of the

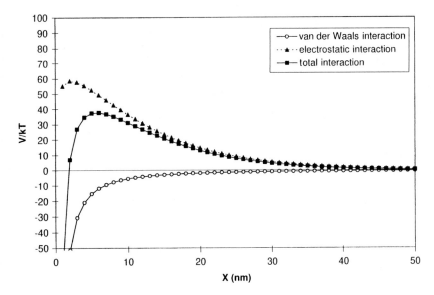

Figure 7 Interaction potential energies relative to the thermal energy of the system. In the situation illustrated, a potential energy barrier to deposition exists. Assumptions are a particle diameter of 0.6 μm, a Hamaker constant of 1.11×10^{-20} J, a particle potential of -20 mV, a wafer potential of -10 mV, an ionic strength of 0.001 M, and a solution temperature of 298 K.

particle diameter, the solution Debye length (via alteration of ionic strength), the wafer potential, or the particle potential can strongly influence the resultant V_T profile.

Figure 8 illustrates the significant impact that ionic strength can have on total interaction energy. Conditions in Fig. 8 are identical to those in Fig. 7 except that a variety of different ionic strengths are explored. Notice that as the ionic strength of the solution increases, both the height of the energy peak and the range of the interaction is reduced. It would be significantly easier for a particle to overcome the energy peak at 0.01 M than the energy peak at 0.0001 M. The likelihood that a repulsive interaction will prevent particle deposition is much stronger when ionic strength is low.

While ionic strength influences both the range of the interaction and the magnitude of the interaction, particle diameter influences only interaction magnitude. Recall from Section III.A.1 that the inverse of the Debye length serves as a good gauge for overall double-layer thickness; because Debye length is not influenced by the properties of the interacting solids, it should not be surprising that particle diameter fails to influence interaction range [Eq. (2)]. Figure 9

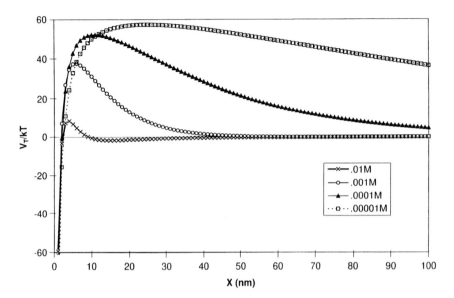

Figure 8   The impact of ionic strength on total interaction energy. Assumptions are a particle diameter of 0.6 μm, a Hamaker constant of 1.11×10^{-20} J, a particle potential of -20 mV, a wafer potential of -10 mV, and a solution temperature of 298 K.

helps to illustrate the influence of particle size on total interaction energy; this figure assumes the same values for key parameters (other than particle size) as used in Figure 7. As is evident in Figure 9, larger particles lead to more significant interaction energies. The likelihood of a repulsive interaction preventing particle deposition shrinks along with particle size.

Finally, consider the role of the wafer and particle potentials. It is somewhat intuitive that as the magnitude of the potentials increases, any existing electrostatic interactions will become more significant. Figure 10 shows the resultant V_T/kT when wafer and particle potential are varied. Figure 10 includes conditions where a repulsion exists (wafer/particle = $-1/-1$, $-10/-10$, $-30/-30$), a situation where no electrostatic interaction exists ($0/-30$), and a condition in which an electrostatic attraction is present ($30/-30$). Notice that if the magnitude of the potentials is small enough ($-1/-1$), an attractive total interaction can exist at all separations regardless of the fact that the two potentials are of the same sign; this occurs when the van der Waals interaction dominates the slight electrostatic repulsion. Figure 10 also illustrates that if one of the two surfaces is at its isoelectric point (potential = 0 mV), there will be no barrier to deposition; in that case, the van der Waals interaction curve defines the total interaction energy. Be aware of the fact that a wafer surface near its IEP can

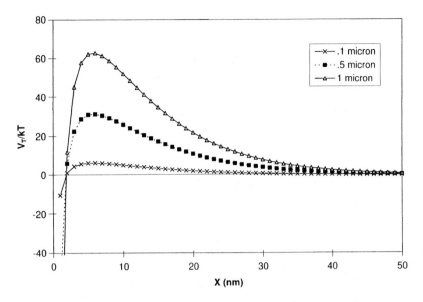

Figure 9 The impact of particle size on total interaction energy. Assmptions are a Hamaker constant of 1.11×10^{-20} J, a particle potential of -20 mV, a wafer potential of -10 mV, an ionic strength of 0.001 M, and a solution temperature of 298 K.

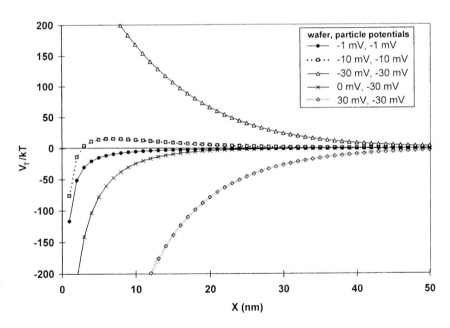

Figure 10 The impact of wafer and particle potential on total interaction energy. Assumptions are a particle diameter of 0.6 μm, a Hamaker constant of 1.11×10^{-20} J, an ionic strength of 0.001 M, and a solution temperature of 298 K.

be extremely sensitive to particulate deposition as it will fail to electrostatically repel both negatively and positively charged particulate matter.

The influence of the electrostatic potential energy curve on resultant deposition can be dramatic. Figure 11, from Riley and Carbonell [46], helps to demonstrate this impact. To gather the data for Figure 11, 100 mm Si wafers were lowered into stirred solutions containing suspended soda-lime glass beads. The suspension media was microelectronics grade DI water. Nominal particle size was approximately 1.6 μm. The wafers were known to have a negative zeta potential in deionized water, and the glass beads in their natural state had a negative zeta potential. Various submersion times were investigated, and it was found that the resultant contamination level was independent of submersion time. This result illustrates the influence of a strong electrostatic repulsion; deposition is prevented from proceeding while the wafers remain submerged in solution. When those same soda-lime glass beads were altered by attachment of aminopropyl surface groups, the beads exhibited a positive zeta potential in water. In all aspects other than the measured zeta potential, the aminopropyl glass beads were identical to the soda-lime beads. Figure 11 shows that the impact of the zeta potential alteration was dramatic. While deposition from the suspension

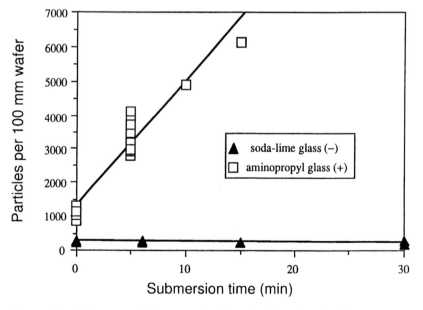

Figure 11   Wafer contamination as a function of submersion time in a contaminated solution with 10,000 particles/mL. The soda-lime glass is repelled by the negatively charged wafer surface, while the aminopropyl glass is attracted to the negatively charged wafer.

media was prohibited in the presence of an electrostatic repulsion (soda-lime glass), it proceeded freely when the repulsion was absent (aminopropyl glass).

When the total interaction energy of a wafer/particle system is repulsive, barriers to deposition can be substantial enough to inhibit particle deposition onto the wafer surface. If an electrostatic barrier does not exist (or is not substantial), deposition can proceed freely at the rate determined by transport limitations. Consideration will be given to transport rates in the next section of this chapter. Using simplified transport models, it is possible to approximate the magnitude of deposition which will occur onto wafer surfaces submerged in a liquid media.

VI. PARTICLE DEPOSITION BEHAVIOR

A. Modeling of Deposition Rate

The rate of particle deposition from cleanroom chemical baths can be controlled by either the transport of the particle toward the wafer surface or by the interaction potential energy of the particle/wafer system. Our emphasis to this point has been on understanding the source of these important wafer/particle interactions. To model deposition rate, however, we must turn attention toward overall mass transport considerations.

In the field of microelectronics, workers seeking to model liquid-based deposition rates onto wafer surfaces have favored a simplified transport model which utilizes the concept of a uniform diffusion boundary layer [44,46]. This simple model has been applied by researchers outside of the microelectronics industry for years [see, for example, Refs. 49–51]; researchers applying the model typically study flow fields in which the boundary layer thickness is well defined (often employing a rotating disc apparatus). Despite the fact that relatively nonuniform flow fields often exist in cleanroom chemical baths, this modeling approach has been found to be highly successful at predicting deposition trends onto wafer surfaces.

In this transport model, contaminant particles are assumed to diffuse through a boundary layer from the bulk liquid to the wafer surface (see Fig. 12). Figure 12 shows that there are actually three "boundary layer" thicknesses of significance in understanding the model. Beyond the hydrodynamic boundary layer δ_H, fluid velocity is defined by bulk liquid motion; within δ_H, velocity will be retarded by the presence of the wafer. Within the diffusion boundary layer δ, particle concentration levels deviate from those found in the bulk solution; beyond the diffusion boundary layer the assumption of a well-mixed solution with uniform particle concentration is considered valid. Finally, δ_E will be referred to as the electrostatic boundary layer thickness; δ_E represents that portion of the diffusion boundary layer for which interaction potential energy

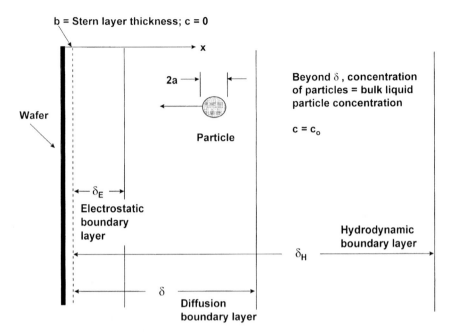

Figure 12 Illustration of the deposition system. Particles diffuse from the bulk liquid to the wafer surface through a diffusion boundary layer, δ. Particle movement is assumed to be unidimensional in the deposition model. b is the Stern layer thickness, a is the particle radius, δ_H is the hydrodynamic boundary layer thickness, and δ_E indicates that portion of the diffusion boundary layer for which interaction potential is significant. Drawing is not to scale.

is significant. Until a particle approaches the electrostatic boundary layer of a wafer, wafer/particle interaction energy fails to influence the particle's behavior.

In the absence of any interaction between the particle and the wafer, steady state particle flux in the direction normal to the surface x would be approximated by the classic Fick's law in which the rate of diffusion is controlled by the particle concentration gradient across the diffusion boundary layer:

$$j = -D\frac{\partial c}{\partial x} \tag{8}$$

The Stokes-Einstein equation is often applied to approximate a diffusion coefficient D relative to the particle radius a:

$$D = \frac{kT}{6\pi\eta a} \tag{9}$$

Equation (8), however, does not adequately describe steady state deposition from a cleanroom solution. If diffusion must occur in the presence of a substantial interaction energy, it is necessary to alter Eq. (8) to reflect the influence of this interaction. Steady state flux to the wafer surface when a substantial interaction energy exists is represented as

$$j = -D\frac{\partial c}{\partial x} - \frac{cD}{kT}\frac{\partial V_T}{\partial x} \tag{10}$$

V_T is the total interaction energy between the particle and the wafer surface. As defined by Eq. (4), this term is approximated by summing the van der Waals interaction energy [Eq. (6)] and the electrostatic double layer interaction [Eq. (7)].

Solving Eq. (10) to find the total particle flux to the wafer surface is relatively straightforward, and details of the solution are available elsewhere [51,52]. If it is assumed that at some value of x very close to the wafer surface ($x = b$) all particles rapidly adhere, then a boundary condition of $c = 0$ at $x = b$ is realized. A second boundary condition can be defined as $c = c_0$ at $x = \delta$ where c_0 is the bulk liquid particle concentration, and δ is the outer edge of the diffusion boundary layer. These boundary conditions yield the following solution to Eq. (10).

$$j = -\frac{c_0 D}{\displaystyle\int_b^{\delta} \exp\left(\frac{V_T}{kT}\right) dx} \tag{11}$$

Note that because the x axis is defined as pointing away from the wafer surface (see Fig. 12), the particle flux to the surface of the wafer will be negative.

If the interaction potential energy in the system is negligible, Eq. (11) reduces to $j \cong c_0 D/\delta$. This is the flux that would be anticipated if diffusion were controlled exclusively by the concentration gradient in the steady state system (no interaction). The exponential term in the denominator of Eq. (11) acts as a retardation (or enhancement) factor with regard to the total deposition that occurs. When the interaction energy becomes significant (especially if a repulsion makes V_T positive), this retardation/enhancement term can dramatically influence the predicted flux.

Readers looking to make deposition predictions based upon Eq. (11) are referred to a discussion by Riley [46] in which such an analysis is carried out. The analysis proposes expressions for the needed limits of integration. The lower limit of integration b is set to the thickness of the Stern layer in Riley's discussion. The expression used to approximate the average diffusion boundary layer thickness (upper integration limit) is found from the expression for uniform flow over a flat plate:

$$\delta = \frac{10}{3}\sqrt{\frac{vL}{u}}\left(\frac{v}{D}\right)^{-1/3} \tag{12}$$

where L is the equivalent length of the plate (wafer), υ is the kinematic viscosity of the liquid, and u is the velocity of bulk liquid flowing past the wafer surface. Notice in Eq. (12) that the size of the diffusing particle influences the thickness of the diffusion boundary layer (through D). Riley [46] suggests an approximate δ of 21 μm for a 1.5 μm particle, and a diffusion boundary layer thickness of 64 μm for a 0.05 μm particle. These predictions were made assuming a 100 mm wafer suspended in a deionized water solution where the bulk liquid velocity is set to 10 cm/s. For these same conditions, the boundary layer thickness for momentum transfer δ_H would be approximately 0.3 cm.

Due to the complex relationship that exists between the total interaction potential energy V_T and the separation distance between the particle and the wafer surface x, it is generally not possible to develop an analytical solution to Eq. (11). While approximate analytical solutions have been considered in the literature [53], numerical integration of the denominator is considered a preferred approach [46].

When conditions lead to large positive values of V_T, the exponential term in Eq. (11) readily grows so large that theoretically predicted particle fluxes approach zero; this corresponds to situations in which the potential energy barrier becomes so large that particles in solution are unable to overcome it. When the interaction potential is negative, however (i.e., a net attraction), the impact of V_T on overall flux is *not* especially dramatic. Why this should be so is most easily understood if Eq. (11) is rewritten as

$$j = \frac{-c_0 D}{\int_b^{\delta_e} \exp\left(\frac{V_T}{kT}\right) dx + \int_{\delta_e}^{\delta} \exp\left(\frac{V_T}{kT}\right) dx} \tag{13}$$

Recall that δ_e represents that portion of δ where interaction potential is significant. Because V_T is effectively zero between δ_e and δ, Eq. (13) can be rewritten as

$$j = \frac{-c_0 D}{\int_b^{\delta_e} \exp\left(\frac{V_T}{kT}\right) dx + \delta - \delta_e} \tag{14}$$

In solutions with an extremely low ionic strength, δ_e can approach 1 or 2 μm, but for higher ionic strengths it rarely exceeds 100 nm; when compared to δ values of 20 μm or more, it is apparent that the quantity $(\delta - \delta_e)$ can often be replaced by δ.

If the interaction potential between a wafer and a particle is large and positive (strong repulsion), the integral in Eq. (14) can have a dramatic influence on the overall flux. If the interaction potential is small or negative (attraction), however, the value of the integral approaches zero. As a result, the denominator in Eq. (14) approaches δ, which is the denominator found if no interaction exists.

The only way to effect an enhancement of particle flux over the situation where no interaction exists is if the ionic strength is so low that δ_e has significance. If δ_e is large enough to influence Eq. (14), it is possible to moderately enhance particle flux; because δ_e is less than 10% of δ even in deionized water, however, the predicted enhancement will not be dramatic. This analysis helps to illustrate that it is more difficult to effect a large enhancement of deposition rate than to effect a large retardation; as particle deposition is rarely a goal in a microelectronics cleanroom, this situation can be considered fortunate. It also helps to illustrate, however, that active deposition does not necessarily indicate that a significant attraction exists between a wafer surface and suspended particulate matter. In reality, active deposition more realistically indicates that a significant barrier to particle deposition is absent. The absence of a significant deposition barrier can be a function of parameters such as particle size (Fig. 9) or ionic strength (Fig. 8); it does not necessarily indicate that the wafer and particle have zeta potentials of opposite sign.

B. The Influence of Solution Chemistry

1. Ionic Strength

As mentioned in Section III.A.1, the Debye length $(1/\kappa)$ is a good relative gauge of double layer thickness; electrostatic potential drops to $\sim$2% of its surface value at a distance $\sim$3/κ. Recall that the Debye length can be directly correlated to ionic strength through the expression

$$\kappa = 3.29\sqrt{I} \tag{15}$$

where κ is in units of nm^{-1}, and I is given in moles/liter. The above expression is applicable for 25°C aqueous solutions in which a 1:1 electrolyte serves as the background electrolyte. Figure 13 illustrates the strong correlation between solution ionic strength and Debye length.

Due to the influence of Debye length on interaction potential energy [see Eq. (7)], it is found that both the strength and the range of electrostatic interaction is significantly reduced as ionic strength is increased. Figure 8 serves to illustrate the strong influence of ionic strength on total interaction energy. As a result of this influence, high ionic strengths tend to dramatically shrink any deposition barriers which might prevent particle deposition. For a system in which the particle and the wafer surface share same-sign zeta potentials, expectation should be for an encouragement of particle deposition as ionic strength increases.

Figure 14 illustrates the predicted impact of ionic strength on particle deposition [46]. For Figure 14, the model of Section VI.A was used to predict deposition onto a 100 mm wafer during a 5 min submersion. The solution was assumed to contain 10,000 particles/mL, and all solid surfaces were assigned

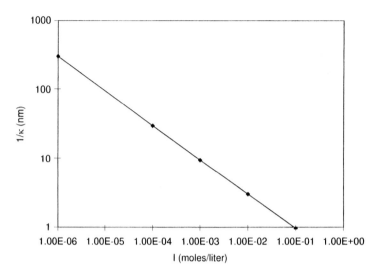

Figure 13 Debye length, $1/\kappa$, as a function of ionic strength.

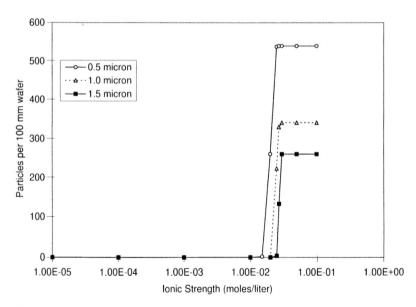

Figure 14 Theoretically predicted bulk deposition from a solution containing 10,000 particles/mL after 5 min of submersion. Both the particle and the wafer surface are assumed to have a negative zeta potential.

negative zeta potentials. Particle diameter (discussed shortly) influences deposition magnitude, and three different particle diameters were considered when creating Fig. 14. At low ionic strengths, the model predicts that a substantial potential energy barrier will prevent deposition. As ionic strength is increased, the barrier eventually shrinks to a point where particles begin to reach the wafer surface. At high ionic strengths, a substantial energy barrier no longer exists, and the rate of diffusion from bulk liquid to the wafer surface defines deposition rate.

Figure 15 is an experimental verification of the predictions shown in Fig. 14 [46]. When Fig. 15 was created, deposition resulting from a 5 min submersion was plotted together with deposition resulting from a rapid dip of a wafer into solution. The data from the rapid dip (labeled "0 minute") serves as a baseline which accounts for deposition from alternative sources (i.e., during wafer passage through the gas/liquid interface, during the drying process, etc.). The soda-lime particles used to create Fig. 15 were not monodisperse; most of the particles were found to be between 0.5 and 1.5 μm in diameter. Figure 15 was created using KCl to adjust ionic strength.

Figures 14 and 15, taken together, serve to illustrate the success of the deposition model in predicting system behavior. The figures also clearly illustrate that increasing ionic strength can cause deposition barriers to collapse. While all surfaces in a system may have negative surface potentials, the shrinking double-layer thicknesses associated with increased ionic strength can enable particle

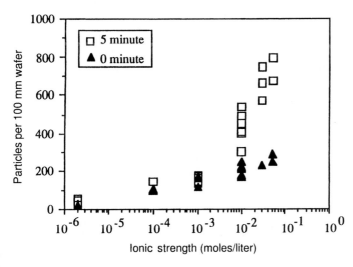

Figure 15 Experimentally observed deposition from a solution containing 10,000 particles/mL. Both the particle and the wafer surface have a negative zeta potential. Each symbol in this figure represents one measurement.

deposition. For the particles depicted here, deposition begins once the ionic strength exceeds 0.01 M.

Figures 14 and 15 address a situation where the wafer and the particle have zeta potentials of the same sign. But what is predicted when opposing signs suggest an attractive electrostatic interaction?

Figure 16 is identical to Figure 14 except that the particles were assigned a positive zeta potential for Figure 16. Figure 17 is Figure 16's experimental counterpart. Figure 17 was created using soda-lime glass beads with attached aminopropyl surface groups; this changes the particle zeta potential from negative to positive at the pH of deionized water [46]. In Figures 16 and 17, it is seen that increasing ionic strength beyond the very low levels in deionized water causes a reduction in deposition. This occurs as the extra ions in solution effectively shield the attractive interaction. Increasing the ionic strength beyond a certain point, however, doesn't encourage a further reduction in deposition. Figures 16 and 17 illustrate that deposition rate is relatively independent of ionic strength when attractive interactions exist except at extremely low ionic strengths. While wafer/particle interactions can prove to be rate limiting when an electrostatic repulsion exists, the absence of a deposition barrier tends to make particle diffusion the rate-limiting factor.

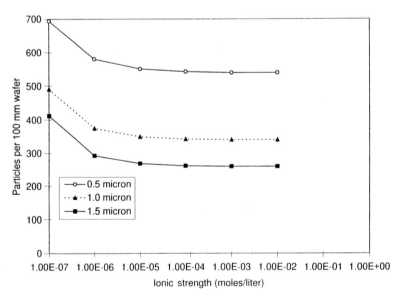

Figure 16   Theoretically predicted bulk deposition from a solution containing 10,000 particles/mL after 5 min of submersion. The wafer surface is assumed to have a negative zeta potential, while the particles are assumed to be positively charged.

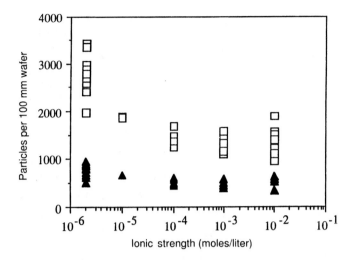

Figure 17 Experimentally observed deposition from a solution containing 10,000 particles/mL. The particle has a positive zeta potential, while the wafer potential is negative.

In summary, the influence of increased ionic strength is a reduction in the influence of electrostatic interaction energy. Electrostatic interactions can significantly influence deposition in deionized water due to the fact that the interactions are relatively long range; reducing the range of the interactions through the addition of ions, however, significantly reduces their relative importance. If a repulsion can prevent deposition at low ionic strengths, moving to a higher ionic strength can induce deposition. If an attraction exists between a particle and a wafer surface, moving away from low ionic strengths may lead to a slight reduction in resultant deposition.

2. pH

When pH is altered in a solution, deposition will be influenced because both the potential of the depositing particles and the potential of the wafer surface are changed. As discussed in Sections II and IV.C, surfaces tend to become more positive (less negative) in acidic solutions and more negative (less positive) in basic solutions.

Each solid surface has a different isoelectric point (IEP); thus the influence of a pH alteration on deposition becomes dependent upon the identity of the materials involved. To deduce the impact of a pH change on deposition, it is necessary to consider the IEPs for the particles/wafer surfaces involved. Consideration of the IEPs allows a determination of whether a pH alteration should increase or decrease the size of the interaction potential energy.

Silicon and silicon dioxide tend to have relatively low isoelectric points (see Fig. 4); as a result, the zeta potentials of wafers in acidic media usually have small absolute values. This means that wafer surfaces will not tend to strongly repel particles in an acid, at the wafer's IEP, there will be no tendency to repel particulate matter regardless of the particle's zeta potential. General expectation should be that the avoidance of particle deposition onto Si and/or SiO2 wafer surfaces is extremely difficult in an acidic environment.

In basic solutions, Si and/or SiO_2 wafer surfaces tend to be strongly negative. As a result, negatively charged contaminant particles may encounter a significant barrier to deposition. Recall, however, that parameters such as ionic strength and particle diameter also influence barrier height; same-sign zeta potentials on particles and wafer surfaces should not be considered a guarantee against particulate deposition. In ascertaining whether a significant potential energy barrier could exist, the specific nature of the primary contaminant particle must be considered. While silicon particles can be significantly repelled by a silicon wafer surface at pH ~8 (all surfaces are negatively charged), alumina particles (IEP ~9) will not respond similarly. At a pH of 8, a negatively charged wafer surface would not repel positively charged alumina particles.

Figure 18 illustrates the impact of pH on particulate deposition using silicon wafers precleaned in a sulfuric-peroxide/dilute hydrofluoric/ammonium hydroxide-peroxide/hydrochloric-peroxide (SPM/HF/APM/HPM) clean sequence [54]. The particles used in the study were polystyrene latex (PSL); these particles had a negative zeta potential over the entire pH range which was studied. At pH values above the IEP of the wafer surface, an electrostatic repulsion prevents particulate deposition. In the vicinity of the IEP (and at pHs below the wafer IEP), deposition occurs as the positively charged or nearly neutral wafer surface no longer repels negatively charged particles. Note in Figure 18 that the smallest particles tested (0.36 μm) deposit significantly at pH 3 while the larger particles don't deposit until the pH level drops further; this particle-size dependence would be anticipated in light of the fact that the interaction energy barrier shrinks along with particle size (Fig. 9).

Finally, consider Figures 19 and 20 to illustrate that deposition response is dependent upon the IEPs of the solids involved [37]. For Figure 19, LPCVD nitride wafers cleaned in SPM were exposed to an aqueous solution contaminated with 0.5 μm PSL (the PSL is negatively charged over the entire pH range). Figure 19 shows the pH dependence of wafer zeta potential along with resultant deposition from a 5 min wafer submersion. Note that deposition onto the wafer surface increases dramatically at pH levels below the wafer's IEP of 3.8. Figure 20 represents an identical study except that the low pressure chemical vapor deposited (LPCVD) wafer was cleaned with SPM then buffered oxide etch (BOE). Jan and Raghavan [37] show that this alternative surface treatment leads to an altered surface potential. In Figure 20, deposition increases dramatically

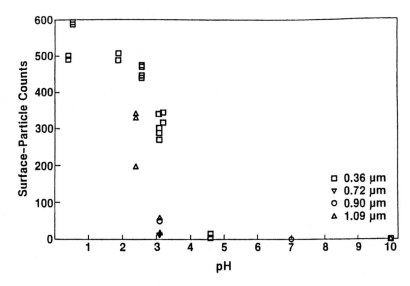

Figure 18 The impact of pH on PSL particle deposition onto silicon wafers precleaned in an SPM/HF/APM/HPM clean sequence. Particle concentration $\sim$ 1000/mL. (From Ref. 54.)

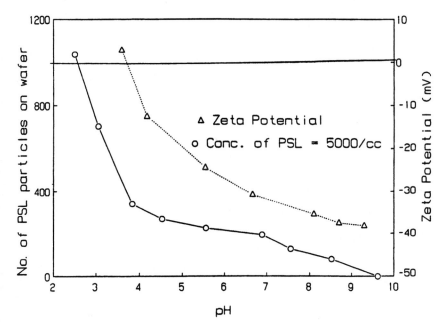

Figure 19 Correlation between zeta potential and PSL particle deposition on LPCVD nitride wafers precleaned with SPM. (From Ref. 37.)

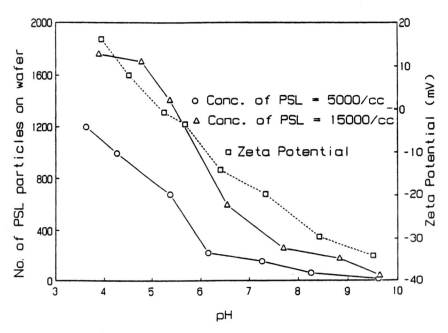

Figure 20   Correlation between zeta potential and PSL particle deposition on LPCVD nitride wafers precleaned with SPM + BOE. (From Ref. 37.)

when the pH drops below this wafer's IEP of 5.3. Note that if both wafer types were submerged into a PSL-contaminated solution at pH 4.5, the deposition resulting over time would differ significantly for the two wafer types.

C.   The Influence of Particle Properties

While the properties of the suspension media can have a dramatic influence on resultant deposition (see Sec. VI.B), the properties of the particle and the wafer surface can also be highly influential. Important parameters that can strongly influence deposition include particle size and particle/wafer surface potential.

1.   Particle Size

As technology advances and the minimum dimension on a chip is reduced, the maximum contaminant particle size which can be tolerated will correspondingly decrease. It is predicted that by the year 2006, particles as small as 0.05 μm will be of concern [55]. While instruments used to detect particles on wafer surfaces cannot yet routinely detect particles that small, modeling predictions can be readily performed to illustrate the impact of shrinking geometries on liquid-based wafer contamination.

From Eq. (11), it becomes apparent that particle diameter can influence deposition in numerous ways; D, V_T, and δ are all dependent upon particle diameter. If the interaction potential energy in the system is negligible (i.e., $V_T \cong 0$), the expression for particle flux to the wafer surface reduces to $j \cong c_0 D / \delta$. Smaller particles have increased diffusion coefficients (tending to increase the flux), but also increased boundary layer thicknesses (tending to decrease the flux). Figure 21 illustrates deposition expected onto a 100 mm silicon wafer when deposition occurs for 5 min, liquid particle concentration is 10,000 particles/ml, and wafer/particle interactions are negligible. The dramatic increase in particle diffusivity as diameter shrinks clearly dominates in this situation.

When interaction energies are considered ($V_T \neq 0$), the relationship between particle size and deposition will be further influenced by the fact that both the van der Waals interaction [Eq. (16)] and electrostatic double-layer interaction [Eq. (7)] are linearly dependent upon particle size. While the results of Fig. 21 will still hold when ionic strength becomes high (and thus interaction energy ceases to be significant), the ionic strength at which this threshold is reached will become a function of the particle diameter being considered.

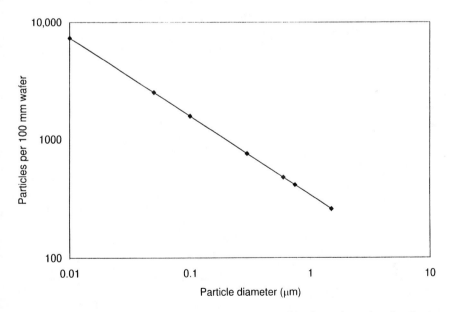

Figure 21 Deposition predicted onto a 100 mm wafer after submersion for 5 min in a solution containing 10,000 particles/mL. Interaction potential energy is assumed negligible in this system.

Figure 22 helps to illustrate the influence of particle diameter on bulk deposition when interaction potential energy is considered. For this figure, the wafer potential was set at -10 mV, and the particle potential was set to -19 mV; the appearance of the figure will be strongly influenced by the choice of potentials. As anticipated from Figure 21, smaller particles deposit more rapidly than larger particles when the ionic strength is high ($V_T = 0$). It is also obvious, however, that the ionic strength at which the repulsive energy barrier is first overcome is significantly reduced as particle diameter shrinks.

For the surface potentials considered to create Figure 22, only particles that are 0.3 μm in diameter or greater will be prevented from depositing in very low ionic strength solution. Even in pure 18 MΩ DI water, deposition of particles 0.1 μm (or smaller) could occur. Figure 22 helps to illustrate that as device geometries shrink and smaller particles become killer defects, deposition from cleanroom liquids will tend to become an increasingly difficult problem. Solutions which don't currently lead to significant particulate deposition (such as DI rinse water) may become problematic as killer defect size is reduced.

2. Surface Potentials

While particle diameter influences Eq. (11) through various terms, the only way that particle surface potential can influence deposition is through its influence on the interaction potential energy V_T. Specifically, the magnitude of surface potential influences the magnitude of the electrostatic potential energy term [Eq. (7)]. When the interaction potential energy is not strongly influencing deposition (for

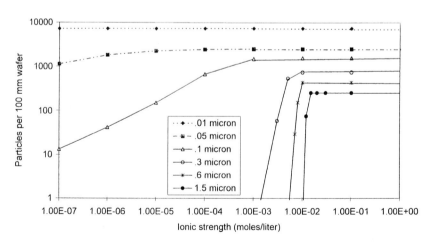

Figure 22 Theoretically predicted deposition as a function of ionic strength for particles of varying size. The wafer potential is set to -10 mV, and the particle potential is -19 mV. The liquid particle concentration is set to 10,000 particles/mL.

instance, when ionic strength is high), the surface potentials of the wafer and the particle surface will have no influence on deposition level.

Figure 23 is identical to Figure 22 except that both wafer and particle potentials are now assumed to equal -35 mV. Under the conditions in Figure 23, only the smallest particles (0.01 μm) would be expected to deposit onto a wafer surface from deionized rinse water; the response to ionic strength depicted in Figure 23 differs notably from the response in Figure 22 where deposition of much larger particles occurs. These figures would suggest that efforts to increase repulsion by altering surface potential could be advantageous as chip dimensions shrink; this alteration of surface potential could theoretically be accomplished through the use of chemical additives. The use of surfactants for this purpose is discussed in Section VI.D. Recall that adjustment of pH is an additional way in which surface potential can be altered.

D. The Influence of Surfactants

Surfactants (i.e., surface-active agents) are substances which can adsorb onto surfaces or interfaces and significantly alter the properties of the interface. These substances have traditionally been used in the microelectronics industry to control/improve a solution's ability to wet the wafer surface (for instance, in BOE solutions). Those wanting to gather additional general information on surfactants are referred to an excellent review by Rosen [56].

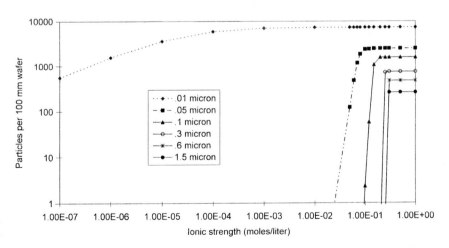

Figure 23 Theoretically predicted deposition as a function of ionic strength for particles of varying size. This figure differs from Figure 22 in that an assumption of -35 mV was made for both wafer and particle potential.

While the industry has made use of surfactants for wettability control for years, recent interest has surfaced in the ability of surfactants to influence particle deposition behavior. Researchers in Japan have put particular emphasis on the impact of surfactants on deposition behavior [57–61].

When surface active agents (surfactants) are placed into an aqueous media with suspended particulate matter, the surface active agent can adsorb to the solid surface and alter its surface potential. As discussed previously in this chapter, the relative zeta potentials of a particle and a wafer surface can dramatically influence the tendency for deposition to occur; it would therefore be anticipated that proper surfactant selection could be used to influence particle deposition behavior.

Investigations by Kezuka et al. [57,59] help to illustrate the impact that surfactants can have on particle deposition. In studies performed by these investigators, an emphasis was placed on particle deposition from acidic media. Si wafer surfaces were used together with suspended polystyrene latex (PSL) particles to study the impact of surfactant usage. Various anionic, cationic, and nonionic surfactants were used in the investigation, with anionic surfactants found to be the most effective at controlling deposition from an acidic media.

The pH used by Kezuka et al. was 3.3, and without surfactant it was determined that the zeta potential of Si in suspension was −32 mV; the PSL particles in the study exhibited a zeta potential of +39 mV. As might be anticipated under these conditions, significant deposition (>10,000 particles on a 4 in. wafer) resulted as a barrier to deposition did not exist. When an anionic surfactant (200 ppm of hydrocarbonic sulfate) was added to solution, both the Si (−32 mV) and the PSL (−67 mV) were found to exhibit negative zeta potentials. When the deposition study was repeated using the anionic surfactant, deposition was found to be dramatically reduced (470 particles on a 4 in. wafer).

Results of Kezuka et al.'s study help to illustrate the possibilities that exist in the use of surface active agents to mediate particulate contamination from liquid suspensions. Some of the potential surfactant applications being studied at this time include the use of surfactants in CMP slurries to reduce residual particle levels [62], and the use of surfactants in cleaning solutions to enhance clean effectiveness [63].

E. The Influence of Particle Transport

Finally, it's worth emphasizing that deposition from a liquid media onto a wafer surface can be controlled by either the wafer/particle interactions in the system or by particle transport considerations. When a significant interaction barrier is not present, deposition will occur as rapidly as permitted by the rate of transport from the bulk solution to the vicinity of the wafer surface. Conditions in a chemical bath which strongly influence this transport to the wafer surface can have a profound effect on wafer contamination levels.

A common chemistry used in microelectronics manufacturing is the HPM formulation comprised of HCl, H_2O_2, and water. Because traditional HPM formulations ($\sim$1 part HCl, 1 part H_2O_2, 5 or 6 parts water) have a low pH and a high ionic strength, the influence of electrostatic interactions tends to be extremely limited for this chemistry. Recall that Si and SiO2 tend to have low IEPs; thus wafer surfaces tend to have low magnitude zeta potentials in acidic media. Because most contaminants tend to be negatively charged, an HPM formulation would have to be adjusted to a pH level greater than the wafer's IEPs (i.e., pH $> \sim$2) to make a wafer/particle repulsive interaction possible. Further, ionic strength would have to be maintained at a relatively low level (see Sec. VI.B.1) to prevent deposition. Hurd et al. [64] have demonstrated that if HCl concentrations are kept below 0.01 M, particle deposition levels can be significantly reduced. The reason for particle control under these conditions is that the wafer/particle repulsive interaction can begin to dominate deposition behavior.

In a study by Riley et al. [65], however, it was shown that less dramatic alterations to the HPM formulation can also have a dramatic impact on particle contamination levels (Fig. 24). In their study, it was found that a 1:1:20 HPM (60°C) formula leads to significantly less wafer contamination than a 1:1:6 for-

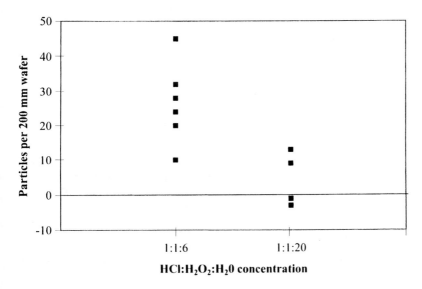

Figure 24 Experimentally observed deposition onto 8 inch wafers from two different formulations of HPM.

mula (85°C). The HCl concentration in the 1:1:20 formula is 0.55 M, which is significantly higher than the range where particle protection from a repulsive potential energy barrier should exist. Despite this fact, significant differences are noted in deposition between the two HPM recipes. Deposition rate with both recipes would be controlled by the rate of particle transport to the vicinity of the wafer surface. If particles are more readily able to reach the wafer surface (or if the number of detectable particles is increased), the number of depositing particles will be higher. While 1:1:6 HPM at 85°C is highly effervescent, the 1:1:20 60°C formulation is not. The violent nature of the effervescent formulation is believed to either increase particle transport rate, increase the number of significant particles in suspension, or both.

A second example of the impact of transport considerations is a study on megasonic-induced deposition by Li et al. [66]. Megasonic energy is commonly used now in industry to aid in removal of particulate contamination from wafer surfaces. In the study by Li et al., however, it was found that the use of megasonic energy can be detrimental under certain circumstances. When contamination levels on wafer surfaces are greater than the relative contamination level of the cleaning solution, megasonic energy is highly beneficial in promoting particle removal. When very clean wafer surfaces are submerged into relatively contaminated cleaning baths, however, the addition of the megasonic energy was actually found to promote particle deposition. The impact of the megasonic energy in this situation would be limited to its impact on particle transport; megasonic energy would not influence the wafer/particle interactions in the system. These observations help to emphasize that particle deposition behavior from cleanroom liquids can be controlled by either particle transport considerations or by wafer/particle interaction energy; the impact that a process change has on wafer cleanliness can be associated with its impact on either mechanism.

REFERENCES

1. Markle, R. J., Dance, D., and Menon, V. B. Role of Equipment in Silicon Defect and Impurity Issues, Proceedings of Microcontamination '91. San Jose, CA, 1991, pp. 609–620.
2. Liu, B. Y. U., and Ahn, K. Particle deposition on semiconductor wafers. Aerosol Sci and Tech 6:215, (1987).
3. Hunter, R. J. Foundations of Colloid Science, vol. 1. Clarendon Press, Oxford, 1986.
4. Hunter, R. J. Zeta Potential in Colloid Science: Principles and Applications. Academic Press, San Diego, CA, 1981.
5. Hiemenz, P. C. Principles of Colloid and Surface Chemistry, 2nd ed., Marcel Dekker, Inc., New York, 1986, chaps. 12 and 13.

6. Shaw, D. J. Introduction to Colloid and Surface Chemistry, 3rd ed., Butterworths, London, 1980.
7. Ross, S., and Morrison, I. D. Colloid Systems and Interfaces. John Wiley and Sons, New York, 1988.
8. Solid/Liquid Dispersons (Th. T Tadros, ed.). Academic Press, London, 1987.
9. Marshall, J. K., and Kitchener, J. A. The deposition of colloidal particles on smooth solids. Colloid Interface Sci 22:342 (1966).
10. Hull, M., and Kitchener, J. A. Interaction of spherical colloidal particles with planar surfaces. Trans Faraday Soc 65:3093 (1969).
11. Davis, A. M. J. Fine particle deposition in laminar flow through parallel plate channels. J Colloid Interface Sci 94:140 (1983).
12. Dabros, T., and van de Ven, T. G. M. A direct method for studying particle deposition onto solid surfaces. Colloid and Polymer Sci. 261:694 (1983).
13. Adamczyk, Z., Zembala, M., Siwek, B., and Czarnecki, J. Kinetics of latex particle deposition from flowing suspensions. J Colloid Interface Sci 110:188 (1986).
14. Christensen, K. HF last performance in centrifugal spray processors, presented at the Electrochemical Society Meeting, Phoenix, AZ, October 1991.
15. MacKinnon, S. Water spot formation on hydrophobic silicon surfaces, Proceedings of Microcontamination '94, San Jose, CA, 1994, pp. 174–184.
16. Riley, D. J., and Carbonell, R. G. Mechanisms of particle deposition from ultrapure chemicals onto semiconductor wafers: deposition from a thin film of drying rinse water. J Colloid Interface Sci 158:274 (1993).
17. Parks, G. A., and deBruyn, P. L. The zero point of charge of oxides. J Phys Chem 66:967 (1962).
18. Furlong, D. N., Yates, D. E., and Healy, T. W. Fundamental Properties of the Oxide/Aqueous Solution Interface. In: Electrodes of Conductive Metallic Oxides, part B (S. Trasatti, ed.), Elsevier Scientific Publishing Co., New York, 1981, p. 368.
19. Zettlemoyer, A. C., and McCafferty, E. Water on oxide surfaces. Croat Chem Acta 45:173 (1973).
20. Lyklema, J. The electrical double layer on oxides. Croat Chem Acta 43:249 (1971).
21. Lyklema, J. Structure of the Solid/Liquid Interface and the Electrical Double Layer. In: Solid/Liquid Dispersions (Th. F Tadros, ed.), Academic Press, London, 1987.
22. Li, H. C., and deBruyn, P. L. Electrokinetic and adsorption studies on quartz. Surface Science 5:203 (1966).
23. Ahmed, S. M. Electrical Double Layer at Metal Oxide-Solution Interfaces. In: Oxides and Oxide Films, vol. 1 (J. W. Diggle, ed.), Marcel Dekker, Inc., New York, 1972, p. 320.
24. Ali, I., Raghavan, S., and Risbud, S. H. Charged particles in process liquids. Semiconductor International: 92 (April 1990).
25. Ali, I. Electrokinetic Characterization of Particulate/Liquid Interfaces and Their Importance in Contamination from Semiconductor Process Liquids. Doctoral Dissertation, University of Arizona 1991.
26. Wiese, G. R., James, R. O., Yates, D. E., and Healy, T. W. Electrochemistry of the Colloid-Water Interface. In: Physical Chemistry Series Two, vol. 6: Electrochemistry (J. O'M Bockris, ed.), Butterworths, London, 1976, p. 53–102.

27. Davis, J. A., James, R. O., and Leckie, J. O. Surface ionization and complexation at the oxide/water interface. J Colloid Interface Sci 63:480 (1978).
28. Yates, D. E., Levine, S., and Healy, T. W. Site binding model of the electrical double layer at the oxide/water interface: J Chem Soc Faraday Trans I 70:1807 (1974).
29. Perram, J. W., Hunter, R. J., and White, H. J. L. Charge and potential at the oxide/solution interface. Chem Phys Lett 23:265 (1973).
30. Verwey, E. J. W., and Overbeek, J. Th. G. Theory of the Stability of Lyophobic Colloids. Elsevier Publishing Co., New York, 1948.
31. Gouy, G. Sur la constitution de la charge electrique a la surface d'un electrolyte. J Phys Thero Appl 9:457 (1910).
32. Chapman, D. L. A contribution to the theory of electrocapillarity. Phil Mag 25:475 (1913).
33. Bijsterbosch, B. H. Electrical Double Layers at Interfaces Between Colloidal Materials and Ionic Solutions. In: Trends in Interfacial Electrochemistry (A. F. Silva, ed.), Reidel Publishing Co., Norwell, MA, 1986, p. 187.
34. Levine, S., and Smith, A. L. Theory of the Differential Capacity of the Oxide/Aqueous Electrolyte Interface. Dis Faraday Soc 52:290 (1971).
35. Blok, L., and deBruyn, P. L. The ionic double layer at the ZnO/solution interface. J Colloid Interface Sci 32:518 (1970).
36. Westall, J., and Hohl, H. Comparison of electrostatic models for the oxide/solution interface. Adv Colloid Interface Sci 12:265 (1980).
37. Jan, D. E., and Raghavan, S. Electrokinetic characteristics of nitride wafers in aqueous solutions and their impact on particulate deposition. J Electrochem Soc 141:2465 (1994).
38. Kitahara, A., and Furusawa, K. Chemistry of Dispersion and Emulsion. 7th ed., Kougakutosyo Ltd., Tokyo, 1979.
39. Parks, G. A. The isoelectric points of solid oxides, solid hydroxides, and aqueous hydroxo complex systems. Chem Revs 65:177 (1975).
40. Israelachvili, J. N. Intermolecular and Surface Forces with Applications to Colloidal and Biological Systems. Academic Press, Orlando, FL, 1985.
41. Krupp, H. Particle adhesion theory and experiment. Adv Colloid Interface Sci 1:111 (1967).
42. Visser, J. Adhesion of Colloidal Particles. In: Surface and Colloid Science, vol. 8 (E. Matijevic, ed.), John Wiley and Sons, New York, 1976, pp. 3–84.
43. Visser, J. On Hamaker constants: a comparison between Hamaker constants and Lifshitz-van der Waals constants. Adv Colloid Interface Sci 3:331 (1972).
44. Saito, A., Ohta, K., and Oka, H. Particle Deposition Mechanism onto Si Wafer, Proceedings of Microcontamination '91, San Jose, CA, 1991, pp. 562–569.
45. Gregory, J. Approximate expressions for retarded van der Waals interactions. J Colloid Interface Sci 83:138 (1981).
46. Riley, D. J., and Carbonell, R. G. Mechanisms of particle deposition from ultrapure chemicals onto semiconductor wafers: deposition from bulk liquid during wafer submersion. J Colloid Interface Sci 158:259 (1993).
47. Hogg, R., Healy, T. W., and Fuerstenau, D. W. Mutual coagulation of colloidal dispersions. Trans Faraday Soc 62:1638 (1966).

48. Weise, C. R., and Healy, T. W. Effect of particle size on colloid stability. Trans Faraday Soc 66:490 (1970).
49. Marshall, J. K., and Kitchener, J. A. The deposition of colloidal particles on smooth solids. J Colloid Interface Sci 22:342 (1966).
50. Hull, M., and Kitchener, J. A. Interaction of spherical colloidal particles with planar surfaces. Trans Faraday Soc 65:3093 (1969).
51. Spielman, L. A., and Friedlander, S. K. Role of the electrical double layer in particle deposition by convective diffusion. J Colloid Interface Sci 46:22 (1974).
52. Riley, D. J. Mechanisms of Particle Deposition from Ultrapure Liquids onto Silicon Wafers, Doctoral Dissertation. North Carolina State University, 1992, p. 205.
53. Ruckenstein, E., and Prieve, D. C. Rate of deposition of Brownian particles under the action of London and double-layer forces. J Chem Soc Faraday Trans 2, 69:1522 (1973).
54. Albaugh, K. B., and Reath, M. Surface Chemistry of PSL Sphere Deposition onto Silicon Surfaces, Proceedings of Microcontamination '91. San Jose, CA, 1991, pp. 603–608.
55. The National Technology Roadmap for Semiconductors Technology Needs. 1997 edition, p. 170.
56. Rosen, M. J. Surfactants and Interfacial Phenomena, 2nd ed., John Wiley and Sons, New York, 1989.
57. Kezuka, T., Ishii, M., Unemoto, T., Itano, M., Kubo, M., and Ohmi, T. Particle Deposition Control for Various Wafer Surfaces in Acidic Solutions with Surfactant, 1994 Proceedings of the IES. Chicago, IL, 1994, pp. 283–288.
58. Saito, A., Ohta, K., Takahara, Y., and Oka, H. Novel Method for Prevention of Particle Deposition in Wet LSI Processes. In: ULSI Science and Technology (J. M. Andrews and G. K. Gellar, eds.), The Electrochemical Society, Pennington, NJ, 1991, pp. 749–754.
59. Kezuka, T., Ishii, M., Hosomi, T., Suyama, M., Maruyama, S., Itano, M., and Kubo, M. The Behavior of Particles in Liquid Chemicals and Their Deposition Control onto Silicon Wafer Surfaces, Chemical Proceedings of the 1993 Semiconductor Pure Water and Chemicals Conference. Santa Clara, CA, 1993, pp. 144–158.
60. Itano, M., Kezuka, T., Ishii, M., Unemoto, T., Kubo, M., and Ohmi, T. Minimization of particle contamination during wet processing of silicon wafers. J Electrochem Soc 142:971 (1995).
61. Saito, A., Ohta, K., Itoh, M., and Oka, H. Prevention of Particle Deposition in HF Solution, Proceedings of the Third International Symposium on Cleaning Technology in Semiconductance Device Manufacturing (J. Ruzyllo and R. Novak, eds.). The Electrochemical Society, Pennington NJ, 1994, pp. 427–433.
62. Free, M. L., and Shah, D. O. Using surfactants in iron-based CMP slurries to minimize residual particles. Micro 16, 5:29 (1998).
63. Ohmi, T. Proposal of Advanced Wet Cleaning of Silicon Wafers, Proceedings of the Fourth International Symposium on Cleaning Technology in Semiconductor Device Manufacturing (R. Novak and J. Ruzyllo, eds.). The Electrochemical Society, Pennington NJ, 1996, pp. 1–12.

64. Hurd, T. Q., Mertens, P. W., Hall, L. H., and Heyns, M. M. Metal removal without particle addition. UCPSS 1994 Proceedings, Acco, Leuven, 1994, p. 41.

65. Riley, D. J., Glick, J. S., Parks, V., and Matamis, G. The impact of temperature and concentration on SC2 cost and performance in a production environment. Mat Res Soc Symp Proc, Vol. 477: Science and Technology of Semiconductor Surface Preparation (G. S. Higashi, M. Hirose, S. Raghavan, and S. Verhaverbeke, eds.), 1997, pp. 519–526.

66. Li, L., Hall, R. M., Jarvis, T., Parry, T., Hawthorne, R. C. An Investigation of Particle and Metallic Deposition in Megasonic Wafer Cleaning, 1996 Proceedings of the IES, Orlando, FL, 1996, pp. 94–99.

8
Deposition of Molecular Contaminants in Gaseous Environments

Allyson L. Hartzell
Analog Devices Incorporated, Cambridge, Massachusetts

I. THE AIR ENVIRONMENT IN A SEMICONDUCTOR FABRICATION FACILITY

The cleanrooms of today's semiconductor manufacturing facilities have unique air environments. The air is highly filtered for particles over both the coarse and fine-size ranges. The number of air exchanges per hour ($\sim$10/min) is much larger than in ordinary office and home environments. Air is forced downwards unidirectionally, entering the cleanroom from the ceiling at a relatively high velocity ($\sim$50 cm/s) compared to air movement in a typical indoor environment. This flow rapidly removes particles generated in the cleanroom by moving the air down and out of the room and through another series of particle filters prior to reintroduction into the "clean" airstream that reenters the cleanroom. What have traditionally not been controlled or even monitored are the molecular contaminants that readily penetrate these high-quality particle filters, although some advanced cleanroom designs in the United States do now include hardware to remove specific families of gaseous species in the air-handling systems. This chapter reviews the types and properties of airborne molecular contaminants found in a cleanroom and their deposition on wafer surfaces.

The cleanroom air chemistry signature is unique to each semiconductor fab line. In addition to contributions from the ambient outside air, multiple sources of air chemistry exist within the cleanroom: volatiles from heated cleaning baths, offgassing of paints and cured materials of construction, offgassing of tool materials, human operators, photoresist chemistry, and solvents. The ratio of outdoor make-up air to recirculated air is much lower in typical fab environments when compared to indoor office environments which can result in a buildup (or increase in concentration) of these internally generated contaminants in the airstream. Outdoor make-up air can introduce combustion products such as nitrogen oxides, ozone, alkenes, alkanes, aldehydes, and sulfur oxides. Nonanthropogenic species such as limonene and alpha-pinene can be present in rural settings. An understanding of how these various species deposit on surfaces and their effect on manufacturing operations, device performance, and reliability are now issues of growing importance to the industry.

II. NTRS REQUIREMENTS

The National Technology Roadmap for Semiconductors [1] has included air chemistry requirements in its 1997 release. Table 1 [1] shows the airborne contaminant concentration thresholds by technology and process step. The real concern is deposition to and adsorption by the wafer. Table 2 [1] presents the wafer surface preparation technology requirements by technology and process area. The required front-end-of-line concentrations of surface metals are much lower than the back-end-of-line concentrations as gate processing with specific metallic contamination can result in 1) reduced breakdown potentials, due to interface trapping states [2,3], and 2) degraded charge retention times, due to decreases carrier lifetimes [3]. The relationship between airborne contaminant concentration and surface deposition is presented in the next section.

III. DEPOSITION VELOCITY

Deposition of molecular contaminants on wafer surfaces is a function of the flux of the species to the surface, their sticking coefficients (or the probability that the species impacting the surface will remain at the surface for a finite residence time), and the concentration of the airborne species. The parameter, deposition velocity, v_d, is defined as the flux of molecules sticking at the surface divided by the airborne concentration of the species beyond the surface boundary layer analogous to the particle deposition velocity discussed in Chapter 6.

Deposition to surfaces is split into three governing resistances [4]:

- Transport to the surface boundary layer (or aerodynamic resistance r_a)
- Diffusive transport through the boundary layer (boundary layer resistance r_{bl})
- Transfer resistance to the surface (r_s)

Table 1 NTRS Air Chemistry Requirements

Year of first product shipment	1997	1999	2001	2003	2006	2009	2012
Technology generation	250 nm	180 nm	150 nm	130 nm	100 nm	70 nm	50 nm
Airborne molecular contaminants (pptM)							
LITHO Bases (as amines)	1000	1000	1000	1000	1000	1000	1000
GATE Metals (as Cu, $\gamma = 2E\text{-}5$)	0.7	0.3	0.3	0.2	0.1	0.07	<0.07
GATE Organics (as MW = 250, $\gamma = 1E\text{-}3$)	300	200	200	100	100	70	50
SAL/CONT Acids (as Cl^-, $\gamma = 1E\text{-}5$)	10	10	10	10	10	10	10
SAL/CONT Bases (as Na^+, $\gamma = 1E\text{-}6$)	80	40	30	20	10	4	<4

Ion indicated is basis for calculation. Exposure time is 60 min, with starting surface concentration of zero. Basis for lithography is defined by lithography roadmap. Gate metals and organics scale as surface preparation roadmap metallics and organics, respectively. Salicidation and contact acids and bases scale as surface preparation BEOL anions and metal, respectively. All airborne molecular contaminants calculated as $J_S = \gamma \langle v \rangle / 4)C$, where J_S is the surface arrival rate (molecules cm^{-2} s^{-1}), γ is the sticking coefficient (between 0 and 1), C is the concentration in the air (molecules/cm^3), and $\langle v \rangle$ is the average thermal velocity (cm/s).

Source: Ref. 1.

Table 2 NTRS Surface Preparation Technology Requirements

Year of first production shipment Technology generation	1997 250 nm	1999 180 nm	2001 150 nm	2003 130 nm	2006 100 nm	2009 70 nm	2012 50 nm
Front End of Line (A)							
Critical metals (atoms/cm^2) (F)	5E9	4E9	3E9	2E9	1E9	<1E9	<1E9
Other metals (atoms/cm^2) (G)	5E10	2.5E10	2E10	1.5E10	1E10	5E9	<5E9
Organics/polymers (C atoms/cm^2) (H)	1E14	7E13	6E13	5E13	3.5E13	2.5E13	1.8E13
Back End of Line (K)							
Metals (atoms/cm^2) (L)	1E11	5E11	4E11	2E11	1E11	<1E9	<1E9
Anions (atoms/cm^2) (M)	1E11	1E11	1E11	1E11	1E11	1E11	1E11
Organics/polymers (C atoms/cm^2) (N)	1E14	7E13	6E13	5E13	3.5E13	2.5E13	1.8E13
Oxide residue (O atoms/cm^2) (N)	1E14	7E13	6E13	5E13	3.5E13	2.5E13	1.8E13

(A) Starting wafer up to deposition of the premetal dielectric.
(F) DRAM requirement for Ca, Co, Cu, Cr, Fe, K, Mo, Mn, Na, Ni, W measured post critical clean for a gettered wafer.
(G) DRAM requirement for Al, Ti, V, Zn (Ba, Sr, and Ta if present in the factory measured post critical clean for a gettered wafer).
(H) Measured post critical clean including pregate, prepoly, premetal, presilicide, precontact, and pretrench fill.
(K) Polysilicide metal dielectric deposition through passivation.
(L) K, Li, Na, measured post critical clean.
(M) Cl, N, P, S, F measured post critical clean. Assumes no fluorinated oxide.
(N) Measured post critical clean of a metallic surface region.
Source: Ref. 1.

The deposition is limited by the highest resistance, or slowest deposition term. The resistances have units of inverse velocity [4]:

$$v_d = (r_a + r_{bl} + r_s)^{-1} \tag{1}$$

Aerodynamic transport to the surface boundary layer by convection is very fast in a cleanroom laminar flow environment and will seldom be the rate-limiting factor in the deposition of molecular contaminants on a wafer surface. Wafers in the cleanroom are typically exposed to cleanroom air in wafer cassettes where the wafers are in close proximity to one another; in this case, diffusion in the space between wafers may make transport to a wafer surface the limiting factor in surface deposition. However, usually the transport mechanisms of most importance to indoor cleanroom surfaces are molecular and turbulent diffusion within the boundary layer and the net surface adsorptive/desorptive rate.

IV. MOLECULAR DIFFUSION

Fick's first law of diffusion relates species flux to the surface, J_D, to the species concentration gradient through a proportionality term, D_g. This constant D_g is the species diffusion coefficient; in cleanroom air this is the diffusivity of the molecular contaminant in air. C is the species concentration in the cleanroom air where y is the distance perpendicular to the wafer surface.

$$J_D = D_g \frac{\partial C}{\partial y} \tag{2}$$

The Hirschfelder equation provides an accurate estimation of the diffusion coefficient of a species in air if empirical data are not available [5–8]:

$$D_g = \frac{BT^{3/2}\sqrt{1/M_1 + 1/M_2}}{PR_{12}^2 I_D} \tag{3}$$

where

D_g = gas diffusivity coefficient (cm^2/s);
$B = 10^4[10.7 - 2.46\sqrt{1/M_1 + 1/M_2}$;
T = temperature in degrees Kelvin, 293.15°K;
M_1, M_2 = molecular weights of air and a molecular contaminant;
$1/M_1 = 0.0345$ (air);
P = pressure at which diffusion takes place, in our case, 1 atm;
$R_{12} = R_1 + R_2/2$ = collision radius, Å.

The collision radius is determined through a summation of atomic volumes, where radius

$$R = 1.18V^{1/3}$$

Element	Atomic volume (Å^3)
Air	29.0
Carbon	14.8
Hydrogen	3.7
Nitrogen in secondary amine	12.0
Silicon (approximation)	20.0

I_D = collision integral for diffusion, a function of kT/ε_{12}

ε_{12} = energy of molecular interaction (ergs)

$\dfrac{\varepsilon}{k} = 1.15T_b = 1.92T_m$, where T_b = boiling point and T_m = melting point; for air, $\varepsilon_1/k = 97°\text{K}$

$$\frac{\varepsilon_{12}}{k} = \sqrt{\frac{\varepsilon_1}{k} * \frac{\varepsilon_2}{k}}$$

k = Boltzmann's constant

A. Diffusivity of HMDS

As an example, the diffusivity in air of hexamethyldisilazane [HMDS, $(CH_3)_3$ SiNHSi$(CH_3)_3$, a wafer-priming agent commonly used in photolithography applications] is calculated:

HMDS atomic volume:

$$V_2 = \sum V_{\text{atomic}} = \sum (V_{6C} + V_{19H} + V_{1N} + V_{2Si}) = 211.1$$

For air,

$$R_1 = 3.62$$

The collision radius of an air molecule and an HMDS molecule,

$$R_{12}^2 = \left(\frac{3.62 + 1.18V_2^{1/3}}{2}\right)^2$$

HMDS parameters:

Boiling point $T_b = 398°\text{K}$

Molecular weight $M_2 = 161.4$ g/mole

$$\frac{kT}{\varepsilon_{12}} = \frac{27.76}{\sqrt{T_b}} = 1.3913 \qquad \text{for HMDS diffusion in air}$$

From Table I [8], the interpolated value of the collision integral I_D as a function of a kT/ε_{12} value of 1.3913 is $I_D = 0.6183$ [8].

Substituting the above expressions into the Hirschfelder equation yields

$$D_g = \frac{22.03 - 5.07\sqrt{.0345 + 1/M_2}\sqrt{.0345 + 1/M_2}}{I_D(3.62 + 1.18V_2^{1/3})^2} * \frac{T^{3/2}}{298.15^{3/2}}$$

$$= 0.0590 \text{ cm}^2/\text{s} \qquad\qquad (4)$$

where $T = 293.15°$K.

For diffusion of HMDS in air (air is the component in higher concentration), $D_g = 0.0590$ cm^2/s.

B. Alternate Calculation for Diffusivity of Small Molecules

For molecular contaminants (assumed spherical) with a diameter much less than the mean free path through air (66 nm at 20°C, 1 atm), the molecular diffusion coefficient can also be approximated by [4]:

$$D_g \approx \frac{2(1.657)\lambda kT}{3\pi \mu d_p^2} \qquad\qquad (5)$$

where k is Boltzmann's constant, T is the temperature, λ is the mean free path, μ is the kinematic viscosity of air ($1.5E^{-5}$ m^2/s at 293 K), and d_p is the gaseous particle diameter. This expression is based on the Stokes-Einstein relationship and is similar to the expression presented in Chapter 6, Eq. (36), for the particle diffusion coefficient in the free molecular regime. The mean free path for a low concentration species in air can be predicted by

$$\lambda = \frac{1}{\pi(1+z)^{1/2}N\delta^2} \qquad\qquad (6)$$

where z is the ratio of the mass of the species under study to the mass of air,

$$z = \frac{M_{\text{species}}}{M_{\text{air}}} \qquad\qquad (7)$$

N is the molecular concentration of air in molecules per cubic centimeter, and δ is the binary collision diameter of the species and air:

$$\delta = \frac{\delta_{\text{species}} + \delta_{\text{air}}}{2} \qquad\qquad (8)$$

Table 3 Empirical Diffusion Coefficients

Compound	D (cm²/s)	Compound	D (cm²/s)	Compound	D (cm²/s)	Compound	D (cm²/s)
pentane	0.0842	formic acid	0.1530	ethylene glycol	0.1005	methylethyl ketone	0.0903
hexane	0.0732	acetic acid	0.1235	propylene glycol	0.0879	dichloromethane	0.1037
octane	0.0616	propionic acid	0.0952	diethylene glycol	0.0730	tetraethylpyrophosphate	0.0475
benzene	0.0932	n-caproic acid	0.0602	triethylene glycol	0.0590	methylpropyl ketone	0.0793
toluene	0.0849	methyl acetate	0.0978	ethylene glycol mono methyl ether	0.0884	1,1,1-trichloroethane	0.0794
o-xylene	0.0727	ethyl acetate	0.0861	ethylene glycol mono ethyl ether	0.0788	mesityl oxide	0.0760
m-xylene	0.0688	methyl-isobutyrate	0.0748	ethylene diamine	0.1009	tetrachloroethylene	0.0797
p-xylene	0.0670	n-propyl acetate	0.0768	n-butyl amine	0.0872	bromine	0.1064
chlorobenzene	0.0747	iso-propyl acetate	0.0770	iso-butyl amine	0.0900	carbon disulfide	0.1045
methyl alcohol	0.1520	ethylene glycol mono ethyl ether acetate	0.0610	diethyl amine	0.0993	benzyl alcohol	0.0712
ethyl alcohol	0.1181	diethyl phthalate	0.0497	triethyl amine	0.0754	benzyl chloride	0.0713
n-propyl alcohol	0.0993	dibutyl phthalate	0.0421	dimethyl formamide	0.0973	aniline	0.0735
iso-propyl alcohol	0.1013	diisooctyl phthalate	0.0370	acrylonitrile	0.1059	isophorone	0.0602
n-butyl alcohol	0.0861	benzyl acetate	0.0600	benzonitrile	0.0710	allyl chloride	0.0975
tert-butyl alcohol	0.0873	carbon tetrachloride	0.0828	triethyl phosphate	0.0552	bromoform	0.0767
acetone	0.1049	chloroform	0.0888	tributyl phosphate	0.0432	mercury	0.1423

Source: Ref. 9.

where

$$\delta = \frac{2(1.18V_2)^{1/3} + 2R_{air}}{2}$$

Empirical data for diffusivity coefficients of some compounds potentially important to cleanroom air are available and listed in Table 3 [9].

V. TURBULENT DIFFUSION

The flux of species through a motion of eddies is expressed in a similar form as the molecular diffusion equation. The difference is in the coefficient of proportionality between the flux due to turbulence, J_T, and the concentration gradient in the y direction from the surface. This is termed the eddy diffusivity, $D_e(y)$, which is a key concept of the mixing length theory. This flux is expressed [10] as

$$J_T = D_e(y)\frac{\partial C}{\partial y} \tag{9}$$

where the eddy diffusivity is

$$D_e(y) = k_0^2 \frac{du}{dy} y^m \tag{10}$$

and y is the distance in the turbulent diffusion zone perpendicular to the surface, k_0 is von Karman's constant ($k_0 \approx 0.4$) and du/dy is the slope of the mean air velocity as taken through the boundary layer [10]. The empirically determined parameter m is in the range of 2–3 [11].

VI. SURFACE FLUX

The impingement rate of a gaseous species per unit area of a surface can be derived using the kinetic theory of gases [12]. The resulting collision flux with the surface depends on the mean kinetic thermal velocity, the average speed at which a molecule of mass M approaches the surface. The collisional flux is

$$J_C = \frac{\langle v \rangle}{4} C \tag{11}$$

where C is the concentration of airborne molecules and $\langle v \rangle$ is the mean thermal velocity of the molecules. The arithmetic mean speed of the molecules can be calculated through statistical mechanics and by equating the kinetic gas pressure with the thermodynamic gas pressure:

$$\langle v \rangle = \sqrt{\frac{8kT}{\pi M}} \tag{12}$$

The collision flux can also be expressed in the following manner [12]:

$$J_C = \frac{P}{\sqrt{2\pi MkT}}$$ (13)

At atmospheric pressure, $P = 10^6$ dynes/cm^2, $T = 300°$K, and $M = 28$ amu (nitrogen dimer), $J_c = $ 3E23 molecules cm^{-2} s^{-1} [12].

The collisional flux is multiplied by the probability of a molecular collision resulting in a trapping event at the surface, γ (the sticking coefficient, which is detailed in a later section). The resulting surface flux expression, J_S, is [10,12]

$$J_S = \gamma J_c = \gamma \frac{\langle v \rangle}{4} C$$ (14)

All three resistance terms are in one conceptual deposition velocity expression, which is an approximation of Eq. (1) [10]:

$$\frac{1}{v_d} \approx \frac{1}{\gamma(\langle v \rangle/4)} + \frac{\chi}{D_g} + \frac{1}{(m-1)k_0^2\, du/dy}\left(\frac{1}{\chi^{(m-1)}} - \frac{1}{\beta^{(m-1)}}\right)$$ (15)

where the boundary layer has been divided into two regimes: one with turbulent transport dominant $(\beta - \chi)$, and one with molecular diffusion dominant (χ), depicted pictorially in Fig. 1. This first term in this deposition velocity expression is the surface flux term. The remaining terms are derived by separating the diffusive boundary layer into molecular and turbulent diffusive regimes and integrating over the transition locations χ and β [10]. The constant m is from Eq. (10).

When the turbulent diffusion term is negligible compared with the molecular diffusion, $\chi \approx \beta$, and Eq. (15) is reduced to [10]:

$$\frac{1}{v_d} \approx \frac{1}{\gamma(\langle v \rangle/4)} + \frac{\chi}{D_g}$$ (16)

The diffusional boundary layer thickness, χ, can be found through modeling the airflow conditions at the wafer surface (Chap. 6).

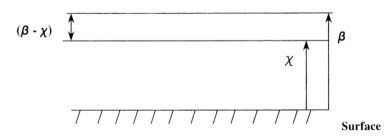

Figure 1 Turbulent and molecular diffusion boundary layers. (From Ref. 10.)

VII. MOLECULAR CLUSTER FORMATION: GAS TO PARTICLE CONVERSION

When does a cluster of molecules become a particle? Particles were defined in Chapter 1 as stable molecular clusters having a diameter in the range of approximately 2 nm to 1 nm. The lower limit is not an exact number because the stability of a molecular cluster varies with composition. Nonetheless, the particle definition implies the existence of a critical size above which molecular clusters will not dissociate but continue to grow into particles. This section relates the critical cluster size to the properties of its constituent molecules.

Molecular cluster growth depends on the chemistry, thermodynamics, and kinetics of each specific gas-phase system. In cleanroom air, the concentration of aerosol particles is low and can be neglected so that the statistics of molecular cluster growth can be estimated by modifying the equilibrium theory for supersaturation [13]. To predict molecular cluster growth of a certain compound, airborne concentration, vapor pressure, and some additional parameters of the compound must be known.

The saturation ratio S is defined as

$$S = \frac{P_1}{P_V} \tag{17}$$

Here, P_1 is the monomer partial pressure and P_V is the saturation vapor pressure above a noncurved surface of the condensed phase of the same compound. A molecular cluster will grow if $S > 1$ and a condensation nucleus (a molecular cluster) of critical size exists in the system. The critical nucleus diameter is [13]

$$d_P = \frac{4\sigma v_m}{RT \ln S} \tag{18}$$

where σ is the surface tension of the condensed phase of the compound, R is the universal gas constant, and v_m is the molar volume of the compound.

An example of particle formation from molecular contaminants is the formation of ammonium chloride particles. Two gaseous species, ammonia and hydrogen chloride, react to form a product with lower vapor pressure [4]:

$$NH_3(g) + HCl(g) \Leftrightarrow NH_4Cl(g) \Leftrightarrow NH_4Cl(s) \tag{19}$$

For homogeneous nucleation (the condensation nucleus is composed of the molecules of the compound itself), the critical cluster size depends on the ratio of the product of partial pressures of ammonia, P_{NH_3}, and hydrochloric acid, P_{HCl}, to the equilibrium partial pressure product, K_p [4], in the following expression, which is derived from the Clausius-Clapeyron equation:

$$\ln K_P = 34.266 - \frac{21,196}{T} \tag{20}$$

Using the following relationship, the critical nucleus diameter d_P can be calculated. [This equation is a rewritten form of Eq. (18).]

$$\ln \frac{P_{NH_3} P_{HCl}}{P_0^2 K_P} = \frac{4\sigma v_m}{RT d_p}$$ (21)

where

P_0 = atmospheric pressure
R = gas constant, 8.314 J K^{-1} mole^{-1}
v_m = molar volume, which is the ratio of the molecular weight to the density of solid ammonium chloride: M_{NH_4Cl} = molecular weight (53.49 g/mole)
ρ = density of NH$_4$Cl (1.527 g/cm^3)
T = temperature, K
σ = surface tension, 150 dynes/cm

$\frac{P_{NH_3} P_{HCl}}{P_0^2 K_P} > 1$ is the equivalent to S > 1, a prerequisite for a critical cluster to form [4]. At 293 K, from Eq. (20), K_p is 2.911E-17. Using the concentrations $[P_{NH_3}/P_0] = 25$ ppb, and $[P_{HCl}/P_0] = 2$ ppb, the product of the concentrations is 5E-17 and therefore S > 1. The critical nucleus size calculated from Eq. (21) under these conditions is $d_P = 16$ nm.

Wafer haze attributed to solid ammonium chloride has been reported in the literature [14], indicating that the kinetics of formation must be favorable in some cleanroom environments. The airborne ammonia and hydrogen chloride concentration values required for this phenomenon to occur are higher than the typical cleanroom ambient, yet some cleanrooms report very high ammonia levels, and elevated airborne hydrogen chloride concentrations due to hydrochloric acid leaks are not uncommon.

VIII. GAS/SURFACE INTERACTIONS

Movement of molecules occurs due to their translational kinetic energy. Above 0 K molecules are in constant random thermal motion and are assumed to travel in straight lines between collisions. In a gaseous mixture, the molecules collide with each other and impact the interaction surface (see Fig. 2). The frequency of the collisions is proportional to the molecular speed of the gas mixture. Surface atoms vibrate at a rate that is negligible to the molecular contaminant velocity, so the surface can safely be assumed to be stationary.

Collision of a molecule with a surface can result in one of three types of electronic interaction with the surface [12].

1. See Figure 3. *Elastic scattering* describes an impact with the surface in which the molecule conserves kinetic energy and directionality (angle

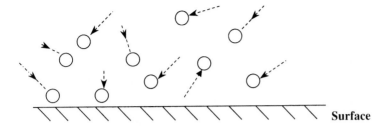

Figure 2 Molecular contaminants as hard spheres colliding with a surface.

of incoming particle = angle of outgoing particle); the surface energy states are therefore not affected. The molecule does not reside at the surface for any significant timeframe.

2. See Figure 4. *Inelastic scattering* results in a weak interaction with the surface. Some internal energy is transferred from the impacting molecule to the surface; therefore, the angle of incident trajectory differs from the angle of emittance from the surface. Again, the molecule doesn't linger at the surface.

3. See Figure 5. *Trapping* could occur, which is an impact resulting in residence of the gaseous molecule within a potential well at the surface; this is due to transfer of the gas molecule's internal energy to the surface. This mode of electronic interaction is necessary for adsorption to occur.

A. Adsorption and Desorption

Interaction of a gas molecule with a surface where the gas resides for a finite time is termed adsorption. Release of the trapped atom or molecule from the surface back into the gas phase is termed desorption. Finite residence times

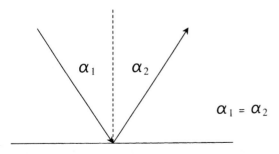

Figure 3 Elastic scattering. (From Ref. 12.)

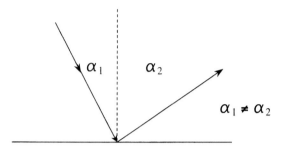

Figure 4   Inelastic scattering. (From Ref. 12.)

can range from 1E-13 seconds—helium physisorbed to a surface—to 1E1100 seconds—O chemisorbed on W [12]. This huge difference in residence time of the adsorbate is due to many interacting factors, including surface electron density, surface structure, kinetic energy of the incoming gas atom or molecule before it hits the surface, temperature of the surface, temperature of the gas phase, interaction between competing species in the gas mix, and chemical structure of the incoming adsorbate.

1. Adsorption

Potential energy is exerted on a gas molecule that approaches a surface. Both repulsive and attractive forces contribute to this potential energy. As the gas molecule travels toward the surface, the repulsive force must increase at a slower rate than the attractive force for a potential energy well to be present. This well or minimum in energy exists an equilibrium distance from the surface r (see Fig. 6). The magnitudes of the repulsive and attractive forces are unique functions of the specific gas-surface interaction. Repulsive forces result primarily from the interaction between filled orbitals of the surface with filled orbitals of

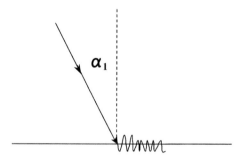

Figure 5   Trapping. (From Ref. 12.)

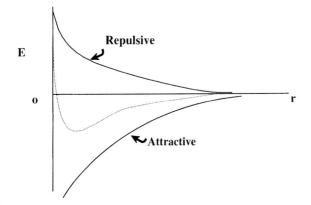

Figure 6   Representation of attractive and repulsive terms that make up potential well.

the approaching gas molecule. The bulk of the attractive forces are due to un-filled surface valence states which exert force on the approaching gas molecule. Adsorption occurs when a molecule becomes trapped within the potential well.

Adsorption is typically grouped into two general categories: physisorption and chemisorption. These are defined quantitatively by the heat of adsorption, which is a measure of the interaction energy between the adsorbate (gas molecule trapped at the surface) and the surface.

Physisorption is characterized by a weak interaction with the surface. The typical heats of adsorption are below 21 kJ/mole (5 kcal/mole) [12,15] [al-though organics physisorb to the range of 42–63 kJ/mole (10–15 kcal/mole)]. The adsorbate-surface interaction is due to van der Waals or dispersion forces—forces that are governed by dipole-dipole interactions. The absence of a true chemical bond is indicative of physical adsorption. A physisorbed species is trapped further away from the surface than a chemisorbed species.

Chemisorbed bonding involves stronger adhesion forces than physisorp-tion. Chemisorbed hydrogen, 84–105 kJ/mole (20–25 kcal/mole), and covalent bonding, 125–627 kJ/mole (30–150 kcal/mole), are examples of chemisorption. These values are illustrative only; specific adsorbate/surface interactions will have unique adsorption and desorption energy values.

The Lennard-Jones model of adsorption is used to illustrate examples of adsorbate/surface interactions [16]. This one-dimensional model is simplified, yet helpful for descriptive purposes. A thorough treatment of this model and its application to airborne molecular contamination is presented by Zhu [17].

1. *Physisorption only.* The interaction between the adsorbate and surface result in a shallow potential well of minimum energy E_p that exists a distance r_p from the surface. This distance is too far to allow chemical bonding, but close

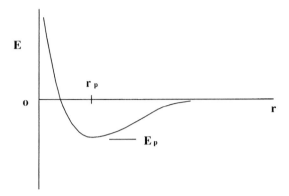

Figure 7 Physisorption.

enough to allow dipole-dipole interaction. The distance r_p is typically 0.4–0.6 nm. (See Fig. 7.)

2. *Chemisorption only.* Again, one potential well exists at minimum energy E_C that is located close enough to the surface to allow chemical interaction of the adsorbate and surface. The depth of the well is larger in energy than a physisorption energy well. This distance is typically on the order of 0.1–0.3 nm. (See Fig. 8.)

Adsorbate molecules with high kinetic energy will likely miss the potential well altogether, bouncing back into the gaseous mixture and resulting in a scattering event. This can occur simply in all cases presented.

3. *Physisorption and chemisorption—nondissociative.* Two energy wells exist at the surface. In this case, a molecule may adsorb in either a physisorbed

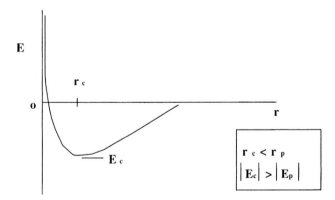

Figure 8 Chemisorption.

or chemisorbed state, and can pass between the states without dissociation. *Dissociation* is the term used when a molecule breaks into fragments and gives off energy in the process. Molecular chemisorption can occur when it costs more energy to break up the molecule than is gained upon adsorption of the fragments. In this case, the molecule will adsorb intact at the surface. (See Fig. 9.)

The physisorbed state is a "precursor" to the chemisorbed state. A physisorbed molecule can chemisorb when it obtains the thermally supplied activation energy to reach the chemisorption potential well. Movement from the physisorbed state into the chemisorbed state is governed by the Boltzmann temperature relationship; it follows that the probability of exceeding the activation energy barrier increases exponentially with temperature.

$$\text{Probability} \propto e^{-E_{act}/kT} \tag{22}$$

4. *Nonactivated chemisorption.* This is a case where the intersection of the attractive and repulsive energy terms create an energy barrier between the precursor and chemisorption energy wells. The crossover of the energy curves for the physisorbed and chemisorbed state occur below zero. Since dissociation of a molecule gives off positive energy, the barrier is more likely to be surmounted when dissociation occurs. These fragments are energetically stabilized during surface bonding. Achievement of chemisorption is more favorable in this case upon dissociation; this is typical for heteronuclear molecules such as NO or CO. (See Fig. 10.)

Molecules that reach the physisorbed state without additional activation energy don't make it over the barrier to the surface; this is termed molecular physisorption.

5. *Activated chemisorption.* Here the crossover of the energy curves for the chemisorbed and physisorbed states results in an energy barrier to chemisorption

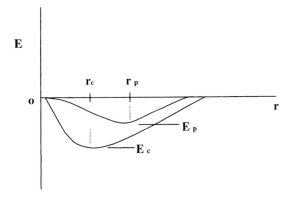

Figure 9 Nondissociative physisorption and chemisorption. (From Ref. 16.)

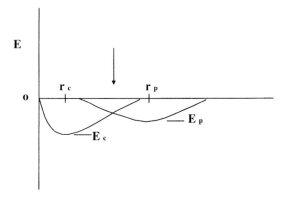

Figure 10 Nonactivated chemisorption. (From Ref. 16.)

that is finite and positive. Dependence on temperature for chemisorption to occur is much more important for activated chemisorption. Again, dissociation of the molecule can provide the energy to surmount the activated energy barrier. This case can occur for homonuclear molecules such as O_2, since homonuclear dissociation gives off so much energy. (See Fig. 11.)

These five cases illustrate some adsorption possibilities but do not cover all possibilities for adsorbate/surface interactions. It must be stressed that each adsorbate/surface interaction is unique, and species noted in cases 4 and 5 are general examples only.

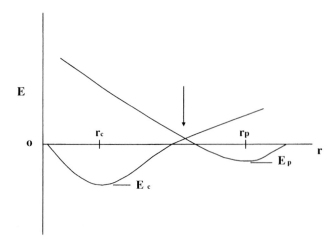

Figure 11 Activated chemisorption. (From Ref. 16.)

2. Desorption

Desorption is the release of the adsorbed atom or molecule from the surface and back into the gas state. The adsorbed atom or molecule sits at the surface, vibrating with the surrounding lattice. Thus, the desorption attempt frequency is typically taken as the atomic lattice vibrational frequency of 1E13/s. To desorb, the adsorbate must achieve the free energy required to get into the desorbed state. The simplest mathematical representation of this is

$$f_{des} = f_0 \exp\left(\frac{-E_{des}}{kT}\right) = \frac{1}{\tau_a} \tag{23}$$

where f_0 is the desorption attempt frequency, E_{des} is the desorption free energy, and f_{des} is the desorption frequency. The lifetime of an adsorbate at the surface, τ_a, is the reciprocal of the desorption frequency.

Figure 12 shows the surface, E_{des} energy hump, and the desorbed state. Adsorption and desorption kinetic models are described in the next section.

3. Sticking Coefficient

The concepts of adsorption and desorption are important to understanding the sticking coefficient. The true definition of sticking coefficient, γ, is *the ratio of the rate of adsorption, r_a, to the rate of impingement, I—the fraction of contaminants striking the surface/unit time that remain on the surface*, but the sticking coefficient can be defined in a number of ways. First, in a homogeneous gas environment an instantaneous view of the sticking coefficient can be taken as the ratio of the number of surface sites filled with an adsorbate, N_A, over

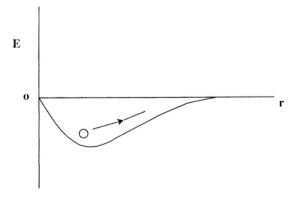

Figure 12 Desorption.

the number of surface sites that are initially available for adsorption, N_0. This instantaneous view of the sticking coefficient is termed the *coverage*, θ.

$$\theta = \frac{N_A}{N_0} \tag{24}$$

This coverage is actually a snapshot in time, and its rate of change is a function of many parameters including the number of currently filled surface sites.

$$\frac{d\theta}{dt} = f(\theta) \tag{25}$$

Models of the adsorption process exist yet many are limited in application. Some simplistic models and one with more detail are discussed in this section. All have limitations and their application to cleanroom air deposition is discussed.

The Langmuir model of the adsorption process limits surface coverage to a single monolayer. Once a surface site is filled with an adsorbed molecule, another molecule cannot also adsorb to that filled site. No interaction between adsorbed molecules is taken into account, adsorbed molecules do not move at the surface once adsorbed, and total adsorbate coverage can never exceed the total number of surface sites. In addition, a molecule striking an occupied site is not adsorbed but is ejected back into the gas phase. This model is most accurate for a homogeneous gas mixture that collides with a homogeneous surface. The Langmuir model is kinetically based; molecular contaminant A adsorbing to surface S is expressed by the kinetic equation:

$$\begin{array}{c} A(\text{absorbed}) \\ | \\ A(g) + S \underset{k_p}{\overset{k_a}{\rightleftharpoons}} S \end{array} \tag{26}$$

where k_a is the adsorption rate coefficient, and k_p is the desorption rate coefficient. The true rates of adsorption and desorption are a function of the surface coverage, and the partial pressure of molecular contaminant A [12].

$\theta_L = $ Langmuir coverage $\quad 0 \leq \theta_L \leq 1$

Rate of adsorption $= k_a p[1 - \theta_L]$

Rate of desorption $= k_p \theta_L$

$p = $ partial pressure of molecular contaminant A

When the rate of adsorption equals the rate of desorption, $k_a p[1 - \theta_L] = k_p \theta_L$, and

$$\theta_L = \frac{k_L p}{1 + k_L p} \tag{27}$$

where

$$k_L = \frac{k_a}{k_p}$$

and

$$k_L p = \frac{J_c * \tau_a}{N_0} \tag{28}$$

For a 2×1 reconstructed Si(100) surface, $N_0 = 2.9E14$ atoms/cm^2.

This model works well for strong chemisorption, where the heat of adsorption is the same for all surface sites. A good application of the Langmuir model is the primed silicon surface with one monolayer of adsorbed hydrogen atoms [18].

The combination of Langmuir adsorption and a rate-limiting unimolecular surface reaction, k_s, can be expressed as

$$k_s \theta_L = \frac{k_s k_L p}{1 + k_L p} \tag{29}$$

Not all gas/surface adsorption is well modeled by the Langmuir model. The Freundlich [19] isotherm is used to model the adsorption of organics to activated carbon:

$$\theta = k[A]^n \tag{30}$$

The Braunauer, Emmet, and Teller (BET) model addresses surface heterogeneity (such as the existence of polar sites and nonpolar sites on the same surface), and allows for adsorption in "islands," where more than one adsorbed layer can exist at the surface [20]. In addition, multiple layers of adsorbate atoms or molecules can grow prior to completion of one monolayer across the entire surface. (See Fig. 13.)

Water on a surface can be described by a BET isotherm model, and is present in large concentrations in ambient air. Air at typical relative humidities equates to an airborne concentration on the order of 10,000 ppm of water at atmospheric pressure. Water adsorbs to polar sites on the surface; water will also want to hydrogen bond to itself at similar bonding energies. When the heat of condensation is equal to or less than the heat of adsorption, condensation or multiple layers will start to form on the surface.

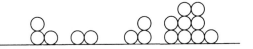

Figure 13 Island formation.

The equivalent of one monolayer of water exists at approximately 40%RH, in islands at the surface. The water to surface bond is rigid, and the next layer of water is relatively fixed as well. As RH levels approach 60%, the islands grow together to form a film of liquid water at the surface [21–23]. The uppermost adsorbed monolayers act as liquid water which solvates ions and allows them to migrate. Ion and electron transfer within an electrolyte are required for metal corrosion to occur. Additional detail on adsorption, desorption, diffusion, and reaction on metal surfaces can be found in Lombardo and Bell [24].

4. Chemisorbed State

Cleanroom air at atmospheric pressure has literally hundreds to thousands of species, all at unique concentrations. When air collides with a surface, the various species compete for surface sites at different rates. The surface of concern in this case is the wafer surface, and the wafer surface can have metal, dielectric, and semiconductor layers exposed to the air. Therefore, the surface is not homogeneous, and these different materials have native oxide and water at their surfaces.

Deposition of molecular contaminants and retention of these compounds at this complex surface is addressed through the precursor model of chemisorption. This model, directly bsed on kinetics of the physisorption and chemisorption processes occuring simultaneously, also accounts for movement of adsorbate molecules on the surface. The model is illustrated for one molecular contaminant only. However, the reader can imagine the expansion of this model to include kinetic terms for all species that interact with the surface, and for differences in reaction rates based on surface inhomogeneities. Add to this interaction between adsorbates and a fully dynamic system that must be solved simultaneously through complex simulation, one can understand the computer power needed to properly predict sticking coefficient.

The example model is based on the precursor model. The kinetics of this model are outlined in Figure 14 [24,25]. An intrinsic precursor to chemisorption

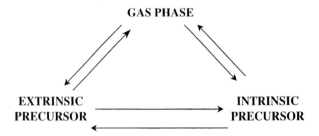

Figure 14   Diagram of precursor states. (From Refs. 24 and 25.)

is an adsorbate that is physisorbed over an otherwise empty state. The adsorbate may not be equilibrated with the surface. An extrinsic precursor is an adsorbate which strikes a filled state at the surface, has a finite lifetime at the surface, and may move to a chemisorption site. The following example is for a nondissociative adsorption process for a homonuclear molecule, such as Cl_2. Assumptions for this model are:

- Chemisorption occurs on a fixed number of sites.
- Only one chemisorbed molecule per site is allowed.
- Desorption of the chemisorbed site back to a precursor state is energetically unfavorable at fixed temperature—the rate constant k_d is negligible.
- Adsorbates randomly occupy sites of adsorption.
- Coverage increases at a rate in direct proportion to the product of the pressure and the number of adsorption sites.
- Weakly adsorbed molecular or precursor species don't behave the same over filled and empty chemisorption sites (intrinsic precursors are not the same as extrinsic precursors).
- The surface lifetime of a precursor is short and its coverage is small; it can be assumed to be at steady state.
- The precursor lifetime is so short that at any instant, an available precursor site is open and movement is not restricted by site occupation; site saturation doesn't occur for precursor sites.

Kinetics:

$$Cl_{2,g} \xrightarrow{I\alpha} Cl_{2,\text{int}}$$

$$Cl_{2,\text{int}} \xrightarrow{k_p} Cl_{2,g}$$

$$Cl_{2,\text{int}} \xrightarrow{k_a} Cl_{2,\text{ads}}$$

$$Cl_{2,\text{int}} \xrightarrow{k_m} Cl_{2,\text{ext}}$$

$$Cl_{2,\text{int}} \xrightarrow{k_m} Cl_{2,\text{int}}$$

$$Cl_{2,\text{ads}} \xrightarrow{k_d} Cl_{2,\text{int}}$$

$$Cl_{2,g} \xrightarrow{I\alpha^*} Cl_{2,\text{ext}}$$

$$Cl_{2,\text{ext}} \xrightarrow{k_p^*} Cl_{2,g}$$

$$Cl_{2,\text{ext}} \xrightarrow{k_m^*} Cl_{2,\text{int}}$$

$$Cl_{2,\text{ext}} \xrightarrow{k_m^*} Cl_{2,\text{ext}}$$

In these reactions, α is the trapping probability from the gas state to the intrinsic state, α^* is the trapping probability from the gas phase to the extrinsic state, and I is the impingement rate. Additional definitions are:

$$Cl_{2,ads} \Rightarrow \text{a chemisorbed } Cl_2$$
$$Cl_{2,int} \Rightarrow \text{a intrinsically physisorbed precursor } Cl_2$$
$$Cl_{2,ext} \Rightarrow \text{a extrinsically physisorbed precursor } Cl_2$$
$$Cl_g \Rightarrow \text{for } Cl_2 \text{ in gas}$$

Here, $[Cl_{2,int}]$ and $[Cl_{2,ext}]$ are separately treated as steady state approximations:

$$\frac{d[Cl_{2,ext}]}{dt} = I\alpha^*\theta[Cl_{2,g}] + k_m\theta[Cl_{2,int}] - k_p^*[Cl_{2,ext}]$$
$$- k_m^*[Cl_{2,ext}](1 - \theta) = 0 \tag{31}$$

$$[Cl_{2,ext}] = \frac{I\alpha^*\theta[Cl_{2,g}] + k_m\theta[Cl_{2,int}]}{k_p^* + k_m^*(1 - \theta)} \tag{32}$$

$$\frac{d[Cl_{2,int}]}{dt} = I\alpha(1 - \theta)[Cl_{2,g}] - k_p[Cl_{2,int}] - k_a[Cl_{2,int}](1 - \theta)$$
$$- k_m\theta[Cl_{2,int}] + k_m^*[Cl_{2,ext}](1 - \theta) = 0 \tag{33}$$

$$[Cl_{2,int}] = \frac{I\alpha(1 - \theta)[Cl_{2,g}] + k_m^*[Cl_{2,ext}](1 - \theta)}{k_p + k_a(1 - \theta) + k_m\theta} \tag{34}$$

$$\frac{d[Cl_{2,ads}]}{dt} = k_a[Cl_{2,int}](1 - \theta) \tag{35}$$

Substituting Eq. (32) into Eq. (34) and then into Eq. (35), and then dividing by I,

$$\frac{d[Cl_{2,ads}]/dt}{I} = \gamma$$
$$= \left\{ \frac{k_a\alpha(1 - \theta)^2[Cl_{2,g}] + k_a k_m^*(1 - \theta)^2\alpha^*\theta[Cl_{2,g}]/[k_p^* + k_m^*(1 - \theta)]}{[1 - k_m^*(1 - \theta)k_m\theta/[k_p^* + k_m^*(1 - \theta)]][k_p + k_a(1 - \theta) + k_m\theta]} \right\} \tag{36}$$

At $\theta = 0$,

$$\gamma_0 = \frac{k_a\alpha[Cl_{2,g}]}{k_p + k_a} \tag{37}$$

Prediction of the sticking coefficient requires knowledge of many parameters, and complexity greatly increases with competition between coadsorbates in a nonhomogeneous gas mixture such as cleanroom air. Zhu has published an example of a theoretical approach to prediction of activation energies and surface kinetics as applied to deposition of airborne molecular contamination [26].

Sticking coefficients are best determined empirically, yet data obtained from high vacuum experimentation is suspect when attempting extrapolation to real indoor air conditions. The collision flux of the nitrogen dimer at typical indoor air conditions is 3E23 molecules $cm^{-2}s^{-1}$ (detailed previously). At ultrahigh vacuum, this flux can be 1E15 times smaller. Experimentation is done with very clean gases, and on cleaned controlled surfaces. More experimentation needs to be performed in real conditions to determine the sticking coefficients of a variety of pertinent surfaces. Deuterated hydrocarbons can be used for this, as well as other isotopes. Atmospheric pressure ionization techniques for surface desorption and detection are important for this work—techniques such as APIMS-MS and IMS-MS.

IX. BASIC SPECIES AND AIRBORNE MOLECULAR CONTAMINATION

Airborne basic compounds have been studied due to their deleterious affects on photoresist mask critical dimensions. Deep-UV (DUV) photoresists contain compounds that, upon exposure to energy at specific DUV wavelengths (typically 248 nm), convert to a super acid with pKa values of -20. The super acids provide a source of protons which catalyze a cleavage reaction of the photoresist organic compound. This cleavage reaction is required to increase the solubility of the exposed photoresist in the developer (for positive resist systems). This is a high-gain reaction that often requires additional energy obtained thermally through a post-exposure-bake (PEB).

Loss of $[H^+]$ at the surface of the photoresist can occur when exposed wafers sit and wait for the post-exposure-bake operation. The waiting is due to wafer lot throughput. Basic species from the air diffuse through the boundary layer and deposit to the wafer, resulting in a chemical reaction where protons are scavenged from the upper layer of photoresist. The resultant decrease in surface proton concentration greatly reduces the cleavage reaction kinetics, and the photoresist mask critical dimensions are poor. A mechanism called t-topping can occur with positive resist systems (see later discussion and Fig. 19). Negative resist systems will require an increase in UV exposure dose.

The reaction shown in Figure 15 is a t-boc/onion salt resist system [27].

Airborne basic compounds in sufficient concentrations are deleterious to photoresist mask critical dimensions. Basic species can be compounds that don't have hydroxyl groups. Lewis bases are electron-donating species, such as amines. Airborne basic species most commonly found in cleanroom air are ammonia and

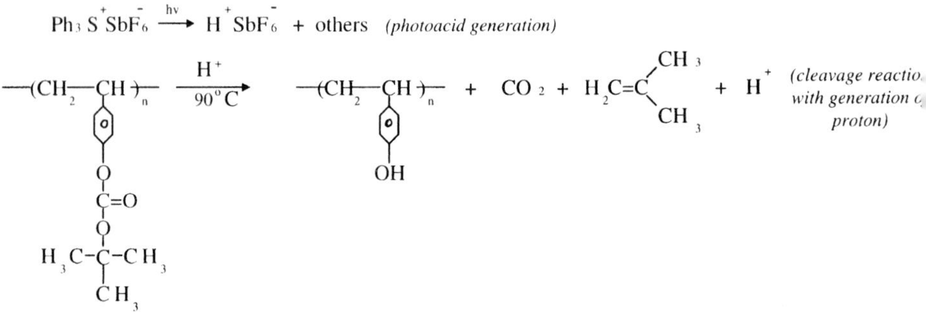

$$Ph_3 \overset{+}{S} \overset{-}{SbF_6} \xrightarrow{hv} \overset{+}{H} \overset{-}{SbF_6} + others \quad (\textit{photoacid generation})$$

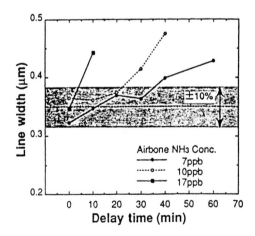

Figure 15 T-boc resist system reaction. (From Ref. 27.)

NMP (photoresist stripping solvent, n-methyl pyrrolidinone). Some concentration thresholds have been reported.

Figure 16 shows the affect of post-exposure delay on linewidth as a function of airborne ammonia concentration. A 10-min delay at 17 ppb NH_3 resulted in a 20% linewidth increase. NMP concentrations of 10 ppb were found to degrade photoresist profiles [29].

Multiple sources of ammonia and amines are present in cleanrooms. HMDS (hexamethyldisilazane) is a common wafer priming agent and gives off ammonia during the surface reaction [30]. Ion Mobility Spectrometry (IMS) monitoring of NH_3 in a wafer track environment at the HMDS priming chamber shows extremely high levels of NH_3 emitted (see Fig. 17). The maximum concentration

Figure 16 Ammonia vs. critical dimensions. (From Ref. 28.)

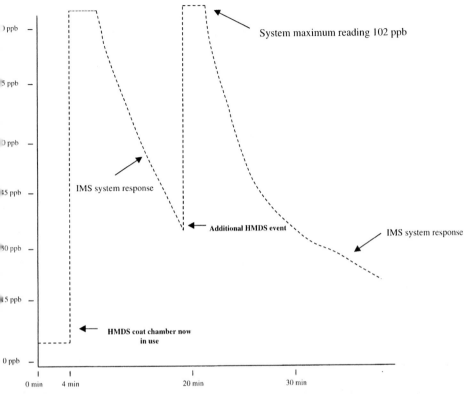

Figure 17 IMS trace of ammonia spike in photoresist track near HMDS priming chamber.

read by the system is 102 ppb. Several events occurred during the following monitoring period.

Typically, simultaneous monitoring at the PEB oven location during HMDS priming events shows low levels of ammonia. However, on occasion the NH_3 level can rise to unacceptable levels, as shown by the following IMS data. This is likely due to poor filtration and lack of laminar air flow through the wafer track. Figure 18 shows this occurrence. Here, the NH_3 level at the PEB oven rose to 73 ppb from an initial concentration of ~3–4 ppb.

Figure 19 is an example of t-topping. Resist footing can also occur on basic surfaces which results similarly from proton scavenging (Fig. 20).

Studies correlating photoresist linewidth changes with solvent-contaminated air and also with offgassing of construction materials have been performed [27,30]. Lewis bases triethylamine, pyridine, *n*-methyl morpholine, and *n,n*-di-

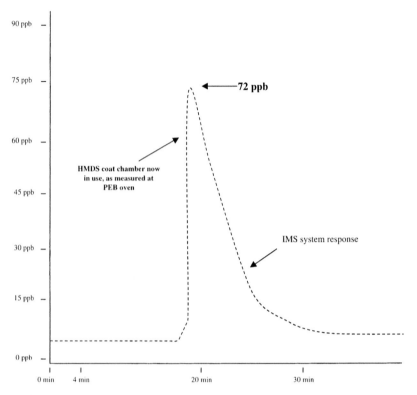

Figure 18 Ammonia concentration measured at post-exposure-bake oven during HMDS event.

methylaniline degrade photoresist performance, while acetone, ethylacetate, and pentane do not.

Amines and amides are electron-donating groups (via resonance effects) while ketones have electron-withdrawing groups and will not act as Lewis bases. Functional groups of aldehydes, nitriles, esters, and carboxylic acids are also electron withdrawing and are likely not to act as Lewis bases to the DUV photoresist chemistry.

Construction materials such as urethane paint, some silicones, PVC glue, and some adhesives have been shown to offgas species that act as Lewis bases to the DUV photoresist, degrading performance [27,31].

Multiple sources of ammonia are present in cleanrooms. HMDS gives off NH_3, as previously discussed. People are a large source of NH_3, as it is a metabolism by-product in expired breath. NH_3 breath levels in the range of 50–1280 ppb were measured in a variety of subjects and studies [32–35]. Resist

Incorrect Photoresist Profile

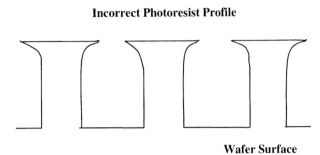

Wafer Surface

Figure 19 Example of t-topping.

developer is often TMAH (tetramethyl ammonium hydroxide) which is an alkyl amine. RCA-SC1 cleans can give off NH_3 [36]. Outdoor air infiltration–rural tropospheric ammonia levels are in the range of 1 ppb; urban areas can range as high as 10 ppb. Also, farm areas have high ammonia levels due to fertilizers and animal metabolic output (which locally results in increased NH_3).

A. Amines

Morpholine has been detected in cleanroom air at approximately 1 ppb [37]. It is an additive to humidification plants and is quite volatile [38]. Cyclohexamine and diethylaminoethanol are also used as humidification solvents [37].

NMP is present in cleanrooms as a photoresist stripper; high-temperature solvent baths are significant contributors to airborne NMP. Table 4 reports NMP concentrations measured with a portable gas chromatograph and an argon ionization/electron capture detector. The specific purpose of the following study was to identify if NMP fumes were escaping through openings in the minienvironment. When wafers were robotically removed from the heated bath and moved into

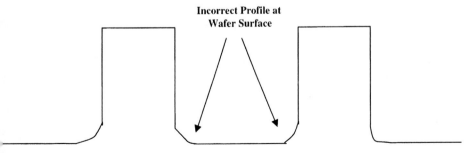

Figure 20 Example of resist footing on nitride.

Table 4 NMP Concentrations Escaping a Minienvironment Enclosing a Photoresist Solvent Sink

Location	Inside or outside minienvironment	Cloud event?	AVE NMP Conc (ppb)
Over tank	Inside	No	424
In hallway	Outside	No	7
Over tank	Inside	Yes	Detector saturated
In hallway	Outside	Yes	837
In hallway	Outside	15 min after event	51

the cooler air above, a cloud event would occur that allowed NMP to escape the minienvironment enclosure. Table 4 documents concentration levels of NMP detected with and without cloud events, inside and outside the minienvironment. Significant levels of NMP escaped the minienvironment during a cloud event. This study was repeated through several cloud events; the results in Table 4 are average levels.

B. Elimination

Elimination of Lewis bases from photo-processing tools and local air is best achieved through a combination of approaches.

- Careful choice of construction materials in the photolithography airstream
- Physical separation of amine and ammonia-containing heated baths through placement in separate airstreams
- Filtration of the DUV tracks and steppers to remove Lewis bases generated during photo processing

Filtration is best performed with a filtration system designed to fit integrally in the photo tools, using an acid-impregnated filtration system [14,39].

Resists are currently being developed that are not as sensitive to proton scavenging and hence will be able to tolerate longer post-exposure-bake delay times.

C. Measurement

Detection of ammonia and amines has been performed by a variety of methods, including

- Ion mobility spectrometry [40]
- Thermal desorption/gas chromatography [40]

- Concentration with annular denuders and subsequent extract analysis with ion chromatography [37]
- Scrubbing through a bubbler or impinger filled with DI water followed by IC
- Concentration on an acid-impregnated sorbent (ORBO 77—Supelco) and analysis of the desorbed extract through IC.

X. ACIDIC SPECIES AND AMC

Airborne acidic compounds are present in fab air due to the presence of acidic etching and cleaning baths in the cleanroom. Increased bath temperature with respect to air temperature results in vaporization of the sulfuric, phosphoric, hydrofluoric, and hydrochloric acids. In the case where exhausted removal of acidic airborne effluents is not wholly efficient, the species can become entrained in fab air. Airborne acidic compounds have high deposition velocities due to their polarity and surface reactivity properties, so deposition to surfaces and subsequent corrosion is the primary concern of this class of airborne contaminants. Corrosion concerns apply to on-wafer metallization, and also to equipment in the cleanroom.

Examples of airborne corrosive acidic species are sulfuric acid and hydrochloric acid. Sulfuric acid can form aerosols with moisture, resulting in airborne "electrolytes." Leaks in HCl lines are not atypical, resulting in increased $HCl(g)$ levels.

A. Metal Corrosion

Metal corrosion occurs in environments at both high and low pH. The region of passivation is in the moderate pH range—where, for example, the aluminum oxide film at the metal-electrolyte interface acts as a protective layer and doesn't allow ionization of the underlying aluminum.

The driving force for corrosion is the potential across an electrochemical interface. Corrosion is suppressed at low electrochemical potential levels and a condition of immunity occurs. The corrosion equilibrium diagram [41] (Fig. 21) for aluminum can be altered by the addition of a small percent of copper in the metal line. Copper is added to decrease the grain boundary surface energy through the precipitation of theta-phase (Al_2Cu) precipitates which reduce the rate of current-density driven electromigration.

However, addition of the copper results in an electrochemical potential difference, or galvanic cell, within the aluminum line itself. The most significant galvanic cell in this system is between the theta phase precipitate formed at the grain boundaries and the Cu-depleted zone adjacent to the grain boundary and

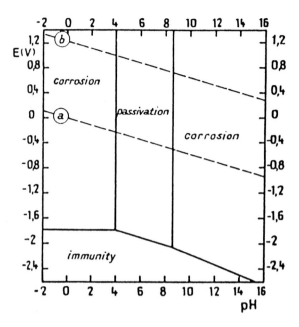

Figure 21   Pourbaix diagram for aluminum in pure water. (From Ref. 41.)

precipitate. The higher open-circuit potential (OCP) of the Al_2Cu precipitate vs. the aluminum matrix OCP makes the Al_2Cu the cathode and the Al the anode.

The aluminum will therefore be preferentially attacked in the proper environment. The gradient of the potential of the system has been shown to be a function of the theta precipitate size [42]. Addition of [Cl^-] or [F^-] to the local environment decreases the pitting potential of the aluminum matrix [43]. If this is lowered to below the OCP of the Al_2Cu precipitate, the aluminum just adjacent to the precipitate will be at a potential above that of the pitting potential, resulting in oxide breakdown and local attack of the underlying metal [46]. A popular theory to describe this mechanism atomically is that the chloride ion penetrates the film through pores, defects, or film density differences easier than other ions [44].

- High current densities increase the rate of Cl ion penetration.
- Once at the surface, the Cl^- increases the rate at which aluminum metal ions enter solution.
- Thus, a critical potential (also called the pitting potential) exists above which breakdown of the aluminum oxide film occurs.

The relationship between pit growth, [Cl^-] or [F^-] in concentration, metallurgy (alloy composition, grain boundary distribution, geometry of feature such as a contact) is not well defined—every metallurgy system will respond uniquely.

An electrolyte must be present for corrosion to occur. A physical model for airborne acid deposition and subsequent metal corrosion has not been fully established. Nonetheless, once a critical concentration of halide (chloride is most active) ion is deposited and the proper conditions exist for an electrolyte to form, corrosion occurs at the aluminum copper surface.

Wafer corrosion defects have been observed in environments with excess airborne [HCl]. Higley [45] reported the elimination of aluminum corrosion on wafers after installation of chemical air filtration to remove the airborne acid (Fig. 22). A correlation of visually evident corrosion defects and an [HCl] airborne concentration of 28 ppb was made.

B. Cleanroom Attack Due to Airborne Acids

Wafers are not the only surfaces susceptible to corrosion due to airborne acids in the fab. Process tools of stainless steel can be attacked by high concentrations of airborne HCl. Stainless steels are resistant to corrosion due to their passivating chromium oxide films, yet these films are particularly susceptible to breakdown by HCl. Removal of HCl from the airstream should eliminate this degradation. Stainless steel corrosion can result in emission of Fe, Cr, and Ni into the airstream.

Air-handling ductwork, if galvanized, can be more readily attacked by acids than stainless steel. Sulfuric acid will attack galvanized coatings. Emission of Zn and Fe into the airstream will result.

HEPA and ULPA filters made of borosilicate glass media can be degraded by excess airborne hydrofluoric acid (HF). Particulate and gaseous contaminants released into the fab airstream have identified HEPA filters as the source. Par-

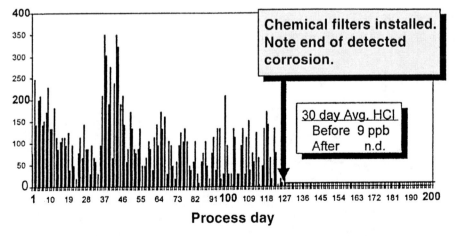

Figure 22 Metallization corrosion and concentration of airborne HCl. (From Ref. 45.)

ticulates from HEPA filters were found to contain Ti, Zn, Al, Ca, Mg, Na, Si, B, K, and Ba [46]. Boron is also emitted in a volatile form from HEPA filters, as it is present at 10–20 wt% of the filter media [47].

C. Sulfuric Acid and Aerosols

Sulfuric acid can be emitted into a cleanroom by heated acid baths, and can also be introduced by infiltration from outdoors. It exists outdoors due to oxidation of SO_2. Infiltration of SO_2 and H_2S can be controlled by filtration in silicon wafer manufacturing. If infiltration of SO_2 into your cleanroom occurs, production of sulfuric acid is a function of NO_x concentrations, hydrocarbon concentrations, and photolytic conditions.

$$SO_2 + OH \rightarrow \text{several . . steps} \rightarrow H_2SO_4$$

Sulfuric acid produced will immediately associate with water molecules to form sulfuric acid aerosol [4]. Sulfuric acid aerosols also increase with increasing relative humidity [48]. (See Fig. 23.)

Figure 24 shows the equilibrium aerosol size as a function of the relative humidity and the number of sulfuric acid moleculs in an aqueous sulfuric acid droplet [49].

Calculations involving equilibrium vapor pressure of sulfuric acid and relative humidity show that homogeneous nucleation of sulfuric acid water aerosols

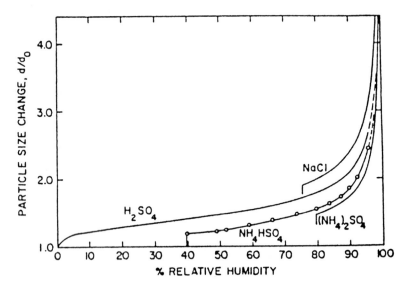

Figure 23 Sulfuric acid aerosol growth. (From Ref. 48.)

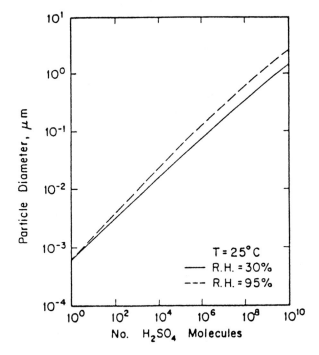

Figure 24   Aerosol size as a function of RH and the molecular concentration of sulfuric acid. (From Ref. 49.)

can occur at a sulfuric acid concentration as low as 13 ppb [50]. This is based on a sulfuric acid/water ratio of 0.01 at RH = 40%.

Typical cleanroom sulfuric acid concentrations are <1 ppb. However, concentrations are likely higher at acid sink locations in the fab, especially if minienvironment airflows are improperly set. One indication of the presence of sulfuric acid aerosols would be high count rates for condensation nuclei counters near sulfuric acid sinks. A witness wafer to measure S on surface with TXRF or VPD-TXRF or airborne monitoring of sulfuric acid can verify this problem.

D.   Monitoring Airborne Acids

Measurement of airborne acids in a cleanroom environment is most commonly performed using impinger or bubbler sampling. The impinger/bubbler solution used is DI water, or DI water with some H_2O_2 [51]. Air is pulled across the impinger/bubbler setup with a constant flow pump at a known flow rate and for a known time (therefore, a known volume of air). After collection at a flow rate of approximately 1 liter/min, the solution is analyzed via ion chromatography using a preconcentrator.

Another popular technique for monitoring airborne acids is use of a sorbent tube technique. The Supelco ORBO-53 sorbent tube has a very low background in acids. The sorbent tube sampling method is much like an impinger or bubbler, yet there are no liquids to spill making this technique more versatile. After a sufficient volume of air is sampled, extraction of the sorbent inside the tube is performed in an appropriate solution, and the extract is analyzed via ion chromatography with a preconcentrator.

Dynamic acid monitoring techniques such as IMS exist, but detection limits are not yet low enough for making measurements without some method of concentrating.

E. Filtration

Filtration is performed with basic compound-impregnated filters. Filtration is required to eliminate the possibility of the corrosion mechanisms discussed previously only if airborne acid concentrations are high. Properly designed and functioning minienvironments over acid sinks will keep acid concentrations low.

XI. AIRBORNE METALS

Airborne metals are of great interest now that the change from aluminum-copper to electroplated or chemical vapor deposited copper metallurgy is becoming more popular. Copper has a very high diffusivity through silicon, as it diffuses interstitially and substitutionally through the silicon lattice. Iron and nickel also diffuse quite rapidly through silicon [52]. (See Fig. 25.)

Airborne copper can deposit to surfaces, including cleaning baths, if it is present in the airstream [53–56]. Copper will tend to plate out on the silicon wafer surface in a contaminated bath, due to the higher electronegativity of copper vs. silicon (see Chap. 10).

Metallic contamination in semiconductor manufacture can result in a variety of failure modes. Minority-carrier lifetime was studied with varying levels of Cu and Fe contamination [57]. The samples were contaminated and a thermal oxide was grown at 950°C. Surface and bulk lifetime components were measured, and different effects were seen with each contaminant. Surface lifetime decreases as the surface copper concentration increased. The bulk component did not change with copper concentration. Iron, however, caused the bulk component to dominate; as surface concentration increases, the bulk lifetime was observed to decrease.

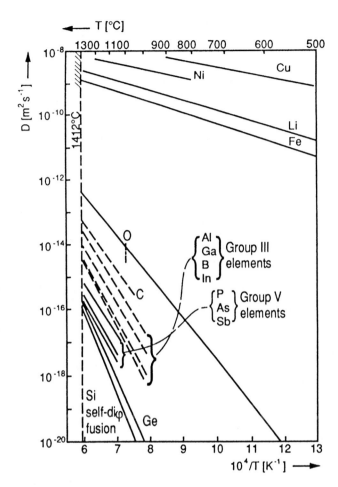

Figure 25 Diffusivities of various metals through silicon. (From Ref. 52.)

DRAMs are particularly sensitive to metallic contamination that results in degraded charge retention times due to decreased carrier lifetimes [3]. Levels in the 1E9 atoms/cm^3 range are detrimental to yield.

Iron surface contamination will be the most harmful to silicon oxide breakdown potentials. Defect density of silicon oxide has been correlated to iron concentration [2]. CMOS transistors with metallic contamination will have interface trapping states that can reduce lifetime through lowered oxide breakdown voltages. Again, levels in the 1E9 atoms/cm^2 range can be detrimental to small geometries [3]. Stacking faults have been linked to metallic impurities in oxide capacitors [58]; a correlation with leakage and breakdown was made with the faults.

A. Empirical Airborne Metals Data

An experiment was performed to determine if airborne Fe would contaminate the CMP area and also other processing areas with shared airstreams. This CMP area had used an iron-nitrate slurry for over one year. No wafer processing occurred during this study.

The experiment consisted of:

- Surface wiping and extraction for metals, using GFAA (Graphite Furnace Atomic Absorption) and ICP-MS (Inductively Coupled Plasma-Mass Spec) (Table 5).
- Teflon impingers with dilute nitric acid at a flow rate of 1 liter/min, sampling for 2–5 days, in the metal-CMP area and also in a shared airstream (Table 6).
- Exposure of witness wafers to the airstream, followed by TXRF (Total Reflectance X-Ray Fluorescence) (Table 7).
- The area was cleaned; surface wiping and airborne metal measurements with impingers were then repeated.

Surface contaminant results are presented in Table 5, in units of ppb metal. The Anticon Gold clean wipes used for surface wiping were extracted in 100 mL 4% HNO_3 solution. Blanks were run on clean wipes that saw all experimental steps except surface wiping; background levels were low. All surface areas sampled showed significant Fe contamination. One location, location 2, had enough Fe contaminant to saturate the detector of the ICP-MS. Measurements repeated

Table 5 Clean Wipe Data—Metal CMP*

Sample	Before equipment cleaning Fe conc (ppb)	After equipment cleaning Fe conc (ppb)
Blank 1	3.56	—
Blank 2	2.12	—
Location 1 (floor)	153.87	18.66
Location 2 (CMP tool A, wafer unloading track)	Saturated	85.36
Location 3 (CMP tool B, wafer unloading track)	211.54	25.96

*Anticon Gold Wipes were used.

Table 6 Airborne Metallic Contaminants in the Metal CMP Area

Metal	Detection limits (ng/liter)	Metal CMP (preclean)	Metal CMP (postclean)	Shared airstream (preclean)	Shared airstream (postclean)
Aluminum	0.0006	0.008	0.17	0.009	0.049
Barium	0.0001	0.003	0.0002	0.0003	0.0003
Boron	0.001	0.071	0.053	0.14	0.067
Calcium	0.001	0.008	0.01	0.001	0.016
Chromium	0.0006	0.009	0.0037	0.004	0.0016
Cobalt	0.0001	0.001	0.0002	0.0001	0.0004
Copper	0.0006	0.003	<0.0006	0.001	0.0014
Iron	**0.0001**	**0.65**	**0.19**	**0.12**	**0.12**
Nickel	0.0006	0.001	0.0032	0.001	0.0018
Potassium	0.001	0.002	0.006	0.001	0.009
Sodium	0.001	0.005	0.004	0.001	0.013
Zinc	0.0006	0.003	0.0053	0.002	0.0027

after cleaning showed that the cleaning reduced the contamination in the room by approximately an order of magnitude.

Airborne testing was performed with preleached teflon impingers containing a 2% HNO_3 solution at 1 liter/min flow rate; results reported in Table 6 are in ng/liter and are blank-subtracted. Typical cleanroom airborne iron levels are <0.02 ng/liter [59,60], yet the airborne levels detected in this experimentation were very high, particularly in the CMP area. The shared airstream showed ele-

Table 7 Witness Wafer Data from the Metal CMP Area (TXRF)

Wafer position	Exposure time	Before cleaning, preexposure Fe Conc	After cleaning, postexposure Fe Conc
Control 1	0 h	<5E9 atoms/cm^2	<5E9 atoms/cm^2
Control 2	0 h	<5E9 atoms/cm^2	<5E9 atoms/cm^2
Metal CMP tool A	2.5 h	Not performed	3.3E10 atoms/cm^2
Metal CMP tool A	24 h	<5E9 atoms/cm^2	7E9 atoms/cm^2
Floor near CMP tool A	2.5 h	<5E9 atoms/cm^2	1.4E10 atoms/cm^2
Floor near CMP tool A	24 h	Not performed	4.4E10 atoms/cm^2
Metrology tool away from CMP tool A	2.5 h	<5E9 atoms/cm^2	<5E9 atoms/cm^2
Metrology tool away from CMP tool A	24 h	<5E9 atoms/cm^2	<5E9 atoms/cm^2

vated levels of iron as well. Cleaning was performed and subsequent repeat sampling showed a reduction in airborne iron concentrations; yet the levels remain higher than typical cleanroom levels in both airstreams after cleaning. Airborne aluminum increased significantly after clean, indicating that the cleaning agent likely contained aluminum.

B. Wafer Exposure Data

Wafers were exposed to the CMP environment for two time periods, 2.5 and 24 h, prior to its cleaning. Some wafers were measured in a minienvironment by TXRF located outside of the fab. Preexposure and postexposure are noted in Table 7. Other wafers were measured only after exposure, and were not removed from the fab environment. CMP tool A is where the Fe-nitrate slurry was used.

Detection limits of the TXRF for iron were 5E9 atoms/cm^2 in this study. Elevated surface iron levels were observed at the CMP tool A and on the floor near the CMP tool A. A metrology tool in the same process area and airstream as CMP tool A was also sampled; this tool was 15 ft across the room from the CMP tool A and no increase in surface iron was observed.

The Fe concentrations measured on wafer that were removed from the fab for the preexposure measurements showed that the removal techniques and minienvironment of the TXRF did not contribute to the Fe levels. An increase in surface iron concentration was observed with increased time of exposure at the floor sampling area, but this pattern was not observed at the metal CMP tool A.

This airborne metals study proves that use of metal-containing CMP slurries can result in increased airborne metal concentrations, and increased surface metal concentrations in the CMP cleanroom area. Increased metal concentration was also observed in a shared airstream which suggests that front-end-of-line process areas should have airstreams isolated from CMP areas. It is likely that the majority of the airborne metals detected in this study were particulate in nature.

C. Airborne Boron in FAB Environment

Airborne boron in cleanrooms originates from two major sources: 1) HEPA filtration media is made up of borosilicate glass. HF fumes escaping into a cleanroom airstream can attack the HEPA media and produce airborne boron; 2) The second source is atmospheric boron present in the range of 20–450 ng/m^3 [61]. Airborne boron can deposit on surfaces and result in increased dopant concentrations. SIMS concentration profiles from a study by Muller show a significant effect of airborne boron deposition when a wafer is exposed to the fab airstream between two polysilicon deposition steps [62].

XII. AIRBORNE ORGANICS

There are multiple sources of airborne organics in a semiconductor cleanroom. Cleaning baths, outgassing of cleanroom paints and plastics, HEPA/ULPA filtration outgassing, adhesives, cassette/pod materials, and other cleanroom construction materials can contribute organic compounds to the air chemistry in the fab.

It is impossible to eliminate all organic species from the cleanroom air, thus, an understanding of the compounds which are destructive to wafer processing is required. The major offenders are higher molecular weight, lower vapor pressure compounds as well as polar species. Deleterious properties of adsorbing organics include a high sticking coefficient to silicon as well as a high energy of desorption.

Organics with vapor pressures in the range of 1E-2 mm or less can be expected to condense to surfaces [63]. However, dissociative desorption of very volatile species such as alkenes on silicon surfaces can also occur, leaving behind carbon-based fragments that can form extremely strong Si–C bonds [64–67]. Organic adsorption varies with specific compounds and organic families, and the properties of the exposed surfaces.

A. Construction Materials

The importance of choosing construction materials for a cleanroom is critical. Table 8 summarizes work by Gutowski, Oikawa, and Kobayashi relating cleanroom construction materials to outgassing compounds [68].

Table 8 Cleanroom Construction Materials and Their Outgassing Compounds

Construction material	Organic compounds outgassed
Flooring materials	Dioctyl phthalate (DOP)
HEPA gel seal	Triethylphosphate (TEP)
Urethane foam sealants for HVAC	TEP and butylated hydroxytoluene (BHT)
Polyurethane adhesives	BHT, amine compounds
Flexible duct connector	Phosphate esters, DOP
Concrete sealing paint	Alkenes, alcohols, amines
Silicon sealant	Cyclic siloxanes
Vinyl materials	DOP, texanol isobutyrate (TXIB), tributyl phosphate (TBP)
Silicon tubing	Siloxanes, dibutyl phosphate (DBP)

Source: Ref. 68.

These outgassing compounds are present in organic construction materials for manufacturing reasons. TEP is typically added as a fire retardant. BHT is a common antioxidant for polymers. Phthalates are plasticizers. Amines are used to enhance cross-linking in polymers. The siloxane structure is the basic building block for silicones.

B. Surface Measurements of Organics

Figure 26 from Fujimoto [51] shows a relationship between the silicon surface and airborne organics.

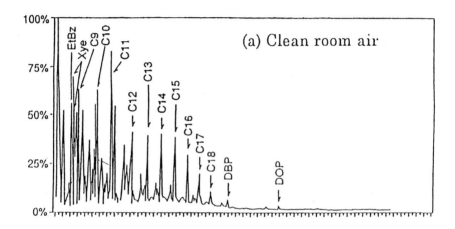

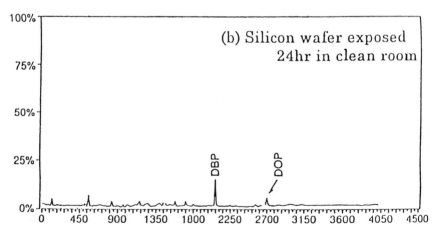

Figure 26   Airborne organics and their affinity to silicon surface deposition. (From Ref. 51.)

C. HEPA Filter Outgassing

Identification of the compounds outgassing from an ULPA filter, presented in Table 9, is based on a standardized technique for outgassing of solids developed by Camenzind and Kumar [73]. A Perkin-Elmer ATD400 Thermal Desorption tool coupled to an H5890A gas chromatograph with an HP5972 mass spec quadrapole detector was used. A variety of species known to be deleterious to wafer surfaces were emitted.

Amines, detrimental to DUV processing, were detected. TEP, a source of phosphorus, was found. BHT, alkenes, and siloxanes were detected. Of interest is NMP adsorbed into the ULPA paper. The likely source is the photoresist stripping bath fumes as this filter was taken from the cleaning bath area of the fab. Phenol and triethylene glycol were also seen in high concentrations.

D. Organophosphorus Contamination

HEPA filter outgassing of organophosphorus compounds has been studied, along with the effects of these compounds when deposited on wafer surfaces. Thermal desorption/GC/MS was performed on wafers exposed to laminar flow from a HEPA filtered module. TEP was present in the cleanroom air and also on the wafer [70]. A correlation between TEP concentration and installation of new HEPA filters were made. When fans above the HEPA filtered module were shut down, airborne TEP concentrations reduced accordingly.

Kodak and MEMC identified wafer doping by organophosphate outgassing of HEPA filtration [71] yet due to a phosphate other than TEP. This study

Table 9 Outgassed Compounds from ULPA Filter Components

Compound	Potting compound (ppmw)	End sealant (ppmw)	ULPA paper (ppmw)
Phenol (C_6H_6O)	169.5	8.1	7.8
Triethylenediamine ($C_6H_{12}N_2$)	72.5	69.2	2.0
C9H12 isomers (alkene)	68.3	nd	nd
Triethyl phosphate ($C_6H_{15}O_4P$)	58.2	16.4	0.4
Xylene isomers	57.5	nd	nd
p-tert-Butyl phenol ($C_{10}H_{14}O$)	42.2	6.8	2.9
BHT ($C_{15}H_{24}O$)	10.4	6.9	0.2
C10H14 isomers (alkene)	5.3	nd	nd
Siloxanes	nd	49.1	1.3
Triethylene glycol	nd	3.9	96.5
1-Methyl-2-pyrrolidinone (C_5H_9NO), NMP	nd	nd	1.1

identified Fyrol PCF [tri(β-chloroisopropyl) phosphate], a fire retardant, as a source of phosphorus dopant. A correlation between unintentional P doping and increased sheet resistance spread was observed.

E. Wafer Carrier Materials

Wafer carriers and pods can outgas and result in organic adsorption to wafers inside. The following work by Budde shows total organics outgassed vs. desorption temperature vs. polymer type [72] using ion mobility spectrometry. Table 10 summarizes the outgassing compounds by polymer type.

Budde used IMS to detect organics adsorbed to a wafer stored for 24 h in a polypropylene carrier [73]. The results are shown in Figure 27. The upper spectrum is a background of a clean wafer showing reactant ions only. IMS utilizes a reaction between reactant ions from the gas mix (typically clean dry air or nitrogen) with desorbed compounds to produce product ions (Chap. 3). The bottom spectrum shows the product ions associated with phthalates, BHT, and a lithography solvent that adsorbed to the wafer during storage in the carrier.

F. Effects of Organics on the Silicon Surface

Adsorption of organics to silicon surfaces can result in surface property changes, for example, from hydrophilic to hydrophobic. This can inhibit the effectiveness of aqueous cleaning baths. Figure 28 represents BHT adsorption to a silicon wafer [72]:

Organic deposition to wafer surfaces can result in the altering of the oxidation kinetics of silicon. Rapid thermal oxidation of the silicon has been shown

Table 10 Outgassing Compounds from Cleanroom Polymers

Polymer	Outgassing compounds
ABS	Benzene, styrene, ethylbenzene, divinylbenzene, diethylbenzene
PP1, PP2, PP3	Butenone, pentenone, hexanone, octanone, heptadienone, octadienone, BHT
PC	Aliphatic compound, chlorobenzene
PVDF	Formic acid ethyl ester, carbonic acid diethyl ester
PFA	4 compounds, not identified
PTFE	Very small amount of one unidentifiable compound

PP1: polypropylene (blue); PP2: polypropylene (antistat); PP3: polypropylene (natural); PC: polycarbonate; PFA: polyfluoralkoxy polymer; PVDF: polyvinylidiene fluoride; ABS: acrylonitrile butadiene styrene copolymer; PTFE: polytetrafluoroethylene.
Source: Ref. 72.

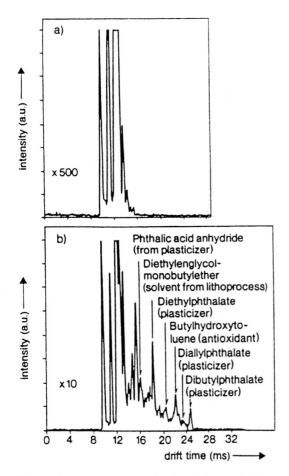

Figure 27 IMS spectra of wafer before and after carrier exposure. Drift spectra of volatiles from wafer surfaces. Upper curve: clean wafer (signals are almost only reactant ions); lower curve: wafer after 24 hr storage (24°C) in a polypropylene box. The values 10 and 500 refer to vertical enhancements. (From Ref. 73.)

to slow down when carbon is incorporated on the surface [74]. Phthalate adsorption from the ambient air has been shown to inhibit room temperature oxidation kinetics [75]. Figure 29, from Licciardello, shows the sequence of a hydroxyl-terminated silicon surface (a), a fluorine-terminated silicon surface after HF dip (b), the adsorption site of a phthalate molecule (c), and the resultant surface after adsorption (d).

For silicon to oxidize with the phthalate at the surface would require oxygen diffusion through the organic surface layer, reducing oxidation kinetics.

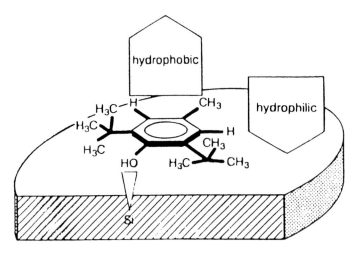

Figure 28   Organic structure of BHT as adsorbed to a silicon wafer. (From Ref. 72.)

Also, if a phthalate-covered surface enters a rapid thermal oxidation system, the total effectiveness of thermal desorption is likely not 100%. CO could be left behind on the surface due to dissociative desorption, creating defects in the silicon oxide layer that can result in a reduction in breakdown voltage.

In a study by Shaneyfelt et al. organic deposition on a preoxidized silicon surface was shown to produce changes in C-V characteristics. The C-V changes are likely due to traps that are a function of the C-V frequency [76]. The authors also found an increase in O-related donors in the silicon substrate. Figure 30 is the C-V plot before and after organic deposition.

XIII. SUMMARY

Deposition of airborne contaminants is not as simple as identifying just the surface flux component to the deposition velocity; sticking coefficients and surface interactions are also important. In many cases, diffusion is the rate-limiting force in deposition. Each adsorbate/surface interaction is unique, and extremely complex. Modeling of this type of interaction is extremely computer intensive, and requires knowledge of many surface kinetics parameters.

Airborne bases have been identified to deteriorate lithography mask CD control when using DUV photo-acid-generated resists. Airborne acids have caused metal corrosion. In the right concentration, sulfuric acids can transform to aqueous aerosols. Metals used in CMP slurries can become airborne and deposit to cleanroom surfaces and wafers. Airborne organics have various sources,

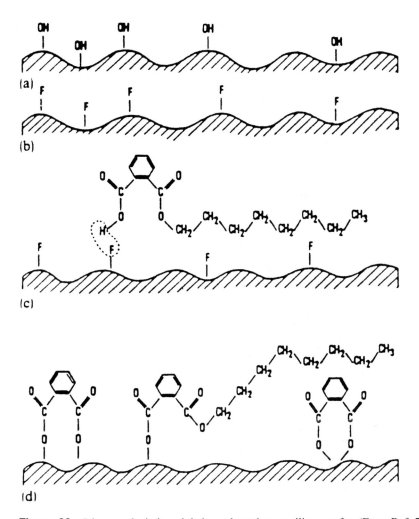

Figure 29   Diagram depicting phthalate adsorption on silicon wafer. (From Ref. 75.)

including cleanroom construction materials. Organophosphates and amines have been shown to outgas from HEPA and ULPA filters.

Of ultimate importance is what stays on the silicon wafer upon adsorption. Phthalates, BHT, and organophosphates have been measured on wafer surfaces and are linked to airborne deposition. Doping profile changes due to P-containing and B-containing compounds in the air have been observed. Oxidation kinetics changes are seen when organics are deposited on silicon surfaces prior to oxidation.

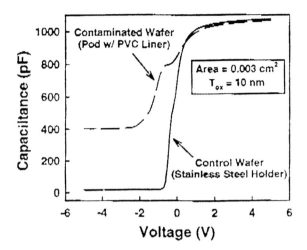

Figure 30 *C-V* changes due to organic deposition preoxidation. (From Ref. 76.)

Removal of airborne molecular contaminants via filtration is important, as is the design of cleanrooms with isolated airstreams, minienvironments, and wafer pods. Understanding of the airstreams in front-end-of-line processing and DUV photolithography, in particular, is important. Elimination of many species can be performed by banning specific construction materials in the fab. Monitoring of species of importance is highly recommended.

ACKNOWLEDGMENTS

The author would like to thank Ben Nowak, Leon Piasecki, Steve Natale, Dave Messier, Craig Seeley, Reg Burt, and Mark Seliger for their contributions to this chapter.

REFERENCES

1. 1997 National Technology Roadmap for Semiconductors.
2. Takizawa, R., Nakanishi, T., Ohsawa, A. Degradation of metal-oxide-semiconductor devices caused by iron impurities on the silicon wafer surface. J Appl Phys 62(12), December 1987.
3. Wauk, M., et al. Controlling heavy metal and dopant contamination during ion implantation. Microcontamination, October 1994.
4. Seinfeld, J. H. Atmospheric Chemistry and Physics of Air Pollution. John Wiley, New York, 1986.

5. Wilkes and Lee. Ind Eng Chem 47:1253 (1955).

6. Hirschfelder, Bird, and Spots. Trans Am Soc Mech Engrs 71:921 (1949).

7. Perry, Ed. Chemical Engineers Handbook.

8. 3M Occupational Health and Safety Division. 3M Organic Vapor Monitor Sampling Rate Validation Protocol, document 70-0701-3308-0 (92.1), September 1992.

9. Lugg, G. A. Analytical Chemistry 40(7):1073, June 1968.

10. Cano-Ruiz, J. A., Kong, D., Balas, R. B., Nazaroff, W. W. Removal of reactive gases at indoor surfaces: combining mass transport and surface kinetics. Atmospheric Environment, vol. 27A, 1993.

11. Chen, B. T., et al. Evaluation of an environmental reaction chamber. Aerosol Sci Tech 17:924 (1992).

12. Hudson, J. B. *Surface Science*, Butterworth-Heinemann, Boston, 1992.

13. Friedlander, S. K. *Smoke, Dust, and Haze.* John Wiley, New York, 1977, chaps. 2–4.

14. Kinkead, D., and Higley, J. Targeting gaseous contaminants in wafer fabs: fugitive amines. *Microcontamination*, June 1993, pp. 37–40.

15. Madou, M. J., and Morrison, S. R. *Chemical Sensing with Solid State Devices.* Academic Press, Boston, 1989.

16. Zangwill, A. Physics at Surfaces. Cambridge University Press, 1992.

17. Zhu, Sheng-Bai. Theoretical study of molecular contamination on silicon wafers. Journal of the IEST, July/August 1998, pp. 30–35.

18. Sinniah, K., Sherman, M. G., Lewis, L. B., Weinberg, H., Yates, J. T., Jr., and Janda, K. Hydrogen desorption from the monohydride phase on Si(100). J Chem Phys 92(9):5700, May 1990.

19. Freundlich, H. *Colloids and Capillary Chemistry*. E. P. Dultons and Company, New York, 1922.

20. Brunauer, S., Emmett, P., and Teller, E. Adsorption of gases in multimolecular layers. J Am Chem Soc 60:309, 1938.

21. Bailey, A. R. Conceptual model of aluminum corrosion of an integrated curcuit. Proceedings of Moisture and Measurement Control Symposia, NBS, 1980. Govt. Doc. C13.10:533.

22. Cvijanovich, G. B. Conductivities and Electrolytic Properties of Adsorbed Layers of Water. Proceedings of the Moisture and Measurement Control Symposia, NBS, 1980. Govt. Doc. C13.10:533.

23. Dante, J. F., and Kelly, R. G. The evolution of the adsorbed solution layer during atmospheric corrosion and its effects on the corrosion rate of copper. J ElectroChem Society 140(7):1890, July 1993.

24. Lombardo, S., and Bell, A. A review of theoretical models of adsorption, diffusion, desorption and reaction of gases on metal surfaces. Surface Science Reports 13: 1–72, 1991.

25. Pfnur H., and Menzel, D. The influence of adsorbate interactions on kinetics and equilibrium for CO on Ru(001). I. Adsorption Kinetics. J Chem Phys 79(5):2400, Sepember 1983.

26. Sheng-Bai Zhu. Theoretical study of molecular contamination on silicon wafers: kinetics. Journal of the IEST, September/October 1998, pp. 36–43.

27. MacDonald, S., et al. Airborne chemical contamination of a chemically amplified resist. SPIE, vol. 1466, 1991.
28. Kawai, Y., et al. The effects of an Organic Base in a Chemically Amplified Resist on Patterning Characteristics Using KrF Lithography. MicroProcess 94, The 7th Annual International MicroProcess Conference, July 11–14, 1994, p. 202.
29. Hinsberg, W. D., and MacDonald, S. A. Quantitation of airborne chemical contamination of chemical amplified resists using radiochemical analysis. SPIE, Advanced in Resist Technology and Processing IX, vol. 1672, pp. 24–32, 1992.
30. Michielson, M., et al. Priming of silicon substrates with trimethylsilyl containing compounds: Microelectric Engineering 11:475–480, 1990.
31. Nalamasu, O., et al. Preliminary lithographic characteristics of an all-organic chemically amplified resist formulation for a single layer deep-UV lithography. SPIE, vol. 1466, 1991.
32. Lovett et al. Real-time analysis of breath using an atmospheric pressure ionization mass spectrometer. Biomedical Mass Spectrometry 6(3), 1979.
33. Larson, et al. A method for continuous measurement of ammonia in respiratory airways. American Physiological Society, 1979.
34. Norwood, et al. Breath ammonia depletion and its relevance to acidic aerosol exposure studies. Archives of Environmental Health 47(4), July/Aug 1992.
35. Hunt, et al. Spectrometric measurement of ammonia in normal human breath. American Laboratory, June 1977.
36. Kern, J. The evolution of silicon wafer cleaning technology. ElectroChem Society 137(6):1887, June 1990.
37. Psota-Kelty, L. A., and Muller, A., et al. Measuring amine concentrations in cleanrooms. IES 1994.
38. Berro, et al. Airborne contamination on semiconductor wafers traced to humidification plant additives. Journal of the IES, November/December 1993, p. 15.
39. Vigil. Contamination control for processing DUV chemically amplified photoresists. SPIE 1994.
40. Dean, K., and Carpio, R. Contamination control for processing DUV chemically amplified photoresists. SPIE 1994.
41. Guy, A. G., et al. Pourbaix diagrams, a firm basis for understanding corrosion, Metal Treatment and Drop Forging, February 1962, p. 42.
42. Scully, J. R. Metastable Pitting Corrosion of Aluminum, Al-Cu, Al-Si Thin Films in Dilute HF Solutions. ECS Symposium, Critical Issues in Localized Corrosion, 1991.
43. Bohni, H., and Uhlig, H. H. Environmental factors affecting the critical pitting potential of aluminum. J Electrochem Society, July 1969.
44. Uhlig, H. H., and Revie, R. W. Corrosion and Corrosion Control. John Wiley, New York, 1985, chap. 5.
45. Higley, J. K. Airborne Molecular Contamination Control in the Cleanroom. Cleanrooms '96 East Proceedings, p. 99.
46. Davis, C., et al. HEPA Filters as a Contamination Source . . . March/April 1981.
47. Stevie, F. A., et al. Boron Contamination of Surfaces in Silicon Microelectronics Processing: Characterization and Causes, October 1991.

48. Tang, I. N., and Munkelwitz, H. R. Aerosol growth studies III: ammonium bisulfate aerosols in a moist atmosphere. J Aerosol Science 8:321–330, 1977.
49. Gelbard, F. The General Dynamic Equation for Aerosols. Ph.D. thesis, California Institute of Technology, Pasadena, CA, 1978.
50. Hartzel, A. Inorganic aerosols: formation and growth. SPWCC 1997, AMC Panel Discussion Presentation.
51. Fujimoto, T., et al. Evaluation of contaminants in the cleanroom atmosphere and on silicon wafer surfaces. 1996 SPWCC Proceedings, p. 325.
52. Shewmon, P. Diffusion in Solids, 2nd ed. The Minerals, Metals and Materials Society, Warrendale, PA, 1989, p. 175.
53. Morinaga, H., Suyama, M., and Ohmi, T. Mechanism of metallic particle growth and metal-induced pitting in Si wafer surface in wet chemical processing. J Electrochem Society 141(10), October 1994.
54. Yoneshige, K., et al. Deposition of copper from a buffered oxide etchant onto silicon wafers. J Electrochem Society 142(2), February 1995.
55. Jeon, J., et al. Electrochemical investigation of copper contamination on silicon wafers from HF solutions. J Electrochem Society 143(9), September 1996.
56. Zhong, L., and Shimura, F. Dependence of lifetime on surface concentration of copper and iron in silicon wafer. Applied Physics Letters 61(9), August 1992.
57. Morinaga, H., Suyama, M., Nose, M., Verhaverbeke, S., and Ohmi, T. Metallic Particle Growth on Si Wafer Surfaces in Wet Chemical Processing and It's Prevention. IES Proceedings, 1994, p. 332.
58. Lin, P., Marcus, R., and Sheng, T. Leakage and breakdown in thin oxide capacitors— correlation with decorated stacking faults. J Electrochem Society: Solid-State Science and Technology, September 1983.
59. Camenzind, M., Liang, H., Fuscko, J., and Balazs, M. How clean is your cleanroom air? Micro, October 1995.
60. Mikulsky, J. Chemically clean air: an emerging issue in the Fab environment, Semiconductor International, September 1996.
61. Fogg, T., Duce, R, and Fasching, J. Sampling and determination of boron in the atmosphere. Analytical Chemistry 55(13), November 1983.
62. Muller, A., et al. Volatile cleanroom contaminants: sources and detection. Solid State Technology, p. 61, September 1994.
63. Muller, A., et al. Concentrations of Organic Vapors and Their Surface Arrival Rates at Surrogate Wafers During Processing in Clean Rooms. In: Semiconductor Cleaning Technology 1989 (J. Ruzyllo and R. E. Novak, eds.), PV 90-9. P. 204, The Electrochemical Society Softbound Proceedings Series, Pennington, NJ, 1989.
64. Bozsco, F., Yates, J. T., et al. Studies of SiC formation on Si (100) by chemical vapor deposition. J Appl Phys 57(8):2771, April 15, 1985.
65. Cheng, C., et al. Hydrocarbon surface chemistry on Si (100). Thin Solid Films 225:196, 1993.
66. Bozack, M., et al. Chemical activity of the C=C Double Bond on Silicon Surfaces Surface. Science 177:L933, 1986.
67. Cheng, et al. Thermal stability of the carbon-carbon bond in ethylene adsorbed on Si(100): an isotopic mixing study surface: Science 231:289, 1990.

68. Gutowski, T., Oikawa, H., and Kobayashi, S. Airborne Molecular Contamination Control of Materials Utilized in the Construction of a Semiconductor Manufacturing Facility. 1997 SPWCC Proceedings, vol. II, p. 143.
69. Camenzind, M., and Kumar, A. Organic Outgassing from Cleanroom Materials Including HEPA/ULPA Filter Components: Standardized Testing Proposal. IES Proceedings, May 1997.
70. Mori, E., et al. Correlating organophosphorus contamination on wafer surfaces with HEPA-filter installation. Microcontamination, November, 1992, p. 35.
71. Lebens, J., et al. Unintentional doping of wafers due to organophosphates in the clean room ambient. J Electrochem Soc 143(9), September 1996.
72. Budde, K., et al. Application of ion mobility spectrometry to semiconductor technology: outgassings of advanced polymers under thermal stress. J Electrochem Soc 142(3), March 1995.
73. Budde, K., et al. Measurement of Organic Contaminants from Silicon Surfaces. IES Proceedings, 1992.
74. Tsuchiaki, M., et al. Carbon Contamination on Silicon Surface. J Electrochem Soc, Vol. 143, No. 9, Sept. 1996.
75. Licciardello, A., et al. Effect of organic contaminants on the oxidation kinetics of silicon at room temperature. Applied Physics Letters 48(1), January 6, 1987.
76. Shaneyfelt, M., et al. Impact of Organic Contamination on Device Performance. Sandia National Labs DOE Contract DE-AC04-94AL85000.

9
Organic Contamination Removal

Steven Verhaverbeke
Applied Materials, Santa Clara, California

Kurt Christenson
FSI International, Chaska, Minnesota

I. INTRODUCTION

A. Background

After the early work of Kern, developing the alkaline and acidic RCA-1 and RCA-2 cleans in 1965 [1,2], little fundamental change occurred in silicon wafer cleaning for 20 years. In the late 1980s, a recognition of the economic impact of contaminants in semiconductor processing drove a renaissance in cleaning technology. This renaissance continues to the present with the added requirement of environmental consciousness.

While most steps in semiconductor processing receive tight quality control on the condition of their incoming material, cleaning serves as a "catchall" or "insurance" process that removes whatever contaminants are present. Advanced lithography requires tight control of incoming resist chemistry and thickness, antireflection layers, adhesion layers, resist bake times and temperatures, development parameters, and hardening steps. Cleaning sequences are expected to remove any existing forms of contamination, either singly or in combination, to levels below their threshold of impact which are often near or below current detection limits. Further, contamination removal must not degrade the device structures that are present on the wafer and must leave the wafer surface in the desired chemical state.

The cleaning processes are further complicated by interaction between individual steps of a given cleaning process. For example, an aqueous-based etch of SiO_2 with HF removes not only the oxide but also metals and particles.

317

Unfortunately, the HF etchant can also deposit metals or particles on the silicon that must be removed by subsequent process steps. "Value added" steps such as the deliberate etching of oxide or metal layers are often integrated into cleaning procedures.

B. Challenges

The contaminants/defects to be removed by the cleaning sequence can be broken into six broad categories: organics, oxides, particles, metals, mobile ions, and crystal damage (see Fig. 1). Organics are typically removed first because they can form nanometer-scale films which mask the cleaning of other contaminants. Oxide films can themselves be a contaminant, and can trap metals and ions. Particles range in size up to thousands of nanometers and can be composed of any material. "A contaminant that can be physically detached from the surface intact" is an alternative working definition of a particle from the viewpoint of contamination removal. Since most metals present on the surface are ionized, the distinction between molecular metals and mobile ions is somewhat arbitrary, based historically on the mobility of ions from column I of the periodic table in Si and SiO_2. In Chapter 10, molecular and ionic contaminants are treated together. Films and molecular contaminants are normally dissolved and removed in a molecular form. Layers of silicon whose crystal structure has been damaged or that contain metals or mobile ions can also be removed by dissolution.

Within these six categories are many individual challenges that cross categories. For example, the via veil present after a reactive ion etch (RIE) metal etch and O_2 ash can contain organics, metallics, and oxides and can be detached from the surface much like a particle. Anion and cation residues present after the cleaning process can be considered a new class of contaminant that is

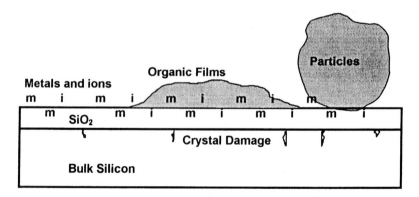

Figure 1   Contamination/defect challenges present on silicon wafers.

receiving more attention as device geometries shrink. Some workers even consider static charge a "contaminant" (Chap. 1).

Once contaminants have been separated from the surface, two further phenomena are necessary for cleaning. First, the contaminants must move by diffusion or other transport mechanisms into a region of bulk fluid flow to be carried away. Finally, for liquid-based processes, the rinse liquid must be dried uniformly and thoroughly.

C. Technologies

The range of challenges, along with the varying demands for compatibility with upstream and downstream processes, have generated a variety of cleaning technologies which can be broken into three categories: liquid, vapor, and vacuum. Liquid, primarily aqueous-based, technologies involve wetting the surface of the wafers. Vapor-based technologies clean with gasses near ambient pressures, volatilizing contaminants. Vacuum technologies clean with gasses, UV, IR, and plasma energy, but at substantially lower pressures than vapor technologies. The distinction between technologies can become blurred—for instance, when vapor technologies use a water rinse to remove soluble, but nonvolatile, contaminants. Vacuum technologies try to avoid the use of water, even in vapor form, so as to remain compatible with vacuum cluster technology.

II. ORGANIC REMOVAL

Organic challenges in wafer cleaning fall into the classes of heavy and light organics. Heavy organic challenges typically consist of photoresist or waxes that are deliberately applied in micrometer-thick layers. As applied, these long chain hydrocarbons are easily removed. But baking, UV exposure, ion implants, and reactive ion etching can toughen the resist. Light organics consist of nanometer-thick scale layers of various carbon compounds that are accidentally deposited onto the wafer during processing, storage, or transport. Light organic compounds commonly detected on wafers include plasticizers, vacuum pump oils, short segments of polymers, siloxanes, and skin oils.

A. Oxygen Plasma Ashing

The dominant method for the removal of photoresist is plasma ashing followed by a wet chemical cleanup of nonvolatile species. In most cases photoresist, even that which could be removed by chemical means, is ashed. In an asher, oxygen gas is excited to form atomic oxygen which directly oxidizes carbon compounds. The atomic oxygen historically has been created with an in situ plasma that either surrounds or is adjacent to the wafers. However, this method carries the risk of: electrostatic discharge (ESD) damage due to exposure to UV light from the plasma; the charging of the wafer leading to ESD; and damage by direct bombardment by hot (high velocity) ions and electrons. In a downstream asher, the atomic oxygen is created in a shielded region remote from the wafers and is transported to the wafers by a carrier gas. Remote production of atomic oxygen eliminates the possibility of damage by UV light and direct bombardment by hot ions and reduces the risk of charging.

B. Piranha and SOM

1. Dissolution and Oxidation of Organics

In the past, front-end resist stripping was performed with inorganic oxidizing mixtures. Inorganic acids such as sulfuric acid (H_2SO_4), nitric acid (HNO_3), chromic acid (H_2CrO_4), phosphoric acid (H_3PO_4), and hydrogen peroxide (H_2O_2) have been used since the early days of semiconductor manufacturing to strip resist layers [3]. Even today, all these chemicals can be found in use for resist stripping. Mixtures of sulfuric and chromic acid, of sulfuric and nitric acid are typically used at 100°C. In contrast, fuming nitric (>95% HNO_3) typically is used at room temperature. Some of these inorganic oxidizing mixtures can also be used in the back end of the line of wafer processing, since many of the concentrated acids are not corrosive to metals at low water concentrations and close to room temperature. Today, as said before, the bulk of the resist is usually removed by ashing and in practice wet stripping is limited to post-ash stripping.

   a. Piranha.   Since 1980, "piranha" wet baths, mixtures of sulfuric acid and hydrogen peroxide, have become the most common method of post-ash resist stripping. Piranha solutions, blends of $H_2SO_4 : H_2O_2$ in the range of 1 : 1 to 10 : 1, have been the primary wet chemical means of removing heavy organics, with the most common mixture being about 4 : 1 H_2SO_4 (>95% wt%) to H_2O_2 (31 wt%). When hydrogen peroxide and sulfuric acid are mixed, "Caro's acid" (i.e., monopersulfuric acid [H_2SO_5]) is formed. Caro's acid is the active etchant in piranha baths.

   Originally, however, the piranha baths consisted of a mixture of concentrated sulfuric acid (>95 wt%) with highly concentrated (85–90 wt%) hydrogen peroxide. Mixing these two chemicals results in the production of Caro's acid [4]:

$$H_2O_2 + HO-(SO_2)-OH \leftrightarrow HO-(SO_2)-O-OH + H_2O \tag{1}$$

Concentrated sulfuric acid is an excellent solvent base for Caro's acid, since Caro's acid decomposes in water. As shown in reaction (1), water is produced in the reaction between hydrogen peroxide and sulfuric acid. The presence of excess water (i.e., as a result of using more dilute reactants) in the mixture actually shifts the equilibrium of the reaction toward the reactants [toward the left in (1)], minimizing the production of Caro's acid. Consequently, using highly concentrated hydrogen peroxide (85–90 wt%) optimizes the production of Caro's acid in piranha baths.

Caro's acid has two significant advantages as a photoresist stripper:

1. It is effective at room temperature.
2. It is noncorrosive toward metals at room temperature in the absence of water.

However, highly concentrated hydrogen peroxide (85–90 wt%) is extremely dangerous; it is a serious fire hazard; it is potentially detonable in the presence of small amounts of organic compounds; and it can cause severe chemical burns. As a result of these safety issues associated with concentrated H_2O_2, the semiconductor industry has generally adopted the use of "laboratory concentrated" H_2O_2 (approximately 31 wt%) for all wet processing, including "piranha" stripping.

The excess water found in 31 wt% H_2O_2 unfortunately shifts the equilibrium in reaction (1) away from the production of H_2SO_5. Additionally, as a result of the heat of dilution of sulfuric acid, the use of dilute H_2O_2 leads to significant heating of the piranha solution when the reactants are mixed. Caro's acid, which is quite heat sensitive, subsequently breaks down, resulting in low equilibrium concentrations of this oxidizing acid. As a result, the mixture of H_2SO_4 and laboratory concentrated H_2O_2 requires very high temperatures (i.e., up to 120°C) in order to be effective in resist stripping. The resulting bath is quite unstable and hydrogen peroxide has to be added periodically. Every time dilute H_2O_2 is added, more water is added and the solution becomes even more unstable. Consequently, piranha baths typically must be changed every 8–12 h.

Thin organic films, waxes, and photoresists dissolve into the sulfuric acid at elevated temperatures and are dehydrated, leaving carbon compounds in solution that are rich in C=C double bonds. Hydrogen peroxide either forms H_2SO_5 (Caro's acid) or decomposes to form two ·OH (hydroxyl radicals). Caro's acid also decomposes to form an ·OH radical and an ·OSO_2−OH radical according to the following reaction:

$$HO-O-SO_2-OH \rightarrow \cdot OH + \cdot OSO_2-OH \qquad (2)$$

The ·OSO_2−OH and the ·OH radicals are active in photoresist stripping:

$$RH + \cdot OSO_2-OH \rightarrow \cdot R \text{ (alkyl radical)} + H_2SO_4 \qquad (3)$$

or

$$R-CH_2-R + \cdot OH \rightarrow \cdot R \text{ (alkyl radical)} + H_2O \qquad (4)$$

The alkyl radicals will react further with any oxidizing species, as follows:

$$\cdot R + \cdot O \rightarrow CO \quad \text{or} \quad \cdot R + \cdot O \rightarrow CO_2 \tag{5}$$

$\cdot O$ represents an oxygen radical that can be produced by any of the oxidizing chemicals found in the sulfuric baths (e.g., H_2O_2, O_3, H_2SO_5).

b. Sulfuric/Ozone Mixtures (SOMs). In many cases, it is possible to replace the hydrogen peroxide oxidant with ozone to form a sulfuric ozone mixture (SOM). The oxidant in the solution converts the carbon compounds to carbon monoxide, carbon dioxide, or small stable organic fragments, effectively "bleaching" the solution.

Sulfuric processing for resist stripping can thus use either hydrogen peroxide or ozone gas as an oxidizing agent. Hydrogen peroxide is expensive and requires high-temperature processing (125°C or greater) and frequent bath change-outs (every 8 h). SOMs can last for at least 4 days of continuous processing.

Table 1 shows reactions of organic compounds that occur in the presence of these various oxidizers.

For hydrogen peroxide (34 g/mole), 100 mL of 31% solutions potentially yields 1 mole of oxygen radicals, while 48 g of ozone gas is required to produce 1 mole of atomic radicals. The action of each oxidant is approximately equivalent on a molar basis, but it is easier to get a mole of oxidant from hydrogen peroxide. A large semiconductor grade O_3 generator can produce 90 g/h or 2 moles/h of O_3. A 200-mL "spike" of hydrogen peroxide can be added to a bath in 4 s.

While convenient and rapid, the use of 31 wt% hydrogen peroxide as an oxidant, as previously noted, has the disadvantage that it contains 69 wt% H_2O. The water gradually builds up in the solution bath with repeated H_2O_2 spikes; the stripping efficiency decreases resulting in film residues; and the bath must be changed. If ozone is used, the only water additions to the bath are from the oxidation of the hydrocarbons (Table 1) and absorption from the atmosphere.

The solubility of gasses in liquids decreases with increasing temperatures. Ozone is more soluble in sulfuric acid at lower temperatures. However, the reaction rate decreases at lower temperatures, so an optimum temperature for

Table 1 Organic Reactions with Oxidizers

Oxidizer	Generic reaction with organic compounds
H_2O_2	$-CH_2- + 3H_2O_2 \rightarrow 4H_2O + CO_2$
H_2SO_5	$-CH_2- + 3H_2SO_5 \rightarrow 3H_2SO_4 + CO_2 + H_2O$
O_3	$-CH_2- + 3O_3 \rightarrow 3O_2 + CO_2 + H_2O$

operation near 85–100°C exists for SOM bath systems. Hydrogen peroxide–based systems typically operate between 120 and 130°C. It is possible to calculate the amount of oxidant required to bleach or "clear" the sulfuric solution. First, determine the mass of the photoresist (area $\times$ thickness $\times$ pattern density $\times$ number of wafers). Assume that the mass is 100%–CH_2–. Every 14 g of –CH_2– (1 mole) requires 3 moles of oxidant to form CO_2 (Table 1). Due to system losses, it is necessary to add 50% more oxidant than the calculated value.

   c. *Species Concentration.*   A conventional 20 liter piranha bath consisting of 4 parts sulfuric acid : 1 part 31% by weight hydrogen peroxide contains 1.24 kg or 36.5 moles of hydrogen peroxide. According to Table 1, this quantity of H_2O_2 corresponds to an equivalent oxidation capacity of 12.16 moles of –CO_2–.

   In the case of an SOM processing system, ozone can be continuously sparged into sulfuric acid. An ozone generation rate of 1.74 g O_3/min (0.036 mole/min) is assumed; this output is available on a particular commercial unit often used for its high output. This production rate corresponds to an equivalent –CH_2– oxidation capacity of 0.012 mole/min (according to Table 1). At a production rate of 1.74 g/min of ozone, 16 h 47 min are required to oxidize an equivalent amount of –CH_2– as the 4 : 1 sulfuric acid/hydrogen peroxide solution for a 20 : 1 bath or vessel. In order to reduce this time, the ozone output can be increased (installing a larger ozone generator) or the bath or vessel volume can be decreased.

   d. *Oxidation Capacity of a Stripping Solution.*   The oxidation capacity of resist stripping solutions can be compared in terms of their ability to oxidize a mole equivalent of methylene (–CH_2–) groups. Clearly, a stripping bath produced with ozone requires many hours before obtaining the same oxidation capacity as a 4 : 1 H_2SO_4/H_2O_2 bath. It is, however, important to calculate how much oxidation capacity is *actually* needed for resist stripping. Typical resist thickness ranges from 1 to 1.5 μm, and the total surface area of fifty 200 mm wafers is 15,700 cm^2. The specific gravity of resist is of the order of 1.2 g/cm^3. Therefore, there is between 1.89 and 2.83 g of photoresist. However, after O_2 plasma ashing, the residue of resist left on a batch of 50 wafers is less than 1% of the original photoresist on the wafers (e.g., about 0.02 g) or less than 15 nm of photoresist. In other words, before ashing about 0.14 mole of equivalent –CH_2– will have to be oxidized, and after ashing about 0.0014 mole of equivalent –CH_2– groups will need to be oxidized. The amount of oxidizers necessary to remove all the photoresist on a batch of 50 or even 100 wafers will differ tremendously depending upon whether or not ashing is completed prior to stripping. Given these quantities of organic compounds present, calculations of bath life (or time required to fully oxidize the organic compounds) can be calculated.

In a 4 : 1 mixture of sulfuric acid/hydrogen peroxide a 20 liter bath initially has 12.16 moles of $-CH_2-$ oxidation capacity present. Therefore, at 100% efficiency, a 4 : 1 bath of sulfuric acid/hydrogen peroxide is good for 86 batches of 50 un-ashed wafers. However, 100% efficiency is clearly unreasonable, as the baths deteriorate rapidly through the degradation of the Caro's acid due to heat and the presence of water (as discussed above). In practice, 20 liter sulfuric baths are rarely used for more than 8 h (or about 20 batches) and are continually spiked with hydrogen peroxide. For 50 ashed wafers, this represents an oxidizing efficiency of only 0.25% of the original peroxide concentration. This reduced efficiency is the result of the fast degradation of Caro's acid in the presence of water and heat, resulting in a competing reaction for the organic oxidation.

In SOM, the production of oxidizing capacity of equivalent $-CH_2-$ groups is 0.012 mole/min. Thus each batch of 50 un-ashed wafers requires a sparging of ozone in the sulfuric acid for 11 min and 40 s at 100% efficiency of ozone use. Obviously, 100% efficiency can never be achieved. After ashing, less than 0.02 g of photoresist will be left on the wafer and SOM can supply ample oxidation capacity. The main advantages of the sulfuric acid/ozone process compared to the sulfuric/peroxide process are that ozone can be added without the addition of water and that the oxidizing power is maintained longer without degradation of performance. As a result of the higher concentration of sulfuric acid (there is no water initially) and the higher oxidation power of ozone vs. hydrogen peroxide, the temperature of the sulfuric/ozone process can be substantially lower than in the case of the sulfuric/hydrogen peroxide process. 85°C has been found to be the optimum temperature for the SOM process.

The quantity of oxidant can be determined empirically by putting no oxidant in the bath, adding the wafers (which turns the solution black) and then adding oxidant until the solution clears.

Both piranha and SOM processes are effective for light organics and both reactive ion etched and lightly ion implanted photoresists. High-dose ion implantation drives off the hydrogen from the photoresist leaving a hardened layer of amorphous carbon dominated by $C-C$ single bonds. This layer is largely resistant to piranha and SOM. Unpatterned areas of hardened resist can often be removed by dissolving the less hardened resist near the wafer surface and undercutting the amorphous carbon layers. But at the edges of the pattern, the carbonaceous layer contacts the wafer forming "fences" of amorphous carbon that are resistant to piranha and SOM. Patterned photoresist that has been implanted at levels above 10^{14}–10^{15} atoms/cm^2 must be ashed prior to wet stripping. In Figure 2, the residues from such "fences" are shown on a wafer with photoresist which has been implanted with As$^+$ at 100 keV and with a dose of

Figure 2 AFM photomicrographs of carbon "fences" on a patterned, resist-coated wafer that was processed two times in a piranha bath at 125°C for 10 min. The picture on the left is a 1 μm by 1 μm scan of the residue area; the picture on the right is the pattern interface where the residues are left. (From Ref. 4.)

1.4e16/cm^2. After processing in SPM at 125°C in two consecutive baths, the residues of these fences still remain. The right AFM scan (20 μm by 20 μm) shows the resist pattern edge and the left AFM scan (1 μm by 1 μm) shows the detail of the residues left.

With an ash, the total organic load is reduced substantially and clearing time is not an issue in SOM processing. Spiking of piranha baths is still necessary due to the breakdown of H_2O_2 in the hot solution with time. Clearing time is also not an issue in systems such as spray acid processors which use chemicals in a one pass or "fresh dispense" mode. Fresh dispense systems do not need to fully oxidize the organic challenge and therefore need less oxidant. The dissolved, dehydrated organics remain in the H_2SO_4 which passes over the wafer once and is then drained. Care has to be taken with these systems to limit the total consumption of H_2SO_4.

2. SPM Haze

Wafer hazing is often observed after an aggressive organic removal sequence using H_2SO_4. After the process, wafers are initially clear and have relatively low particle counts. On extended exposure to air, large numbers of particles form in a period from hours to weeks. These particles are sulfur rich, are very water soluble, and once rinsed away do not reform [5]. Evidence to date is consistent with the formation of an insoluble sulfate compound during the exposure to H_2SO_4. It is likely that the surface is terminated partially with SO_4^-. This surface termination reacts with ammonia vapors in the fab air, coalesces, and forms water-soluble particles, by etching the top monolayer of the terminated oxide. TOF-SIMS measurements indicate that these particles are comprised of SO_x^- and NH_4^+ [6].

A number of methods for its suppression or elimination have been developed empirically.

- The final number of particles can be reduced by the use of an extended hot DI rinse using megasonics [6].
- The growth can be eliminated by exposure to either dilute HF or SC-1 ($NH_4OH : H_2O_2 : H_2O$) after the strip. The SC-1 exposure also serves to remove other particles on the wafer.
- Adding small amounts of ammonium hydroxide to the post piranha rinse bath can eliminate the growth [6].
- The growth can be eliminated by the addition of a small amount of HF to the strip chemistry [7,8].

The addition of small amounts of ammonium hydroxide (e.g., sufficient to achieve pH10) to the post piranha rinse bath has been found to be effective in reducing the surface concentration of sulfur, as well as mitigating the piranha-induced particle growth. Sulfur concentration, measured by total reflectance x-ray fluorescence (TXRF), is shown in Figure 3 for both basic rinsing and

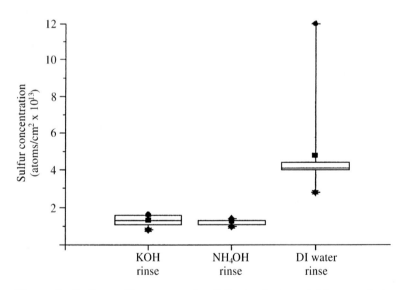

Figure 3 Surface sulfur concentration following basic and DI water rinsing. (From Ref. 6.)

rinsing in deionized water. The hydroxyl ion is clearly the active species with respect to sulfur removal.

By adding small amounts of HF to the sulfuric acid, a fluorine (F) passivated surface is obtained that is hydrophobic. This F passivation displaces the otherwise obtained SO_4^- termination. During subsequent rinsing in DI water, the F passivation is replaced again with OH^- hydroxyl termination.

C. Alkaline Solutions

The SC-1 solution, $NH_4OH : H_2O_2 : H_2O$ in a $1:1:5$ ratio, was designed to remove organic contaminants that are attacked by both the solvating action of the ammonium hydroxide and the powerful oxidizing action of the peroxide. In basic solutions, H_2O_2 breaks down into two hydroxyl radicals. Hydroxyl radicals have a very high oxidation potential. Kern's original process of a 10 min exposure in $1:1:5$ SC-1 at 80°C is sufficient to remove most light organic species. Note that SC-1 has a significant number of other effects on wafers such as Si, SiO_2 and metal etching, metal deposition and removal and roughening. Hossain et al. showed that, in many cases, after resist ashing, a heated SC1 process can replace the commonly used sulfuric/peroxide or sulfuric/ozone process [9].

D. Solvents

Both acidic and alkaline chemistries can attack the metal interconnect wiring used in integrated circuits. Therefore, solvents are used in the BEOL (back end of line) to remove organics to improve metal adhesion and for post photoresist plasma ash cleanup, whenever metals are exposed. N-methyl-2-pyrrolidone (NMP) is the primary component of most solvent mixtures at present and has replaced the flammable acetone blends used earlier. Chemical manufacturers blend in various proprietary additives which improve NMP's performance. Although NMP is relatively nontoxic and environmentally benign, it is still a target of the strong desire to eliminate all solvent usage in semiconductor fabrication.

The dissolution rates for solvents increase rapidly with temperature. NMP is often used near its flash point for maximum efficiency. In some cases, it is used at temperatures above the flash point with a nitrogen blanket to prevent ignition. The boiling point (T_b), flash point (T_{FP}), autoignition point (T_{AI}), and threshold limiting value (TLV) for acetone, IPA, and NMP are shown in Table 2.

Chemical analogs of NMP with similar solvating characteristics and higher flash points are becoming available and will reduce the potential of combustion. Chemical manufacturers have also developed aqueous proprietary stripping chemistries that have shown promise, but need to be tested in each target process.

Table 2 Physical Characteristics of Commonly Used Organic Solvents

Solvent	T_b (°C)	T_{FP} (°C)	T_{AI} (°C)	TLV (ppm)
Acetone	56	−17	464	750
IPA	82	12	399	400
NMP	202	96	346	10

Three main classes of new generation strippers are now becoming available for accomplishing one of the following functions:

1. Removing positive resist and organic residue
2. Removing multicomponent etch residues
3. Removing multicomponent etch residues with a controlled substrate etch

Most of these newer generation strippers no longer contain NMP. They all function mainly by means of solvation:

Step 1. Penetration (this is the primary function of the stripper)
Step 2. Swelling
Step 3. Dissolution

The first class of new generation strippers consists of:

• Polar organic solvent (the major component)
• Added amine (a minor component)
• Corrosion inhibitor

The second class of new generation strippers consists of:

• Basic amine (the major component; it has a high penetration ability and is basic to attack acidic resist)
• Polar solvent (a minor component)
• Corrosion inhibitor
• Optionally water

The third class of new generation strippers consists of:

• Polar solvent
• Tertiary amine (optional)
• Water (optional)
• Fluoride (to achieve a controlled substrate and oxidized residue removal)

E. $O_3 : H_2O$

Ozone gas from current ozone generators can be dissolved in deionized (DI) water at atmospheric pressure to concentrations from 10 to 50 ppm by weight to form a DI : O_3 mixture (DIO3). By using higher pressure and lower temperature, concentrations of up to 100 ppm are possible. In water, the ozone decomposes to oxygen and hydroxyl radicals ($\cdot$O and $\cdot$OH). Hydroxyl radicals are very reactive, rapidly attacking C=C and C=O bonds and slowly attacking C−C bonds. DIO3 is similar in "oxidizing power" to piranha and SOM mixtures and has a unique set of advantages and disadvantages.

DIO3's advantages center around its low overall cost. While there is a significant initial investment for DIO3 generation hardware, the operating costs for raw oxygen, DI water, and electricity are far below the purchase, disposal, and environmental costs of sulfuric acid mixtures.

DIO3's disadvantages are associated with mass transfer—getting the hydroxyl radicals to the wafer surface. In H_2SO_4-based chemistries, the organic is dissolved from the wafer and dispersed throughout the solution. Oxidant molecules throughout the solution have a significant chance of interacting with an organic molecule. In DIO3, radicals near the surface of the wafer react with the photoresist, cleaving the long polymer chains into smaller, water-soluble fragments. These fragments then disperse and can be oxidized by radicals throughout the solution. But the initial dissolution process in DIO3 is much slower than in the sulfuric process due to the extremely low concentration of ozone in the DIO3.

The reaction rate is also affected by the low solubility of ozone. While DIO3 typically contains 10–50 ppm of dissolved ozone, a 4 : 1 piranha mixture initially contains 60,000 ppm H_2O_2. The mass transfer can be improved by cooling the water and thereby increasing the solubility of ozone. Unfortunately, cooling also reduces the rates of the chemical reactions in the oxidation process.

The speed of DIO3 cleaning is ultimately limited by the total quantity of organics to be consumed. Light organic contamination is typically thinner than 10 nm—over 1,000 times less than photoresist challenges. DIO3 light organic cleaning times are therefore substantially shorter than DIO3 strip times. DIO3 has been shown to remove the outgassing products from wafer shipping containers in 60 s [10]. The organics need not be completely oxidized, but only oxidized to the point that the fragments are slightly soluble in water. The dissolved organic fragments are then washed from the wafer surface and out of the system. This washing will be of the greatest value when removing photoresist, waxes, or other gross organic challenge. The reduction in organic contamination can be seen from the reduction of the $-(CH_2)_n-$ and C−CH_3 hydrogen absorption peaks. Even at such low concentrations, ozonated water can reduce the hydrocarbon contamination in less than 60 s.

F. UV/Oxygen

UV exposure for cleaning has been practiced since the early 1970s [11]. UV light can produce ozone from oxygen and Vig [12] showed that UV/oxygen is very effective at removing light organic contamination. UV/ozone has been used for photoresist stripping as an alternative to oxygen/plasma ashing. Historically, the wafer throughput of a UV/ozone process has been lower than that of oxygen/plasma ashing. UV single-wafer technology has not been very successful in the marketplace.

G. Laser Ablation

There has been some initial "evaluation level" tool offerings that remove organics using pulsed eximer lasers. A number of possible organic removal mechanisms are being explored and utilized singly or in combination:

- Evaporation due to the local heating of the substrate by the beam
- Direct breaking of bonds in the organic by UV photons
- Chemical reactions with gaseous species such as ozone that are created by the laser
- Ablation of bulk organics

None of these techniques is presently in production. But the large parameter space available using pulse laser excitation (wavelength, pulse energy, pulse duration, repetition rate, geometrical optical configuration, composition of gaseous ambient) makes this a technology worthy of further research.

REFERENCES

1. WA Kern and DA Puotinen. "Cleaning solutions based on hydrogen peroxide for use in silicon semiconductor technology," RCA Rev 31, pp. 187–206, 1970.
2. WA Kern. "Radiochemical study of semiconductor surface contamination," RCA Rev 31, pp. 207–265, 1970.
3. LH Haplan and BK Bergin. "Residues from wet processing of positive resists," J Electrochem Soc 127, p. 386, 1980.
4. J Wei, G Smolinski, S Verhaverbeke, and J Parker. "Ozone use for post-ashing resist stripping: mechanisms and recent findings," SPWCC 1997, Santa Clara, CA, March 3–7, 1996. In Semiconductor Pure Water and Chemical Conference 1997, (M. K. Balazs, ed.) (Balazs Analytical Laboratory, Santa Clara, CA, 1996), vol. II, p. 81.
5. PJ Clews, GC Nelson, NC Korbe, PJ Resnick, and CLJ Adkins. Sulfuric Acid/ Hydrogen Peroxide Rinsing Study: Cleaning technology in semiconductor device manufacturing IV (J. Ruzyllo and R. E. Novak, eds.), PV 95-20 (The Electrochemical Society, Inc., Pennington, NJ), 1996, p. 66.

6. P Resnick, C Adkins, D Kittelson, T Kuehn, R Gouk, Y Wu, P Clews, and C Matlock. Proceedings of the Third International Symposium on Ultra Clean Processing of Silicon Surfaces. (Acco, Leuven, Belgium, 1996), p. 195.
7. M Fleming, Jr., W Syverson, and E White. "Reduction of foreign particulate matter on semiconductor wafers," U.S. Patent 5,294,570, March 15, 1994.
8. S Verhaverbeke, R Messoussi, and T Ohmi. "Improved rinsing efficiency after SPM (H2SO4/H2O2) by adding HF," Second International Symposium on Ultra Clean Processing of Silicon Surfaces, September 19–21, 1994, Brugge, Belgium. In Proceedings of the Second International Symposium on Ultra-Clean Processing of Silicon Surfaces (Acco, Leuven, Belgium, 1994), p. 201.
9. SD Hossain and MF Pas. "Comparison of post ash cleaning processing," Third International Symposium on Cleaning Technology in Semiconductor Device Manufacturing, 184th Electrochemical Society Meeting, New Orleans, Louisiana, October 10–15, 1993. In Proceedings of the Third International symposium on Cleaning Technology in Semiconductor Device Manufacturing (The Electrochemical Society, Pennington, NJ) PV 94-7, 1994, p. 111.
10. S Ojima, K Kubo, M Kato, M Toda, and T Ohmi. "Megasonic excited ozonized water for the cleaning of silicon surfaces." J Electrochem Soc 144(4), April 1997, p. 1482.
11. RR Sowell, RE Cuthrell, DM Mattox, and RD Bland. J Vac Sci Technol 11(1), 1974, pp. 474–475.
12. JR Vig. J Vac Sci Technol A3, 1985, p. 1027.

10
Deposition of Metallic Contaminants from Liquids and Their Removal

Steven Verhaverbeke*
Applied Materials, Santa Clara, California

I. INTRODUCTION

In Chapter 9, we saw that contaminants can be broken up into six broad categories: organics, oxides, particles, metals, mobile ions, and crystal damage. Whereas Chapter 9 dealt with organics, this chapter focuses on the deposition and removal of metallic and mobile ions. Most metals present in the solution are ionized, and therefore this chapter is applicable to any ionic contamination, be it metallic ions, mobile ions, or even anions. However, because of the technological importance of metallic ions and metallic impurities, we will mainly refer to ionic contamination as metallic contamination.

The removal and cleaning of metallic impurities are treated together with the deposition of metallic impurities; since we are usually dealing with equilibrium reactions, the deposition and removal are determined by the same mechanisms. It would be better to refer to the interaction of metallic impurities with solid surfaces. As the title of the chapter clearly points out, we will limit ourselves in this chapter to the interaction of metallic impurities and solid surfaces in liquids. This is because of the technological importance of this subject. There are some gas phase processes for removing metallic impurities from semiconductor surfaces, such as high-temperature oxidation with chlorine added to the gas ambient [1] and UV-excited Cl_2 cleaning [2]. Others exist as well, but none are as generic as the liquid-based cleaning technique. Most of the gas phase cleaning processes can be very useful for a specific contaminant but don't have the

---

*This chapter was prepared by the author while at CFM Technologies, West Chester, PA.

Table 1 Metals Analysis on City Water Contaminated Wafers: Values Averaged Over All Measured Points $*1e10$ atoms/cm^2

	TOFSIMS						TXRF							
	Li	B	Na	Mg	Al	K	Ca	Ti	Cr	Mn	Fe	Ni	Cu	Zn
Virgin wafer	1	840	80	8.6	12.4	<	2.9	<	<	<	15	<	10.36	<
City water, contaminated wafer	20600	1260	40000	71000	740	2567	42000	33	17	0.33	5	5.7	567	1100
SOM—HF/HCl	2.4	2.6	360	0.8	0.6	<	<	<	<	<	<	<	<	<
SC-1—HF/HCl	0.2	16.4	2.8	1.8	1.2	<	2.7	1.7	<	<	1	<	<	<

<, Below detection limits; SOM, sulfuric and ozone mixture (see Chap. 9).
Source: Ref. 3.

overall efficiency of the liquid-based techniques. Table 1 shows typical metallic impurity levels on virgin wafers and also on wafers after contamination with city water. The values shown in the table are values averaged over all the measurement points. When a measurement point fell below the detection limit, its value was assumed to be 0. It is clear that city water exposure contaminates wafers with predominantly Li, B, Na, Mg, Al, K, Ca, Cu and Zn (up to concentrations as high as 7×10^{14}/cm^2 for Mg). B and Na are detected on both virgin wafers (row 1) and after cleaning (rows 3 and 4) and may originate from airborne contamination, since the analysis was not performed in a cleanroom. Na values on virgin wafers were always in 1×10^{11}/cm^2 range and very uncontrollable. Even when using just electronic grade chemicals, as in these measurements, almost any wet chemistry sequence geared at removing metallic contamination removes all of the metallic contamination with efficiencies >99.8%. This is unparalleled by any of the gas phase technologies. There are a large number of controllable parameters available to wet cleaning technology. These include concentration; temperature; time; transport process parameters, such as agitation or spinning; other physical ways of energy transfer, such as megasonics or light (including UV); chemistry of the reactants; type of impurities, including the materials of construction; isotropic conditions, such as a homogeneous reagent and temperature distribution.

In this chapter we will look mainly at the fundamental principles governing the deposition and removal of the metallic impurities from surfaces, relatively independent of the tool construction itself. We will therefore limit ourselves to the chemistry of the reactants, the impurities, and the surface condition itself. It has to be stressed here that the other parameters also influence the total result and therefore the selection of the right tool is also of prime importance.

II. ADSORPTION FORCES, TYPES OF SURFACES, SURFACE TERMINATION, AND REMOVAL MECHANISMS

A. Adsorption Forces

There are four different mechanisms by which metallic impurities can adhere to the wafer surface

- *Physisorption* is characterized by very weak forces, such as electrostatic forces due to the dipolar attraction, Van der Waals forces, and hydrogen bonding. These forces are important for a polymer surface such as a photoresist surface.
- *Chemisorption* is characterized by much stronger bonds, such as covalent bonding (sharing the same electron pair) or strong electrostatic

(or ionic) bonds. An example is the chemisorption of transition metals
to an oxide surface.

- *Metallic bonds* are characterized by delocalized electrons and are formed
  primarily at high temperatures. A typical example is the silicide formation.
- *Displacement plating* is a form of chemisorption, which is particularly
  important, since it can take place on bare Si surfaces.

B. Types of Surfaces and Surface Termination

1. Types of Surfaces

In wafer processing, there are a multitude of different materials which can be in
contact with cleaning liquids. All of these surfaces can interact differently with
metallic impurities. Therefore it is important to specify the surface type being
cleaned. Surfaces encountered in semiconductor manufacturing include:

- Silicon surfaces, both monocrystalline, such as bare wafers or epi sur-
  faces, and polycrystalline silicon. (The doping levels vary widely and
  can impact the metallic impurity adsorption.)
- Dielectric films, such as oxide, nitride, oxynitride, doped oxides (e.g.,
  BPSG), spin-on oxides
- Metallic films, such as Al, Ti, TiW, Cu, TiN, Ta, TaN
- Polymers, such as photoresist and low-K materials

In semiconductor manufacturing the removal and adhesion of metallic
impurities are most critical in the front end of the process sequence, since at
this stage the active regions of the device can be exposed to the contamination.
During the back-end processing the device is sealed from any contamination and
only interconnections are exposed. It is for this reason that the materials used in
the front end are most important technologically. Polymers in the front end are
limited to photoresist layers and are completely removed from the surface of the
wafer (see Chap. 9). After removing the photoresist, a metallic cleanup is usually
performed. Therefore, if we restrict ourselves to surfaces of practical importance
in semiconductor manufacturing, we can limit ourselves to understanding silicon
surfaces and dielectric films.

2. Surface Termination

The phenomena describing adsorption/desorption of metallic impurities on the
surface of a wafer are so surface dependent that the nature of the underlying
structure is not usually important. In the front of the line of the semiconductor
manufacturing process, there are four main surface terminations:

- Oxide termination: $-Si-O-Si-O-$
- Hydroxide termination (silanol groups): $-Si-OH$

- Hydride termination: $-Si-H$
- Fluoride termination: $-Si-F$

Almost all materials are terminated with one of these four groups typically obtained after the following processes*:

- Oxide termination: After gas phase oxidation, this surface is hydrophobic, i.e., it has a contact angle with DI water of about 40–50° [5].
- Hydroxide termination: After most wet cleaning solutions, such as the RCA-1 [6] (or SC-1) cleaning solution.
- Hydride termination: After a dilute HF etch or BHF etch down to the bare Si surface.
- Fluoride termination: Partially present after a dilute HF or BHF without any rinse; also present after anhydrous gas phase HF [7] on a bare Si surface and after sulfuric acid mixed with HF on an oxide surface [8].

When oxide surfaces are immersed in DI water, both the oxide termination and the fluoride termination convert to a hydroxide-terminated surface [8]. Both the oxide termination and the fluoride termination are hydrophobic, whereas the hydroxide termination is hydrophilic. The conversion to a hydroxide termination can easily be measured by the decrease in contact angle during rinsing. When a silicon wafer comes out of the oxidation furnace, its oxide surface will be hydrophobic. However, after 30 min rinsing, it will be turned into a hydrophilic surface. This can be seen in Figure 1.

At high temperature and at high pH, such as in an SC-1 solution, this conversion goes much faster and is more complete (lower final contact angles). On silicon surfaces, the fluoride termination will also transform to a hydroxide termination during a rinse. The fluoride termination is very polar and very reactive. It is hydrophobic, like the hydride termination. It occurs mainly as an intermediate step in the etching of silicon dioxide, but is not stable. In HF solution, the fluoride termination will transform into a hydride termination. Raghavachari et al. [9] proved this conversion of a fluoride-terminated surface into a hydride-terminated surface based on first principles. The reaction is schematically shown in Figure 2.

Therefore, the only two terminations which are important for understanding the metallic impurities deposition/removal immersed in aqueous media are the hydroxide and the hydride terminations. In the rest of this chapter, we will

---

*Recently, it has been shown that a methoxy-terminated surface, $-Si-O-CH_3$, can be obtained [4]: This termination was obtained under very special conditions, involving immersion in pure methanol spiked with I_2 to act as a catalyst and only under UV irradiation. Only when all of these conditions were present was the methoxy termination obtained. However, it is not clear how stable this termination is, especially when immersed into an aqueous cleaning solution. It is most likely that this methoxy termination will be exchanged with a hydroxide termination during a DI water rinse.

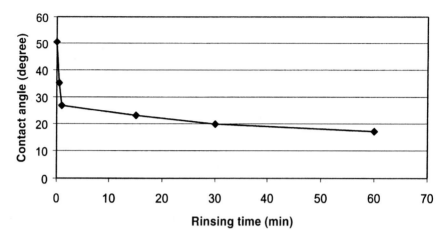

Figure 1 Contact angle of DI-water on oxide as a function of rinsing time in DI-water at 20°C (wafer removed from oxidation furnace at time = 0 min). (From Ref. 8.)

show the underlying principles governing metallic impurity deposition/removal on these two surfaces:

The silanol group (Si terminated with hydroxide) will occur on dielectric surfaces immersed in aqueous environments and is:

- A weak acid group.
- A complex-forming anion which forms salts and acts as an ion exchange site.
- Dipolar: May have different electrostatic charges, depending on the solution pH, may attract oppositely charged particles.
- Hydrophilic.

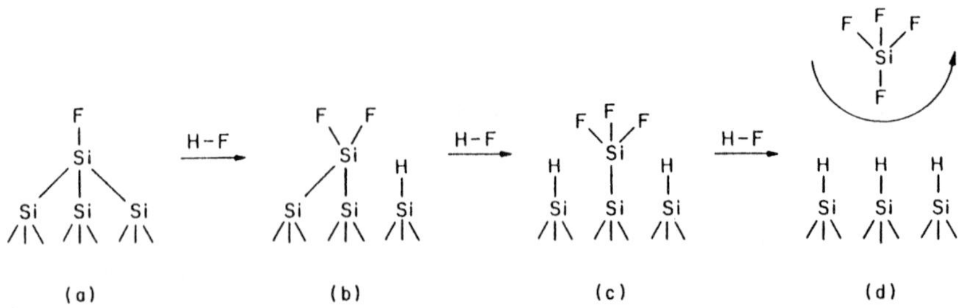

Figure 2 Schematic representation of the mechanism of H-passivation. (From Ref. 9.)

The hydride termination will occur on bare Si surfaces after an HF etch and is

- Nonionic. The Si—H bond is a covalent bond in which the H is not acidic.
- Quasi-apolar. It is very weakly polar.
- A good organic adsorber.
- Hydrophobic. The contact angle is roughly 72° [5].
- Not stable in water and in the atmosphere. Eventually it will convert into a hydroxide termination.

Even though the hydride termination is not stable, it takes up to 2 weeks to convert half the hydride sites into hydroxide sites [10] in a clean room environment and therefore this termination will occur frequently on semiconductor surfaces in the front end of the line.

C. Removal Mechanisms

Now that we understand the surfaces we will be dealing with and understand the adhesion forces, we can list the available removal mechanisms:

- Surface charge change with acids, bases
- Ion competition
- Etching of the surface layer
- Oxidation or decomposition of the impurity
- Physical desorption by solvents and surfactants
- Other physical methods: surface ablation, gas phase reaction, RIE, plasma processes, etc.

The first four mechanisms are important for removing chemically bound metallic impurities in an aqueous immersion environment and will be the focus of the remainder of this chapter. The first two mechanisms are most important on hydroxide-terminated surfaces, and the third and fourth mechanisms (etching of the surface layer and oxidation) are most important on hydride terminated surfaces.

Physical desorption takes place in solvents and by the action of surfactants, but it also occurs during rinsing and therefore is an important mechanism. Physical desorption is very mass-transfer determined, and diffusion and bulk transport mechanisms determine the success of physical desorption. These mechanisms are quite different from the mechanisms for chemically bound impurities. In this chapter we will limit ourselves to the chemically bound impurities, since many diffusion and bulk transport mechanisms have already been discussed at length elsewhere in this book and are very tool dependent.

III. STATE OF THE ELEMENTS OR IMPURITIES IN SOLUTION

The state of the impurities in solution is one of the most fundamental characteristics. There are three different states:

- Solid substances, e.g., $Fe(OH)_2$, $Fe(OH)_3$, $CaCO_3$, Au
- Dissolved substances, e.g., Fe^{2+}, Fe^{3+}, Ca^{2+}, H_2O_2
- Gaseous substances, e.g., H_2, O_2, HCl, HF, SiH_4

The state of the elements in solution depends on the pH, redox potential, and interacting anions in the solution. One of the key requirements of a cleaning solution is that the impurity has to be soluble in the cleaning solution. As we will see later, this requirement is not enough by itself to constitute a good cleaning solution, but it is definitely the first requirement. The different states of an element can be represented in a *Pourbaix diagram*. An atlas with a Pourbaix diagram for every element was published by Pourbaix [11] and is a very useful reference document. In this diagram the different states of an element are represented as a function of pH and electrochemical potential. As an example, Figure 3 shows the Pourbaix diagram for Cu.

It is important to understand which interactions are taken into account when reading a particular Pourbaix diagram. A Pourbaix diagram always lists the reactions which are taken into account before each diagram. It is important to review this list and to understand that, in a specific solution of interest, interactions not listed in the *Atlas* [11] may be of importance. In its simplest form, only the interactions of pH and electrochemical potential are taken into account. When using these diagrams, it is good practice to start with these simple interactions and then add more complex but relevant interactions to them, such as the interactions with other particular anions in the solutions. When anions form complexes with a particular element, the state of the element in the Pourbaix diagram may change from dissolved to solid, or vice versa. A typical example is the diagram for Cu when ammonia is present in the solution, published by Norga et al. [13]. Norga et al. clearly show the changes introduced into the diagram when a simple anion such as NH_4^+ is added to the interactions list. Commercial software exists to construct these diagrams based on a database of interactions. Nevertheless, the Pourbaix *Atlas* remains a very useful reference tool, since the diagrams can be looked up immediately without having to construct them for each case (even with the help of software). But again, their limitations must be understood. Since the state of an element in solution is dependent upon the pH and the electrochemical potential, it is important to know the characteristics of a cleaning solution. In Figure 4 several common solutions are plotted on a Pourbaix diagram.

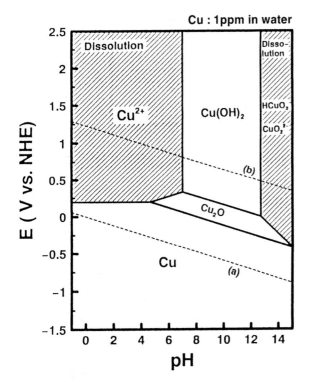

Figure 3 pH-electrochemical potential diagram for Cu. Line (a) represents the reduction equilibrium of water according to the reaction, $H_2 \leftrightarrow 2H^+ + 2e^-$ [$E_a = 0.0 - 0.591$ pH (p$H_2 = 1$ atm)]; Line (b) represents the oxidation equilibrium of water according to the reaction, $2H_2O \leftrightarrow O_2 + 4H^+ + 4e^-$ [$E_b = 1.229 - 0.591$ pH (p$O_2 = 1$ atm)]. (From Ref. 12.)

Two important lines on this diagram are the lines that plot the pH dependence of the electrochemical potential of an aqueous solution saturated with (a) H_2 at 1 atm and (b) O_2 at 1 atm. A solution at pH = 0 and saturated with H_2 ([H^+] = 1 mol/L) will have an electrochemical potential arbitrarily defined as 0 V. This is the definition of the hydrogen electrode. A solution at pH = 0 saturated with O_2 has an electrochemical potential of 1.229 V. This is the standard oxidation/reduction potential of O_2. The potential of solutions saturated with O_2 or H_2 decreases 60 mV/pH unit. Ultrapure water (UPW), when saturated with air (without CO_2), will have a pH of 7, and an electrochemical potential of roughly 0.7 V. When pure O_2 at 1 atm is bubbled through this solution (UPW, high O_2), the potential will increase toward the theoretical potential of a solution saturated with O_2, which is 0.809 V at pH = 7. When the solution is N_2 blanketed and

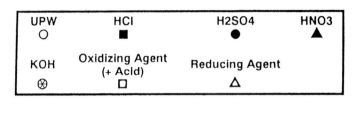

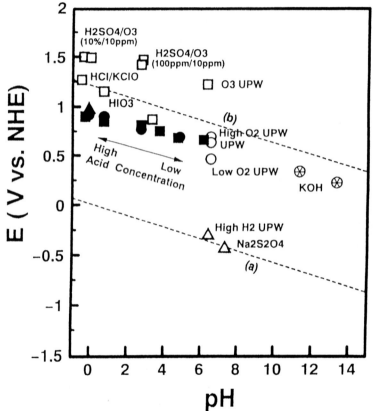

Figure 4 pH and electrochemical potential of typical cleaning solutions. Line (a) represents the reduction equilibrium of water according to the reaction, $H_2 \leftrightarrow 2H^+ + 2e^-$ [$E_a = 0.0 - 0.591$ pH (pH$_2$ = 1 atm)]; Line (b) represents the oxidation equilibrium of water according to the reaction, $2H_2O \leftrightarrow O_2 + 4H^+ + 4e^-$ [$E_b = 1.229 - 0.591$ pH (pO$_2$ = 1 atm)]. (After Verhaverbeke et al. [12].)

degassified, the potential drops (UPW, low O_2). When H_2 is bubbled through the solution at 1 atm, the potential will drop towards the H_2 saturated line (line a). The potential of a H_2-saturated solution at pH 7 is 0 V–0.060 × 7 = −0.42 V.

When an acid such as HCl, H_2SO_4, or HNO_3 is added to UPW saturated with air, the potential increases by 60 mV for each unit of pH decrease. Finally, if a stronger oxidizer than O_2, such as O_3, HClO, or HIO_3, is added to the solution, then the potential of the solution can be higher than the O_2 saturation line. If a stronger reducing agent than H_2, such as $Na_2S_2O_4$ ($E^\circ = -0.056$ V), is added to the solution, then the potential of the solution can be lower than the H_2 saturation line.

As an alternative to looking up the Pourbaix diagram for a certain element, one may look up the standard reduction potential table. The standard reduction potential table lists the potential of different reactions at pH $= 0$ ($[H^+] = 1$ mol/L). Even though this table is a one-dimensional cut of the Pourbaix diagram (pH is fixed and is 0), it is still very instructive. It lists all reactions in one table (a one-dimensional list), whereas the Pourbaix diagram represents only a limited subset of the reactions. Moreover, most metal removal solutions are acid solutions and therefore the tabulated potential values will be relatively accurate representations for the real potentials, since most acid solutions will be used at pH values not too different from pH $= 0$. However, care should be taken in using these values for basic solutions, since this would assume that the pH dependence of the listed reactions is similar. Since the pH dependence can vary from 0 to 60 mV/pH unit to 120 mV/pH unit or be even more complicated, the potentials of basic solutions are likely to vary significantly from the values for solutions at pH $= 0$.

Another tool available is the solubility product. The solubility product represents the heterogeneous equilibrium constant for the reaction between a slightly soluble substance and its ions in a saturated solution. For example, for the reaction

$$Ag_2CrO_4 \rightleftharpoons 2Ag^+ + CrO_4^{2-} \tag{1}$$

the solubility product is given by

$$K_{sp} = [Ag^+]^2[CrO_4^{2-}]$$

where $[Ag^+]$, $[CrO_4^{2-}]$ are the saturation concentrations, in moles per liter, of Ag^+ and CrO_4^{2-} respectively.

The solubility product for metallic ions is well characterized and is tabulated in most chemical handbooks (see, for example, the *CRC Handbook of Chemistry and Physics* [14]). This constant therefore provides a very easy and quick first check on the state of an element in solution. The solubility of the hydroxide is important in aqueous solutions. From the solubility product constant, a solubility of the ion in the presence of $[OH^-]$ can be calculated [15] and is shown in Table 2. Obviously, other reactions, other anions, and the wafer surface itself can change this solubility. Therefore, this table has to be used with the utmost caution and only as a first, quick reference. A typical example where this table fails is the solubility of Al at high pH, since, at high pH, Al will be in solution in the form of $[Al(OH)_4]^-$, whereas the table is only valid for Al in solution as $[Al^{3+}]$.

The published Pourbaix diagrams usually take into account these different reactions and are therefore usually a more complete set of data to determine the state of an element in solution. Consider the Pourbaix diagram for Al shown in Figure 5. At low pH, Al^{3+} is thermodynamically the most stable form, i.e., Al will dissolve as Al^{3+}. At intermediate pH, Al is stable as $Al_2O_3 \cdot 3H_2O$ (passivation region) and, at high pH, Al will dissolve as $Al(OH)_4^-$. Metallic Al exists only at E values below -1.663 V. The deposition of Al onto a hydroxide terminated oxide surface is shown in Figure 6.

As previously noted, interactions with other anions and with the silicon wafer itself can change the Pourbaix diagram. It was indicated before that the hydroxide termination can act as a weak acid group and, therefore, it can interact with the metallic impurities in solution and can change their state. This will be the subject of the next section.

It is possible to include the interactions with the wafer surface into a Pourbaix diagram, but little effort has been undertaken to construct these diagrams. The previously cited behavior of Al is a typical example where the omission

Table 2 Solubility of Selected Metal Hydroxides at pH = 7 and pH = 10

Metal	Solubility (ppb) at pH = 7	Solubility (ppb) at pH = 10
Al^{3+}	5×10^{-4}	5×10^{-13}
Ca^{2+}	Very high	Very high
Cu^{2+}	1×10^3	1×10^{-3}
Co^{2+}	6×10^6	6
Fe^{2+}	2×10^5	0.2
Fe^{3+}	1×10^{-10}	1×10^{-19}
Pb^{2+}	300	3×10^{-4}
Mg^{2+}	Very high	1×10^4
Mn^{2+}	Very high	1×10^3
Hg^{2+}	6×10^{-4}	6×10^{-10}
Ni^{2+}	3×10^6	3
Sn^{2+}	6×10^{-5}	6×10^{-11}
Zn^{2+}	4×10^5	0.4

Source: Ref. 15.

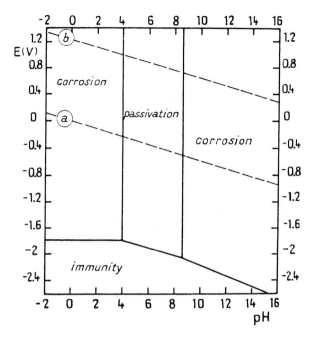

Figure 5 Pourbaix diagram for Al. Line (a) represents the reduction equilibrium of water according to the reaction, $H_2 \leftrightarrow 2H^+ + 2e^-$ [$E_a = 0.0 - 0.591$ pH (pH$_2$ = 1 atm)]; Line (b) represents the oxidation equilibrium of water according to the reaction, $2H_2O \leftrightarrow O_2 + 4H^+ + 4e^-$ [$E_b = 1.229 - 0.591$ pH (pO$_2$ = 1 atm)]. (After Pourbaix et al. [11].)

of these surface interactions leads to completely different conclusions. Even though Al, in the absence of a silicon wafer, becomes highly soluble at high pH (in accordance with most published Pourbaix diagrams for Al; Fig. 5), Al will deposit on the hydroxide silicon surface at high pH very efficiently through its reaction with the weak acid Si−OH group. When a hydroxide-terminated oxide surface is present, Al will react with the surface and form (SiO)$_2$Al(OH) or SiOAl(OH)$_2$ on the surface. This shows clearly that the presence of a wafer in the solution can change the state of the elements from that shown in a Pourbaix diagram if that diagram does not take the interactions with the silicon wafer into account. Not surprisingly, as we will see later, the simple consideration of the hydroxide solubility, as in Table 2, leads to a more accurate prediction of the Al behavior on oxide-covered silicon wafers than the more complete Pourbaix diagram.

Finally, it must be noted that when a metal ion is in an aqueous solution, it will stabilize its positive charge by surrounding itself with a hydration sphere

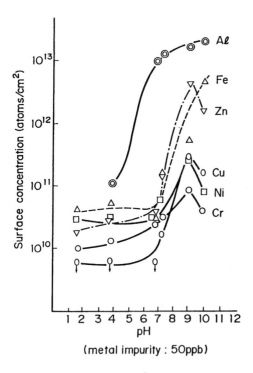

(metal impurity : 50ppb)

Figure 6 Deposition of Al^{3+} on hydroxide-terminated oxide surfaces. (From Ref. 16.)

of 6 water molecules [17]. If other species are present in the solution, especially Lewis bases or anions, the metal can preferentially coordinate with the ligands, thereby displacing the H_2O coordinated to the metal:

$$M(H_2O)_6^{x+} + yL^- \rightleftharpoons ML_y(H_2O)_{6-y}^{x-y+} \qquad (2)$$

where M = a metal and L = a ligand.

However, this mere fact does not change the further treatment of metallic ion chemisorption and desorption and therefore will not be considered here. Moreover, the exact state of metallic ions in a solution is still very controversial. For all practical purposes it can be represented as an isolated metallic ion in solution. Similarly, strictly speaking, wherever H^+ is being used in the equations, it should be understood as H_3O^+, as a proton in solution will immediately be coordinated with a water molecule. For the purposes of this book, we do not have to take the exact state of the ions in solution into account and can simplify this to the more common

$$M^{x+} + yL^- \rightleftharpoons ML_y^{x-y+} \qquad (3)$$

IV. CHEMISORPTION/DESORPTION OF METALLIC IMPURITIES ON OXIDE SURFACES

A. Introduction

As indicated above, there are two surfaces of major importance: the oxide surface and the silicon surface. We will first look at the chemisorption/desorption of metallic impurities on oxide surfaces. Then we will look at some technological implications of those mechanisms, and in Section V we will look at the chemisorption/desorption of metallic impurities on silicon surfaces. In Section VI we will consider the chemisorption/desorption of metallic impurities on thin oxide surfaces, a special case of the oxide surface discussed here in Section IV.

B. Chemisorption Mechanisms

The oxide surface can behave as an acid or a base depending on the pH of the system. As said before, immersed in an aqueous solution, the oxide surface will convert to a hydroxide-terminated surface (hydrophilic). In a system that involves only water, the three possible surface species on a hydroxide terminated oxide surface are $Si-OH$ (neutral), $SiOH_2^+$ (positive), and SiO^- (negative). The equilibrium equations and constants between those three species are [18]:

$$Si-O-H \rightleftharpoons SiO^- + H^+ \qquad \log K_1 = -7.5 \qquad (4)$$

$$Si-O-H + H^+ \rightleftharpoons Si-O-H_2^+ \qquad \log K_2 = \text{(not available)} \qquad (5)$$

As can be seen from the equilibrium constant K_1, the silanol surface termination is only weakly acidic. The interaction of metal ions in solution with the silanol surface group can be described in the same way as a weak acid ion exchange resin. Even though the H^+ is bound fairly strongly (the silanol is weakly acidic), it can be substituted by a positive metal ion, as shown in the following reaction:

$$M^{x+} + y\,(Si-OH) \rightleftharpoons (Si-O)_y M^{(x-y)+} + yH^+ \qquad (6)$$

In fact, the hydroxide-terminated oxide surface behaves very much like silica gel. The interaction of silica gel with ions in the solution is well characterized and its equilibrium constants are tabulated in the literature. In general, $y \leq 2$. Most likely $y > 2$ is prohibited by steric hindrance [19]. Therefore, this means that there are only two reactions to be considered when describing metallic ion adsorption onto oxide surfaces:

$$-Si-OH + M^{x+} \rightleftharpoons -Si-O-M^{(x-1)+} + H^+ \qquad (7)$$

$$2-Si-OH + M^{x+} \rightleftharpoons (-Si-O-)_2 M^{(x-2)+} + 2H^+ \qquad (8)$$

Several general conclusions can be drawn from the equations cited above describing metal adsorption on oxide surfaces. The first is that, based on this model, it would never be possible to have levels of metal adsorption larger than the surface concentration of Si$-$OH which is 4.6–5.5 $\times$ 10^{14}/cm^2 [19–21]. The second is that this suggests two ways to reduce the surface adsorption. One is to increase [H$^+$]; the other is to decrease the free metal ions in the solution [M^{x+}]. Hence the often reported and measured pH and metal concentration dependence of metallic impurity adsorption. For example, the adsorption species and the estimated equilibrium constants of Fe^{3+}, Cu^{2+}, and Mg^{2+} on silica gel are [22]

$$\text{SiO}-\text{H} + \text{Fe}^{3+} \rightleftharpoons \text{SiO}-\text{Fe}^{2+} + \text{H}^+ \qquad \log K_{1(\text{int})} = -1.77 \qquad (9)$$

$$\text{SiO}-\text{H} + \text{Cu}^{2+} \rightleftharpoons \text{SiO}-\text{Cu}^+ + \text{H}^+ \qquad \log K_{1(\text{int})} = -5.52 \qquad (10)$$

$$\text{SiO}-\text{H} + \text{Mg}^{2+} \rightleftharpoons \text{SiO}-\text{Mg}^+ + \text{H}^+ \qquad \log K_{1(\text{int})} = -7.7 \qquad (11)$$

$$2\text{SiO}-\text{H} + \text{Fe}^{3+} \rightleftharpoons (\text{SiO})_2\text{Fe}^+ + 2\text{H}^+ \qquad \log K_{2(\text{int})} = -4.22 \qquad (12)$$

$$2\text{SiO}-\text{H} + \text{Cu}^{2+} \rightleftharpoons (\text{SiO})_2\text{Cu} + 2\text{H}^+ \qquad \log K_{2(\text{int})} = -11.19 \qquad (13)$$

$$2\text{SiO}-\text{H} + \text{Mg}^{2+} \rightleftharpoons (\text{SiO})_2\text{Mg} + 2\text{H}^+ \qquad \log K_{2(\text{int})} = -17.15 \qquad (14)$$

Figure 7 shows the modeling results on the surface species distribution for those metals ions on silica gel. The relative contributions of reactions (9) to (13) to the overall adsorption process as a function of pH are clearly illustrated.

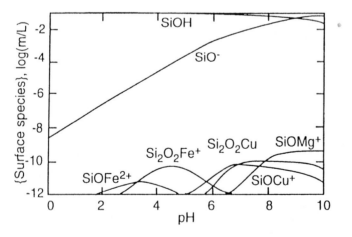

Figure 7 Distribution of surface species for Fe, Cu, and Mg adsorptions onto silica at 25°C as a function of pH. SiO$_2$ = 1M, [Fe^{3+}] = [Cu^{2+}] = [Mg^{2+}] = 10 ppb. (From Ref. 22.)

Reaction (14), which occurs at pH higher than 10, is not shown in Figure 7. Fe^{3+} starts to adsorb near pH 2, Cu^{2+} at pH 6, and Mg^{2+} at pH 8. From these reactions it is clear that Fe^{3+} adsorption increases with Fe^{3+} concentration in the solution, even at acidic pH values. This was experimentally confirmed by Rotondaro et al. [23] and is shown in Figure 8.

Using the equilibrium constants for silica gel in the case of oxide surfaces provides us with tabulated values from the literature. As shown above, using these values in the case of Fe^{3+} would predict the adsorption of Fe^{3+} to start around pH 2. This was confirmed experimentally by Hurd et al. [19]. Figure 9 shows the experimental result. In this figure, the surface adsorption of Fe^{3+} on oxide surfaces is shown as a function of pH. It can be clearly seen that Fe^{3+} starts to adsorb on oxide surfaces at pH 2. This agrees very well with the

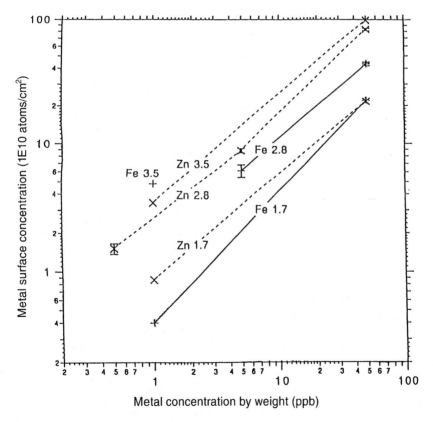

Figure 8 Fe and Zn adsorption on oxide surfaces as a function of contamination in the solution and as a function of pH. The pH values are adjusted by spiking with HNO_3 and are varied between 1.7 and 3.5. (From Ref. 23.)

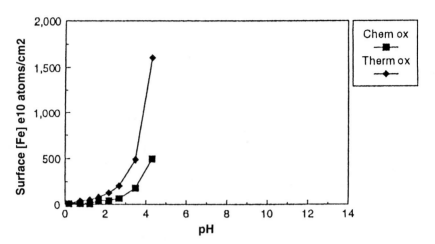

Figure 9 Iron concentration remaining on chemical and thermal oxide surfaces following a 10 min immersion in 100 ppb spiked HCl solutions of varying pH. (From Ref. 19.)

equilibrium constants that are tabulated for Fe^{3+} on silica gel. Moreover, from Figure 9 it can be seen that there are only minor differences between different sources of oxide.

As stated before, the adsorption of metallic ions onto oxide surfaces, for all practical purposes, can be adequately described by reactions (7) and (8) above. The stability constants for both these surface complexes that are formed can be defined as follows:

$$K_{1(\text{int})} = \frac{\{\equiv \text{SiOM}^{x-1+}\}[\text{H}^+]}{\{\text{SiOH}\}[\text{M}^{x+}]} \quad \text{and} \quad K_{2(\text{int})} = \frac{\{(\equiv \text{SiO})_2\text{M}^{x-2+}\}[\text{H}^+]^2}{\{\text{SiOH}\}^2[\text{M}^{x+}]}$$

where pressure, temperature, and ionic concentration (P, T, I) are held constant, { } indicates surface concentration, and [ ] indicates a solution concentration.

Even though the exact stability constants for an oxide surface on a silicon wafer are usually unknown, the tabulated values for silica gel, which are widely available in the literature, can be used to predict and model the adsorption behavior. However, it can happen that the stability constant of silica gel with a particular metallic ion of interest cannot be found. In such a case, the behavior of that metal with the oxide surface can still be predicted based on an analogous mechanism known as the hydrolysis of metallic ions. It has been noted [17] that the $-\text{Si}-\text{OH}$ groups covering the silica surface are chemically similar to the $\text{H}-\text{OH}$ of water and thus a valuable guide as to how aggressively a metal will react with the silanol groups on the silica surface is provided by the metal's hydrolysis behavior. Because of this similarity, the more readily hydrolyzable a metal is the more readily it will bond with the oxide surface. Since very simple rules exist to predict the hydrolysis, we can use these rules to predict the adsorption of metals to oxide surfaces. This is especially useful in those cases where the stability constants of the particular metal with silica gel are not available.

Hydrolysis can simply be described by the coordination of a metallic ion in solution with water while ejecting a proton:

$$M^{x+} + y\text{OH-H} \rightleftharpoons M(\text{OH})_y^{x-y+} + y\text{H}^+ \tag{15}$$

y can be larger than x. For example, Al^{3+}, Zn^{2+}, and Cr^{3+} will form $[Al(OH)_4]^-$, $[Zn(OH)_4]^{2-}$, and $[Cr(OH)_4]^-$. Sometimes this will be written as a metallic ion surrounded with 6 water molecules. However, this is not really relevant for the description of metallic ion adsorption on to oxide surfaces, as noted earlier.

The analogy of oxide surface adsorption with hydrolysis provides us with two important instruments to use. First, if we cannot find the stability constant of the metal with silica gel in the literature, we can look for the hydrolysis constant of the metallic ion and if this is available in the literature, we can derive the stability constant of the silica surface with the metal as follows [24]:

$$\log K_s = -0.09 + 0.62 \log K_h$$

where K_h is the hydrolysis constant of the metal ion and K_s is the stability constant of the metal ion with silica gel. As mentioned earlier, the stability constant of a metal ion with silica gel is a good approximation of its stability constant with the oxide surface on a silicon wafer.

Additionally, if the hydrolysis constant cannot be found, we can use a simple rule to predict the relative value of the hydrolysis and therefore of the surface complex stability constant. The charge density (ionic charge/ionic radius) of a metallic ion gives a rough guide to determine how readily a metal ion will hydrolyze [25]. The higher this number, the more easily an ion will hydrolyze; i.e., small ions with multiple charges will easily hydrolyze, whereas large single charged ions will hardly undergo hydrolysis. For example, Al^{3+}, which is a small ion triple charged, will easily be hydrolyzed. This is shown in Figure 10.

Loewenstein and Mertens [21] confirmed this for metallic ion adsorption to oxides on silicon wafers. They found a good correlation between the metal coverage on oxide layers on silicon wafers and the ratio ionic charge/ionic radius. The solubility of the hydroxides as shown in Table 2 follows the same dependency and follows a very analogous reaction path, since it is a particular form of hydrolysis. This explains why the solubility table initially gave such a good prediction of the metallic impurity adsorption on oxide surfaces [5]. It has the same pH dependency and follows the same relative order between different metallic ions. Thus it provides us with an additional engineering tool. The solubility table can be used to get a quick idea of how likely a metallic

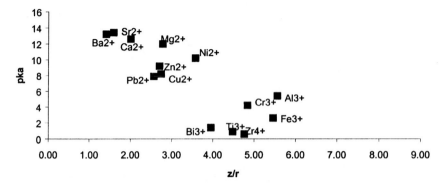

Figure 10 Hydrolysis stability constant as a function of charge density (ionic charge/ionic radius). (From Ref. 26.)

ion will chemisorb on oxide surfaces and it can even provide an indication around which pH values the chemisorption will start. Similar information is found on a Pourbaix diagram. Indeed, the hydrolysis behavior of a metallic ion can also be read from a Pourbaix diagram. This explains the success of using a Pourbaix diagram in predicting the metallic ion chemisorption on oxide surfaces, since the same tendencies can be observed. However, it is difficult to explain the difference between oxide and silicon surfaces if precipitation were the main contamination mechanisms. Moreover, a thorough rinsing should be able to remove most precipitates.

Finally, it must be noted that when dealing with solutions of pH > 9, oxide is being etched at an appreciable rate and as such the surface is changing continuously. However, Loewenstein (21) showed that equilibrium conditions occur almost instantaneously and, therefore, there is almost no time and no etching rate dependence.

Loewenstein (21) also showed that at high pH, the solution contains other cations in addition to metallic impurity cations and H^+ ions. In the case of a standard SC-1 solution, the majority of cations will be NH_4^+. These cations react in a similar manner as the metallic cations and therefore will terminate the silanol groups according to the following reaction:

$$-Si-O^- + NH_4^+ \rightleftharpoons -Si-O-NH_4 \tag{16}$$

This means that less silanol groups are available to react with the metallic ions in the solution, since silanol groups terminated with $NH4^+$ are inaccessible for metallic ions. This predicts that the higher the NH_4OH concentration in an SC1 solution, the lower the metallic impurity concentration, which has been observed (21).

Now, we can understand the chemisorption of cations to a silanol-terminated surface as a competitive chemisorption between the various cations in the solution, H^+, NH_4^+ and the metallic cation impurities.

$$-Si-O^- + H^+ \rightleftharpoons -Si-O-H \tag{17}$$

$$-Si-O^- + NH_4^+ \rightleftharpoons -Si-O-NH_4 \tag{18}$$

$$-Si-O^- + M^{x+} \rightleftharpoons -Si-O-M^{(x-1)+} \tag{19}$$

The chemisorption reaction of metallic impurities on the silanol-terminated surface shown earlier as chemical reaction (7) is actually a combination of the reactions (17) and (19). From reactions (17), (18), and (19) a more general conclusion can be drawn: in order to reduce the metallic impurity adsorption onto silicon oxide surfaces, one can:

1. Increase the H^+ concentration (making the solution acidic) and/or
2. Increase the NH^+ concentration (adding NH_4OH or an ammonium salt)
3. Decrease the M^{x+} concentration (adding a chelating ligand).

This may offer the explanation for the sometimes reported observation that BHF resulted in higher yields on LOCOS structures than dilute HF (higher NH_4^+ concentration in BHF).

C. Desorption Mechanisms

The first and foremost requirement in designing a cleaning solution is that the metallic ions to be removed must be soluble in the solution. The second requirement is that the solution must exchange the chemisorbed surface metallic ions with H^+. Alternatively, NH_4^+ can be used to terminate the surface, or the surface can be left as $Si-O^-$ and metallic impurities can be bound in the solution.

As explained earlier in this chapter, in order to determine whether a metallic ion is soluble in the cleaning solution, one can use the Pourbaix diagrams or the Standard Reduction Potential tables. The Pourbaix diagrams show the pH dependency, but often do not include the interactions with all the anions in the solution. That is, a standard Pourbaix diagram as published in the *Atlas* [11] will not show the effect of NH_4OH when studying a SC-1 (or APM) solution as indicated earlier. However, as also shown earlier, the effect of NH_4OH can easily be added to the diagram. Strictly speaking, standard reduction tables are only applicable at pH = 0, but when describing acid solutions, the relative order of the reactions in the table usually are still valid. The advantage of this approach is that more reactions can be considered simultaneously.

The second requirement states that the solution exchanges the surface metallic ions, which are chemisorbed to the oxide surface, with H^+. Technologically, this is the most common approach. Later we will see that we can do this also with NH_4^+ and chelating agents. From the previous section, we saw that one way to drive the surface concentration down is to increase $[H^+]$, i.e., make the solution more acidic; the other way is to drive down the $[M^+]$. $[M^+]$ is the concentration of the free metal ion. One way to drive the free metal ion concentration down is to add a competing ligand to the solution. If the ligand-metal complex is more stable than the oxide-metal complex and if that complex is also soluble, then the ligand can effectively prevent the metal from adsorbing on the surface and from precipitating out of solution. A comparison of ligand stability constants with the metal-silica gel complex constants will indicate how effectively a ligand will be in preventing metal adsorption on oxide surfaces.

We will now study a couple examples of these general guidelines. Metals are always soluble in their ionized state. Other soluble forms exist as well, but if one wants to design a solution in which a multitude of metallic ions are soluble, one has to make the solution such that the most noble metal which is of interest is still ionized. For acid solutions with pH values close to 0, this can easily be seen from the Standard Reduction Potential series (see Table 3).

In Table 3, the standard reduction potentials of some common elements are listed. The most common oxidizing species is dissolved oxygen. This is always

Table 3 Standard Reduction Potential of Selected Elements

	E^0 (V vs. NHE)
$O_3 + 2H^+ + 2e^- \Leftrightarrow O_2 + H_2O$	2.07
$H_2O_2 + 2H^+ + 2e^- \Leftrightarrow 2H_2O$	1.776
$Au^{3+} + 3e^- \Leftrightarrow Au$	1.50
$O_2 + 4H^+ + 4e^- \Leftrightarrow 2H_2O$	1.229
$Ag^+ + e^- \Leftrightarrow Ag$	0.799
$Cu^+ + e^- \Leftrightarrow Cu$	0.521
$Cu^{2+} + e^- \Leftrightarrow Cu$	0.337
$2H^+ + 2e^- \Leftrightarrow H_2$	0.000
$Pb^{2+} + 2e^- \Leftrightarrow Pb$	−0.126
$Ni^{2+} + 2e^- \Leftrightarrow Ni$	−0.257
$Fe^{2+} + 2e^- \Leftrightarrow Fe$	−0.440
$SiO_2 + 4H^+ + 4e^- \Leftrightarrow Si + 2H_2O$	−0.857
$A^{3+} + 3e^- \Leftrightarrow Al$	−1.662
$Mg^{2+} + 2e^- \Leftrightarrow Mg$	−2.37
$Na^+ + e^- \Leftrightarrow Na$	−2.714

Source: Ref. 12.

present in aqueous solutions. If no other oxidizing species is used in the solution, the oxidation potential of the solution will be determined by O_2. At pH $= 0$ and saturated with 1 atm of O_2, the potential of dissolved O_2 is 1.229 V. As can be seen from the table, this is enough to oxidize most of the common contaminants encountered, including Cu^{2+}. Therefore, acidic solutions that are saturated with O_2 are generally good solutions for removing metallic ions from oxide surfaces, since they meet both requirements for most common metals. This has been known for over 30 years, as is shown in Figure 11, after Kern and Puotinen [6]. The most common choice for the acid is HCl. Other acids may be used as well. The choice of the acid has to be such that the anion introduced (Cl^- in the case of HCl) does not form any insoluble precipitates with common metallic impurities. On top of that, the acid itself should be relatively inert to oxidation, since, as we will see further, usually these acid solutions are used in the presence of an oxidizer. Finally, the anion introduced should not form chemical bonds with the surface to be cleaned. As an example, H_2CO_3 is convenient, since it

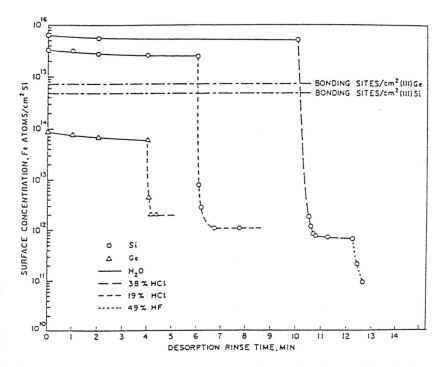

Figure 11 Efficiency of water, HCl (38% and 19%) and HF for desorbing adsorbates from silicon and germanium. Fe deposits from hot NaOH were used. All rinsing treatments were conducted at 23°C. Number of bonding sites for Si and Ge are indicated for reference. (From Ref. 6.)

can be introduced by bubbling CO_2 gas into DI-water, but, while acidifying, it also introduces the HCO_3^- and CO_3^{2-} ions into solution which form insoluble precipitates with many metallic ions, including the important contaminant Ca^{2+} [27,28]. Therefore, this acid is not preferred for metallic impurity removal. Good choices for the acid include HNO_3, H_2SO_4, and HCl. The most common choice is HCl. This was originally driven by the fact that HCl is a dissolved gas and as such it is expected to leave little or no residues behind. Disadvantages of HCl include the fact that it can be oxidized by both H_2O_2 and O_3 to Cl_2, HClO, and $HClO_2$ and the fact that the Cl^- ion adheres easily to solid surfaces. The oxidation of Cl^- can be observed in SC-2 solutions as evidenced by the rapid decomposition of the H_2O_2. Nevertheless, the continued use of SC-2 solutions in the industry for over 30 years proves that these disadvantages can be dealt with in a manufacturing line. This is mainly due to the fact that the exact amount of oxidizing agent and acid is not so important. The removal of metallic impurities from an oxide surface will be effective over a wide process window of H_2O_2 and HCl concentrations.

More recently, Glick [29] showed similar results. He compared the efficiency of dilute HCl with that of SC2 (HCl with H_2O_2 added as an oxidizer) in removing both iron and zinc (Fig. 12). From Figure 12, we can see that the addition of H_2O_2 does not help in the removal of iron or zinc from oxide surfaces. These results were confirmed by Rotondaro et al. [23] and Hurd et al. [19]. Hurd et al. showed similar efficiencies for other elements such as Ca, Mn and Ni (Table 4).

The effect of the $[H^+]$ concentration on the desorption can also easily be measured. Hurd et al. [19] published the efficiency of the removal of Fe, Ca, and Zn from oxide surfaces as a function of the dilution of the HCl which was used (Fig. 13).

As stated previously, the Standard Reduction Potential table (Table 3) should only be used when dealing with acid solutions with pH values close to 0. When using very dilute solutions such as in Figure 13, the pH will be substantially different from 0 and the use of the Standard Reduction Potential table becomes less accurate. In this case a Pourbaix diagram should be used. In Figure 14, the Pourbaix diagram for Fe is shown with several cleaning solutions. As can be seen from this diagram, the solubility of Fe is almost equal for concentrated HCl or SC2 (or even HF) solutions. However, when using dilute HCl solutions, the solubility will be substantially reduced. Even so, the measured reduction in removal efficiency for Fe when using dilute HCl solutions (Fig. 13) is most likely a result of the reduced ion exchange at the surface of the oxide at a reduced $[H^+]$ concentration rather than being caused by the reduced solubility of Fe^{3+} in the solution.

A particular element of interest is Cu. Cu is being used extensively in the semiconductor manufacturing lines as an impurity element in Al alloys or as

IRON REMOVAL

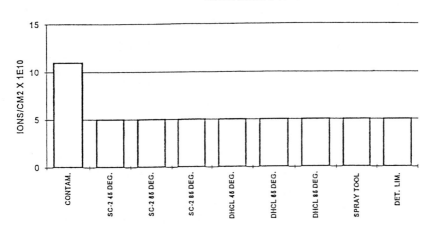

ZINC REMOVAL

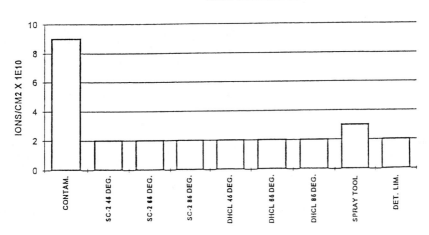

Figure 12 Iron and zinc removal with dilute HCl and SC2 solutions at varying temperatures. (From Ref. 29.)

a conductor material itself. Moreover, Cu is often the most noble element [or highest on the Standard Reduction Potential table (Table 3)] of the contamination elements to be removed. From Table 3, we can already see that Cu will be oxidized by dissolved O_2. This can also be seen from the Pourbaix diagram shown in Figure 15.

Table 4 Selected Metallic Ion Removal with Dilute HCl and SC2 Solutions

	Ca	Mn	Fe	Ni	Zn
Cleaning			$(10^{10}$ atoms/cm^2)		
Initial contamination (SC1)	9.2	0.9	357	2.1	37
SC2 (1/1/5, HCl/H$_2$O$_2$/H$_2$O)	0.6	<0.1	1.4	0.7	0.6
HCl (1/6, HCl/H$_2$O)	1.2	<0.3	1.1	1.8	1.3

Source: Ref. 19.

In Figure 15, it is clearly shown that both SC-2 solutions and HCl solutions fall into the soluble Cu^{2+} region. This proves that the addition of H$_2$O$_2$ to HCl is not necessary to remove Cu from oxide surfaces, contrary to common belief. This has been confirmed experimentally by Hurd et al. [17].

However, the addition of H$_2$O$_2$ to HCl to yield a general metallic impurity cleaning solution is not completely useless. In fact, it is a good practice. In state-of-the-art fabs, degassified DI water is commonly used. This is usually used to prevent any unwanted oxidation of exposed bare Si. However, when using degassified DI water, one cannot rely anymore on the presence of dissolved O$_2$ to keep the ions oxidized in solution and precipitation may occur. If the fab uses a central degassification stage in the DI water production, the level of dissolved

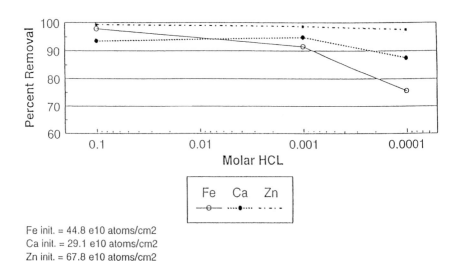

Fe init. = 44.8 e10 atoms/cm2
Ca init. = 29.1 e10 atoms/cm2
Zn init. = 67.8 e10 atoms/cm2

Figure 13 Removal of metallic impurities with dilute HCl at 25°C as a function of dilution. (From Ref. 19.)

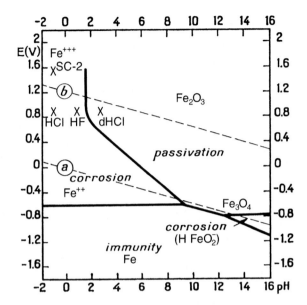

Figure 14 Pourbaix diagram for Fe with the following solutions listed: SC2, HCl (concentrated), HF, and dilute HCl (dHCl). Line (a) represents the reduction equilibrium of water according to the reaction, $H_2 \leftrightarrow 2H^+ + 2e^-$ [$E_a = 0.0 - 0.591$ pH (pH$_2$ = 1 atm)]; Line (b) represents the oxidation equilibrium of water according to the reaction, $2H_2O \leftrightarrow O_2 + 4H^+ + 4e^-$ [$E_b = 1.229 - 0.591$ pH (pO$_2$ = 1 atm)]. (After Verhaverbeke et al. [12].)

oxygen at point of use is relatively uncertain and depends on any contact with air in between the degassification stage and the point of use. This may vary between different points of use in the fab. When using pure HCl solutions to remove metallic contamination, one specifically relies on the oxidizing action of this uncertain dissolved oxygen to keep exchanged metals oxidized and therefore to keep them from precipitating. In order to remove this uncertainty, it is good practice to add some H_2O_2 to the solution to assure the oxidation power of the solution for all elements of concern.

When H_2O_2 is added to the solution, the oxidation power of the solution will be dominated by the H_2O_2 reduction half-reaction [12]:

$$H_2O_2 + 2H^+ + 2e^- \rightleftharpoons 2H_2O \tag{20}$$

The oxidation potential of this reaction is

$$E = E^\circ + 0.0347 \log [H_2O_2] = 1.78 - \frac{RT}{F} \text{pH} + 0.0347 \log [H_2O_2]$$

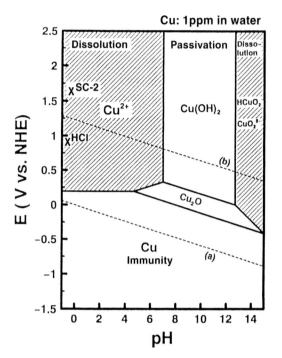

Figure 15   Pourbaix diagram for Cu showing both SC2 and HCl solution in the soluble Cu^{2+} region. Line (a) represents the reduction equilibrium of water according to the reaction, $H_2 \leftrightarrow 2H^+ + 2e^-$ [$E_a = 0.0 - 0.591$ pH ($pH_2 = 1$ atm)]; Line (b) represents the oxidation equilibrium of water according to the reaction, $2H_2O \leftrightarrow O_2 + 4H^+ + 4e^-$ [$E_b = 1.229 - 0.591$ pH ($pO_2 = 1$ atm)]. (From Ref. 12.)

In 1 M HCl solutions (pH = 0), this gives

$[H_2O_2] = 1$ mol/L $\Rightarrow$ $E = 1.78$ V

$[H_2O_2] = 0.26 \times 10^{-3}$ mol/L (10 ppm) $\Rightarrow$ $E = 1.66$ V

Clearly, the oxidation potential is very weakly dependent on the H_2O_2 concentration, and therefore H_2O_2 should only be added in low concentrations. As a comparison, one has to keep in mind that a solution, which is saturated with O_2, contains dissolved O_2 of the order of 10 ppm.

In addition to the assurance of oxidation power in the solution, when dealing with degassified water, some metallic ions have a higher oxidation potential than oxygen. Au has a standard reduction potential of 1.5 V and will not be oxidized by O_2. In order to oxidize Au, a stronger oxidizer is necessary. This can be seen from the Standard Reduction Potential series (Table 3). This was the original reason why Kern and Puotinen [6] added H_2O_2 to HCl in order to create the "universal" metallic impurity cleaning solution. We can also see this from a Pourbaix diagram for Au. The Pourbaix diagram for Au is shown in Figure 16. It is clear that Au is not soluble in an HCl solution, but will be dissolved in an SC-2 solution. Experimentally Kern and Puotinen confirmed that hot SC-2 solutions were able to remove Au from oxide surfaces. This is shown in Figure 17.

Alternatively to H_2O_2, O_3 can be added to the solution. O_3, as can be seen from Table 3, will provide an even greater oxidation power than H_2O_2. Even though the oxidation power of H_2O_2 is more than sufficient, the use of O_3 can be justified by cheaper consumables. O_3 can be generated at point of use from O_2. Recently, this practice is becoming more accepted (see Chap. 9 and panel discussion at the Balazs 2000 SPWCC). However, it is not clear if the dissolution at point of use of simple O_2 instead of O_3 wouldn't give the same

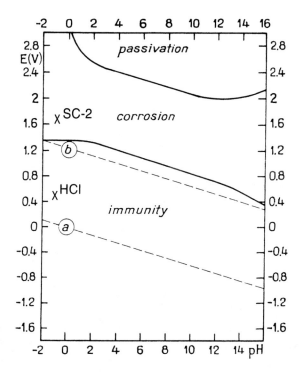

Figure 16 Pourbaix diagram for Au showing SC-2 solutions in the soluble region and HCl solutions in the insoluble region. Line (a) represents the reduction equilibrium of water according to the reaction, $H_2 \leftrightarrow 2H^+ + 2e^-$ [$E_a = 0.0 - 0.591$ pH (p$H_2 = 1$ atm)]; Line (b) represents the oxidation equilibrium of water according to the reaction, $2H_2O \leftrightarrow O_2 + 4H^+ + 4e^-$ [$E_b = 1.229 - 0.591$ pH (p$O_2 = 1$ atm)]. (From Ref. 12.)

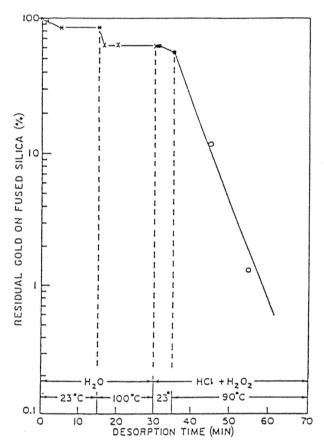

Substrates:

Fused quartz etched in 49% HF + Au^{195}; 100% = 1.56 × 10^{12} Au atoms/cm^2

Solutions:

* Deionized distilled H$_2$O, 23°C
× Deionized and distilled H$_2$O, 100°C
● 1 vol 30% H$_2$O$_2$ + 1 vol 1N HCl + 8 vol H$_2$O, 23°C
○ 1 vol 30% H$_2$O$_2$ + 1 vol 1N HCl + 8 vol H$_2$O, 90°C

Figure 17 Desorption efficiency for gold labeled with Au^{195} from fused quartz under various conditions. (From Ref. 6.)

results. In most cases the uncertainty of the level of dissolved oxygen in state-of-the-art fabs can be removed by purposely dissolving O_2 and even saturating metallic removal solutions with O_2 at 1 atm.

Finally, the discussion related to the use of an oxidizer in metallic removal solutions is not only applicable to these cleaning solutions, but is equally applicable to the subsequent rinsing solution. This was clearly pointed out by Loewenstein and Mertens [21]. Rinsing solutions which follow metallic removal solutions and which are used on oxide surfaces need to contain an oxidizer to prevent any metallic impurity precipitation in the subsequent rinsing solution. The easiest way is to saturate the rinsing solution with O_2. Even though this practice is not very widespread, isolated incidents have been reported when fabs introduced central degassification as part of their DI water plants. Most of these incidents can be traced back to metal outplating (see the next section) or precipitation in the absence of O_2 in the rinse water.

D. Summary and Conclusions

In Section IV we showed the mechanisms and the tools we have to manipulate the desorption mechanism (surface ion exchange) and the solubility of the dissolved ion. We have seen that the interaction of metal ions in solution with the silanol surface group can be described in the same way as a weakly acid ion exchange resin. The equilibrium reaction governing chemisorption/desorption was described by the Equation (6) repeated here:

$$M^{x+} + y(Si-OH) \rightleftharpoons (Si-O)_y M^{(x-y)+} + yH^+ \tag{6}$$

From this equation, one can see that there are two ways to desorb a metallic ion from the oxide surface. The first way is to increase $[H^+]$. Acidifying the solution produces at the same time a solution in which most common metallic ions are soluble, provided that a suitable oxidizer, such as dissolved O_2, H_2O_2, or O_3, is present in the solution to prevent any reduction, especially of such metallic ions as Cu^{2+}. This is the basis of the most common metallic impurity removing solution, i.e., SC-2.

However, the chemisorption/desorption equilibrium equation provides us with another mechanism for metallic impurity removal, i.e., decreasing the free metal ion concentration in the solution $[M^{x+}]$. If this is achieved by binding the free metal with a ligand so that the combined complex remains soluble, we have created the same conditions for metallic impurity removal as the common SC-2 solution. These ligands are commonly referred to as chelating, complexing, or sequestering agents. Typical chelating agents are polyacrylates, carbonates, phosphonates, and gluconates. The ethylenediaminetetraacetic acid (EDTA) family is very popular. In 1991, Verhaverbeke et al. [27] showed the benefit of EDTA as a chelating agent in SC-1 solutions. More recently, this approach has gained more popularity and presently several companies are offering chemicals with appropriate chelating agents added. One of the problems encountered with many chelating agents, including EDTA, is that the metal complexes are being sent through the waste treatment systems and into the environment. However, chelating agents exist which are both effective and can be treated appropriately in the waste treatment system, e.g., EDDS.

The advantage of this approach to metallic impurity removal is that the acid environment is not necessary anymore. Using chelating agents, metallic impurities can be removed from oxide surfaces at alkaline pH values, which is commonly used to remove particles such as the ubiquitous SC-1 solution [6] (an aqueous mixture of NH_4OH and H_2O_2). This opens the door for an all-in-one, universal cleaning solution. The effectiveness in removing metals of such a solution is shown in Figure 18 for Fe, Al, Cu, and Ca.

The surface metallic contamination remaining can be described by the following formula:

$$\{\equiv SiOM^{x-1+}\} = K_{1(int)} * \{SiOH\} * \frac{[M^{x+}]}{[H^+]}$$

Therefore, when raising the pH, if $[M^{x+}]$ can be reduced by binding the metallic impurities with a ligand to the same extent that $[H^+]$ is reduced by raising the pH, the same final metallic impurity concentration on the oxide surface will result. When comparing SC-1 solutions with SC-2 solutions, the free metal ion concentration has to be reduced by 10^{10} through binding with a chelating agent to achieve a similar metallic removal efficiency. Loewenstein

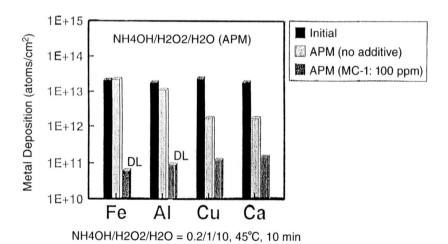

Figure 18 Removal of metallic contamination (Fe, Al, Cu, and Ca) in SC-1 (or APM) cleaning solutions with and without a chelating agent (MC-1). (From Ref. 30.)

(21) has shown that this is a worst case, since from experimental observations, the equation looks on the average like:

$$\{\equiv SiOM^{x-1+}\} = K_{1(int)} * \{SiOH\} * \frac{[M^{x+}]^{0.54}}{[H^+]^{0.3}}$$

The reason for the change in exponents, is the competition between NH_4^+, H^+ and metallic ions for available chemisorption sites. The components are slightly different for different metallic cations. This makes the effect of increasing the pH much less than the binding of the free metallic ions. According to this equation, a similar metallic impurity removal can be expected if the free metal ion concentration can be reduced by 10^6. In the particular case of Fe^{+++}, the reduction in free ions of roughly 10^3, has a similar effect as a reduction in pH from the alkaline SC-1 to the acid SC-2 solution.

V. CHEMISORPTION/DESORPTION OF METALLIC IMPURITIES ON BARE SILICON SURFACES

A. Introduction

In this section we will look at the chemisorption/desorption of metallic impurities on bare silicon surfaces. The mechanisms taking part in chemisorption/desorption on bare silicon surfaces are substantially different from the same mechanisms on oxide surfaces. Whereas the surface changes with acids, bases, and the ion competition were the most important mechanisms for chemisorption/adsorption on oxide surfaces, the adsorption of metallic impurities on silicon surfaces will be mainly determined by outplating and physical adsorption. The removal of metallic impurities from the silicon surface will be mainly by oxidation or decomposition of the impurity (necessarily together with the underlying silicon) and by etching of the surface layer. Obviously when oxidizing, an oxide surface will be formed; therefore the same mechanisms discussed in Section IV for the oxide surface come into play once the entire surface is covered with an oxide. For physically adsorbed impurities, physical desorption by solvents (such as rinsing water) and surfactants will be possible together with decomposition of precipitates, if precipitation on the surface is occurring. The bare silicon surface of interest here is the hydride-terminated surface for reasons explained earlier. Finally, once dissolved, the same guidelines and rules for the solubility or the state of metallic ion in solution, as in Section IV, still apply here.

B. Adhesion Mechanisms on Bare Si Surfaces

The three main mechanisms by which metallic impurities contaminate a bare hydride-passivated Si surface are 1) electrochemical deposition, 2) precipitation, and 3) physisorption.

Precipitation is a specific contamination case by physisorption and takes place whenever the concentration of the contaminant in the solution exceeds its solubility. A typical example of this is the adhesion of Ca and Mg to a bare silicon surface in HF solutions as CaF_2 and MgF_2 precipitates:

$$Ca^{2+} + 2F^- \rightleftharpoons CaF_2 \tag{21}$$

CaF_2 is only sparingly soluble in HF solutions. Exchanging the F^- ions with other anions can easily prevent precipitation. Indeed, when an acid stronger than HF is added to a dilute hydrofluoric solution, no more CaF_2 or MgF_2 precipitation will occur [28]. At medium and high pH, metal-hydroxide precipitation is possible with several common contaminants. At high pH, bare silicon is being etched quite rapidly and this, of course, complicates the precipitate settling.

Even though physisorption and precipitation is the only contamination mechanism for most impurity elements on a bare silicon surface, the remainder of Section V will deal with the electrochemical deposition mechanism. Even though the electrochemical deposition mechanism applies to only a limited subset of the metallic impurities on a bare silicon surface, the resulting contamination levels are several orders of magnitude higher than from the physisorption mechanism. Because of these high contamination levels, it is important to understand and control electrochemical deposition. Moreover, one of the elements which can contaminate bare silicon by this mechanism is Cu. Cu is being used not only in an alloy with Al, but more and more frequently as the primary metallic conductor on a silicon chip. This widespread usage will increase the chances of Cu contamination on silicon wafers.

C. Electrochemical Metal Deposition on Bare Silicon

Metals which exhibit a higher electronegativity or a standard reduction potential higher than Si can potentially undergo electrochemical metal deposition by exchanging electrons with the Si surface [31,32]. Table 5 shows the redox potential (standard reduction potential) of major metals.

As can be seen from Table 5, quite a few common metallic impurities have a higher standard reduction potential than silicon, e.g., Cu^{2+}, Ni^{2+}, and Fe^{2+} to name a few. Other common metallic impurities have a lower standard reduction potential, such as Al^{3+}, Ca^{2+}, Na^+, and K^+. These metal ions can only deposit on the bare silicon surface by precipitation or physical adsorption. However, in order to determine if electrochemical deposition can occur, as will later become clear, when bare silicon is passivated with hydrogen, the actual mixed potential or the corrosion potential of the silicon surface has to be considered and not just the standard reduction potential. Passivated silicon has the characteristic of an irreversible potential electrode—it acts as a highly polarizable electrode. The actual corrosion potential of the silicon can be substantially higher than the

Table 5 Standard Reduction Potential of Major Metals and Hydrogen

Element	Standard reduction potential E^o (V vs. NHE)	
Au^+/Au	1.692	
Au^{3+}/Au	1.5	
Pt^{2+}/Pt	1.188	
Pd^{2+}/Pd	0.951	
Ag^+/Ag	0.799	
Hg^{2+}/Hg	0.797	
Cu^{2+}/Cu	**0.337**	
Ge^{2+}/Ge	0.24	
W^{3+}/W	0.1	
H^+/H_2 (N HE)	0	These metals can potentially and thermodynamically react with bare silicon and plate out.
Pb^{2+}/Pb	−0.126	
Sn^{2+}/Sn	−0.1375	
In^+/In	−0.14	
Mo^{3+}/Mo	−0.200	
Ga^+/Ga	−0.2	
Ni^{2+}/Ni	**−0.257**	
Co^{2+}/Co	−0.28	
Cd^{2+}/Cd	−0.403	
Fe^{2+}/Fe	**−0.440**	
Ta^{3+}/Ta	−0.6	
Cr^{3+}/Cr	−0.744	
Zn^{2+}/Zn	−0.762	
SiO_2/Si	**−0.857**	
Al^{3+}/Al	−1.662	
Mg^{2+}/Mg	−2.37	
Ca^{2+}/Ca	−2.866	
Na^+/Na	−2.714	
K^+/K	−2.924	

Source: Ref. 14.

reversible standard reduction potential, which is −0.857 V vs. NHE (Table 5). The mixed potential of the silicon surface is the standard reduction potential plus the overpotential. This overpotential is determined by the Tafel slope of the silicon oxidation reaction and the concentration and nature of the oxidizing species oxidizing the silicon. At any time in any aqueous solution, an oxidizer for silicon will be present. In pure water the only two oxidizers present are dissolved O_2 and H^+. The corrosion potential of the silicon surface will always be in between the standard reduction potential of the silicon (-0.857 V) and the standard reduction potential of the oxidizer, which is 1.229 V for oxygen and 0.00 V for H^+. Exactly where in between these two numbers the mixed potential of the silicon surface really is depends on the overpotential for the oxidation of silicon and the overpotential for the reduction of either oxygen or H^+, whichever is dominant. It is sometimes assumed that metals with a negative redox potential can never be deposited on Si, since the H_2 evolution will always dominate. However, in the absence of oxygen, the corrosion potential of the silicon surface will be between the hydrogen evolution and the silicon oxidation. Since the hydrogen evolution can exhibit a very high overpotential on some electrodes, including silicon, the actual corrosion potential on the silicon surface can be significantly less than the standard reduction potential of the H_2 evolution. The silicon corrosion potential can therefore under certain circumstances be even more negative than, say, Ni^{2+}/Ni (-0.250 V) and Fe^{2+}/Fe (-0.440 V). Under these circumstances, electrochemical deposition can potentially occur with Ni^{2+} and Fe^{2+}. As we will see later, this can occur in completely degassified solutions. Indeed, whenever the mixed potential of the silicon surface is below the standard reduction potential of the particular metallic impurity, that impurity can deposit on the silicon surface. Nevertheless, the likelihood of electrochemical deposition of a metallic impurity will decrease with decreasing standard reduction potential and whether at all possible will be dependent on the dissolved oxidizers, such as oxygen, in the solution. As an example, Figure 19 shows how much metal deposits on the Si surface in DHF solutions and ultrapure water spiked with 1 ppm of Cu, Fe, and Ni [33].

It is clear from Figure 19 that the higher the standard reduction potential, the more electrochemical deposition that will occur. Electrochemical deposition will follow the following order: $Cu^{2+} > Ni^{2+} > Fe^{2+}$. Therefore, for the remainder of this section, Cu^{2+} will be used as the example element, since the effects will be most pronounced for Cu^{2+}. There are contaminants even more likely than Cu^{2+} to deposit onto the bare silicon, such as Hg^{2+}, Ag^+, Pd^{2+}, Pt^{2+}, and Au^+, but these contaminants are less common than Cu^{2+}.

Cu^{2+} ions deposit onto the bare silicon surface in the form of metallic particles [33] (see Fig. 20). Moreover, in HF solutions, Cu^{2+} deposition causes pits in the silicon surface. The number and diameter of the pits agree very well with the number and size of metallic Cu particles. The electrochemical reduction of Cu^{2+} ions follows the following half-cell reaction:

$$Cu^{2+} + 2e^- \rightleftharpoons Cu \qquad E^0 = 0.337 \text{ (V vs. NHE)} \qquad (22)$$

where E^0 = standard reduction potential (redox potential in a standard state).

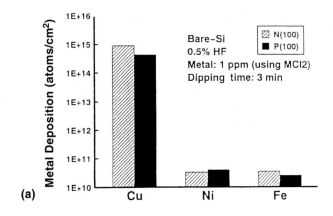

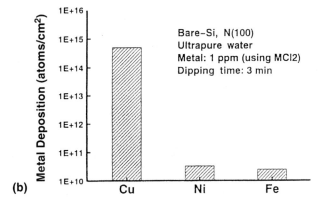

Figure 19 Metal deposition on a Si surface from 1 ppm solutions of Cu, Fe and Ni in (a) DHF solutions and (b) ultrapure water. (From Ref. 33.)

The other half-cell reaction is the oxidation of Si:

$$SiO_2 + 4H^+ + 4e^- \rightleftharpoons Si + 2H_2O \qquad E^o = -0.857 \text{ (V vs. NHE)} \qquad (23)$$

The potential of the silicon oxidation half-cell reaction is pH dependent, i.e. it decreases with 60 mV/unit pH increase. In any case, the Cu^{2+} ion always has a much higher redox potential than Si. While depositing Cu, SiO_2 is formed and if this is happening in an HF solution, HF will immediately etch the formed oxide. This causes the formation of a pit. Figure 20 shows SEM images of the Si surface with Cu deposition from: 1) 0.5% DHF; 2) ultrapure water; and

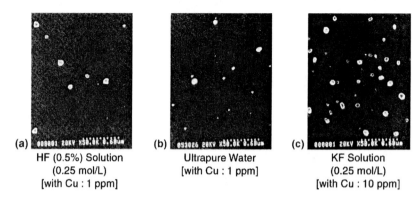

(a) HF (0.5%) Solution
(0.25 mol/L)
[with Cu : 1 ppm]

(b) Ultrapure Water
[with Cu : 1 ppm]

(c) KF Solution
(0.25 mol/L)
[with Cu : 10 ppm]

Figure 20 SEM images of a Si surface after Cu deposition from (a) DHF (0.5%) solution, (b) ultrapure water and (c) KF solution, each spiked with Cu using $CuCl_2$ (dipping time = 3 min). (From Ref. 33.)

3) a KF solution. Figure 21 shows the AFM images of the Si surface after Cu particles are removed by SPM cleaning (10 min).

As can be seen in Figure 21, many pits are detected on the Si surface on which Cu was deposited from the DHF solution. On the other hand, no pits are detected on the Si surface on which Cu^{2+} was deposited from either ultrapure water or the KF solution.

Figure 22 shows the mechanism of Cu^{2+} deposition onto Si surfaces in solutions. Firstly Cu^{2+} ions in the vicinity of the Si surface withdraw electrons from the Si and are deposited in a form of metallic Cu. This is how a nucleus of a Cu particle is formed. Subsequently, it becomes easier to exchange electrons with the silicon through this Cu electrode, and therefore more Cu^{2+} will deposit

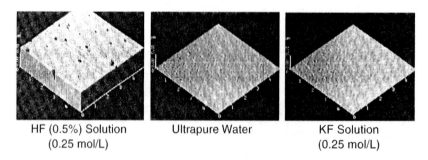

HF (0.5%) Solution
(0.25 mol/L)

Ultrapure Water

KF Solution
(0.25 mol/L)

Figure 21 AFM images of the Si surface after Cu particles are removed by SPM cleaning (10 min). Cu was deposited in either DHF (0.5%) solution, ultrapure water or a KF solution. (See Fig. 20.) (From Ref. 33.)

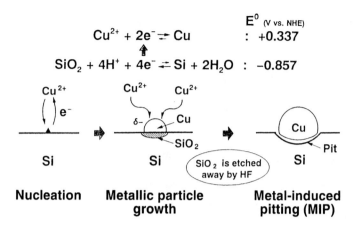

$$Cu^{2+} + 2e^- \rightleftharpoons Cu \quad : \ +0.337$$

$$SiO_2 + 4H^+ + 4e^- \rightleftharpoons Si + 2H_2O \quad : \ -0.857$$

E⁰ (V vs. NHE)

Nucleation   Metallic particle   Metal-induced
              growth   pitting (MIP)

Figure 22 Mechanism of Cu^{2+} deposition onto Si surfaces in solutions. (From Ref. 33.)

on the now formed electrode. Accordingly, the Si surface around the Cu particle must be oxidized to SiO_2. In HF solutions, this oxide is continuously etched away and a pit is formed. In ultrapure water and in KF solution, a local oxide is grown which is incorporated into a chemical oxide subsequently formed when immersing the silicon in an SPM solution. When etching this entire chemical oxide away, no residual damage can be found on the silicon surface. The Cu nucleus is started where the Si surface is electrically active. The silicon surface is terminated with a stable hydrogen passivation in DHF solutions. Electron exchange can hardly take place. In other words, a large overpotential is necessary to oxidize the silicon and the oxidation rate will be extremely slow. However, Cl^- or Br^- ions or OH^- ions can replace the H passivation locally, especially at kink sites, such as steps on the surface. If any of these anions are present in the solution, electrochemical deposition of Cu^{2+} on silicon will be greatly catalyzed. In DHF solutions, trace amounts of Cl^- and Br^-, 1 ppm and 10 ppb, respectively, have been reported to be enough to start the electrochemical deposition [33]. OH^- ions are abundant in ultrapure water (pH = 7) and this explains why Cu deposits so easily in ultrapure water on silicon surfaces, whereas the deposition is much less in HF solutions. Figure 23 shows how much Cu^{2+} deposits on p-type Si surface in ultrapure water and DHF solution, both spiked with Cu salts of various anion species.

The catalytic enhancement of Cu^{2+} deposition on a silicon surface due to Cl^-, Br^-, or OH^- ions in solutions can mathematically be described by a lowering of the Tafel slope of the silicon half-cell oxidation reaction. Cu^{2+} will only be deposited onto the silicon surface if the corrosion potential of the silicon

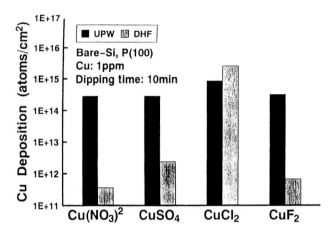

Figure 23 Cu^{2+} deposition on a Si surface in ultrapure water and DHF solutions with Cu^{2+} salts of various anion species added. (From Ref. 33.)

surface is lower than the potential of the Cu^{2+} reduction. If the Tafel slope is large, the overpotential will be high even for very low corrosion currents. This means that very small oxidation rates will cause large swings in the corrosion potential. The oxidation of the silicon, and therefore the corrosion potential of the silicon surface, can be controlled by changing (1) the oxidation species in the solution, (2) the pH (some redox potentials are pH dependent), and (3) the concentration of the oxidation species. Changing the Tafel slope of the silicon surface itself can also control silicon oxidation. The redox potential of the oxidation species in solution at the particular pH and concentration can be measured with a simple Pt electrode. This will be further referred to as the redox potential of the solution. It is in fact the redox potential of the actual oxidizing species measured at open circuit ($I = 0$) vs. the NHE. If we consider a similar silicon surface in all cases, then the actual corrosion potential of the silicon surface will directly depend on the open circuit potential of the oxidizing species, unless there is a substantial difference in Tafel slopes between the different oxidizing species. Figure 24 shows the effect of redox potential of the solution on the concentration of Cu^{2+} deposition. Cu^{2+} deposition is significantly suppressed when the redox potential of the oxidizing species exceeds 0.75 V vs. NHE.

It is clear that increasing the redox potential of the solution can prevent Cu^{2+} deposition onto bare silicon. However, in doing so, a corrosion current is generated, and hence silicon is being oxidized. Similar to ultrapure water, increasing the redox potential of a DHF solution can also prevent Cu deposition. In this case, however, the silicon oxide will be continuously etched. As long as Cu^{2+} deposition is prevented, no pitting will occur, since the oxide will be

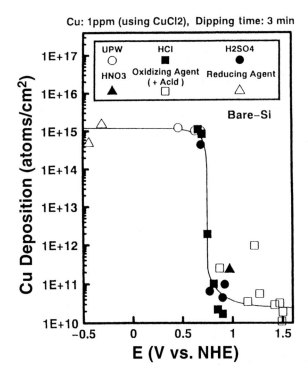

Figure 24 The effect of solution redox potential (ultrapure water with various chemicals added) on Cu^{2+} deposition onto a Si surface. (From Ref. 33.)

uniformly formed by oxidation due to such oxidizing species as O_2, HClO, and O_3 and immediately etched in HF. Figure 25 shows Cu^{2+} deposition onto a Si surface from various DHF solutions as a function of the redox potential of the solution.

In all previous cases, it was indicated that the corrosion potential of the silicon surface is of major importance in the deposition of Cu^{2+} from ultrapure water and from DHF solutions. The corrosion potential can be manipulated by changing the oxidizing species, their concentration, and the pH. As was the case with metallic impurity adhesion on oxides, the deposition of Cu^{2+} on a silicon surface can also be manipulated by binding the free Cu^{2+} ions in solution. Figure 26 demonstrates that adding a chelating agent, such as EDTA [12], can prevent Cu^{2+} deposition onto the bare Si surface from ultrapure water containing 1 ppm of Cu^{2+}.

Finally, Cu^{2+} deposition in solution can also be suppressed by the injection of surfactants [34]. Figure 27 shows how Cu^{2+} deposition from a DHF

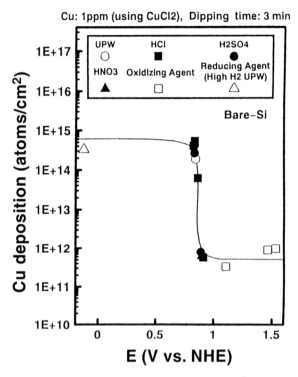

Figure 25 The effect of redox potential on Cu^{2+} deposition onto a Si surface in DHF solutions containing various added chemicals. (From Ref. 33.)

solution with 1 ppm of Cu^{2+} added, is affected, when an oxidizer (10% H_2O_2) or a surfactant (0.1% anion type) is injected. It is clear that the surfactant is as effective in preventing the deposition of Cu^{2+} as the oxidizing agent. The surfactant will increase the Tafel slope of the silicon oxidation by covering the surface. This increases the overpotential for a given oxidation rate.

Since the chemisorption of Cu^{2+} onto bare silicon is well understood, it is easy to give some general guidelines to prevent Cu^{2+} deposition on to bare silicon:

1. *The case of ultrapure water.* If oxidation is allowed, add an oxidizer such as O_3. If oxidation has to be minimal, but can be nonzero, saturate the water with O_2 and acidify the rinse water with H_2SO_4. Do NOT acidify the rinse water with HCl, HBr, or HNO_3. HCl and HBr will catalyze the Cu^{2+} deposition and HNO_3 might oxidize the silicon more than necessary. Alternatively, a surfactant can be added. This can minimize the oxidation even further. If absolutely no oxidation

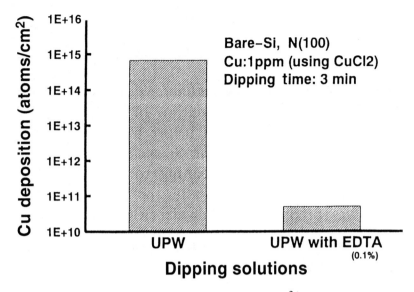

Figure 26 The effect of chelating agent injection on Cu^{2+} deposition onto a Si surface in ultrapure water. (From Ref. 12.)

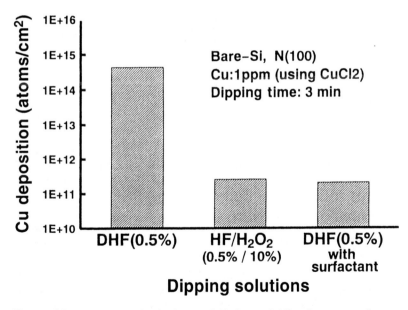

Figure 27 The effect of injecting 10% H_2O_2 or 0.1% anion-type surfactant on Cu deposition onto a Si surface in a DHF solution. (From Ref. 12.)

can be tolerated, degassify the water, even saturate the water with H_2 and add a chelating agent.

2. *The case of HF solutions.* If etching is allowed, add an oxidizer such as H_2O_2 or O_3. If minimal etching can be tolerated, saturate the HF solution with O_2, and acidify the HF with H_2SO_4 or trichloroacetic acid. Avoid HCl, HBr, HNO_3. (HCl and HBr catalyze the Cu^{2+} deposition; HNO_3 is an oxidizer). Similar to the water case, a surfactant can be added to reduce the etching even further. If absolutely no surface etching can be tolerated, degassify the solution, even saturate the solution with H_2—and add a chelating agent.

Obviously in both cases (ultrapure water and DHF), Cu^{2+} deposition can be prevented by removing all Cu^{2+} from the water and the chemicals. This is the most common practice in most fabs: Make the process as clean as possible without changing the chemistry to avoid deposition. However, in this section we showed that changing the chemistry slightly can avoid deposition even when Cu^{2+} contamination is present.

D. Electrochemical Metal Desorption from Bare Silicon

For the same reasons as stated above, we will consider only the case of Cu as an example of metallic impurities to be removed from bare silicon. This is because of the difficulty in removing Cu from the silicon surface. Most other metals are only physisorbed to the bare silicon surface, such as precipitates, and can be removed by simple acidified rinsing (acidified to create a high solubility in the rinsing water). A typical example is CaF_2. Cu electroplated on bare silicon is much more difficult to remove. Therefore we will use Cu again as an example for removing electroplated metals from a bare silicon surface. Figure 28 compares the conventional cleaning methods (SPM and HPM or SC-2), typical acids, and typical oxidizing agents in terms of Cu removal efficiency. Diluted HCl and diluted H_2SO_4 at room temperature have been shown to be as effective in removing Cu as SPM and HPM at high temperature. This is because the HCl and the H_2SO_4 solutions were saturated with O_2 and therefore, their redox potential is high enough to remove Cu. It turns out that the same redox potential, which is necessary to prevent Cu^{2+} outplating on the Si, is sufficient to remove Cu from the Si. Moreover, it is clear that even ozonated DI water, without any chemicals added, is effective in removing Cu.

Acid solutions and oxidizing agents with a redox potential of over 0.75 V are effective in removing Cu, just as they are effective in preventing Cu outplating. As can be seen from Figure 28, however, they are not able to reduce the Cu concentration on a Si surface to the blank level. This is because some of the Cu is included in the grown oxide on the Si surface. The metallic im-

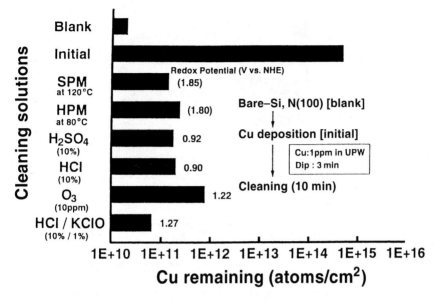

Figure 28 Comparison of conventional cleaning methods (SPM and HPM), typical acids, and typical oxidizing agents in terms of Cu removal efficiency. (From Ref. 33.)

purities included in the oxide can only be removed by etching. Of course, the etching solution used to etch the oxide must also be a solution which prevents redeposition. This is shown in Figure 29.

Figure 29 shows the effect of various HF cleanings to remove the oxide following an SPM cleaning. The Cu concentration on the Si surface hardly decreased when the wafer was treated with a 0.5% DHF solution following the SPM cleaning. However, when an HF/H_2O_2 solution or an HF/surfactant solution was used to etch the chemical oxide, the Cu concentration was reduced to the blank level. Of course, the oxidizing and etching step can be performed in a one-step chemical solution such as HF/H_2O_2. This will result in similar removal efficiency as a two-step SPM + HF/H_2O_2 cleaning [33]. However, this is not recommended, since an HF/H_2O_2 solution will remove Cu while locally roughening the surface. In the absence of Cu contamination on the silicon surface, however, HF/H_2O_2 solutions will not roughen the surface and, therefore, there is no problem using it in a two-step process such as SPM + HF/H_2O_2 or O_3 + HF/H_2O_2. In these two-step processes, when the bare silicon is exposed to HF/H_2O_2, no more Cu will be left on the surface and no roughening results. This is shown in Table 6.

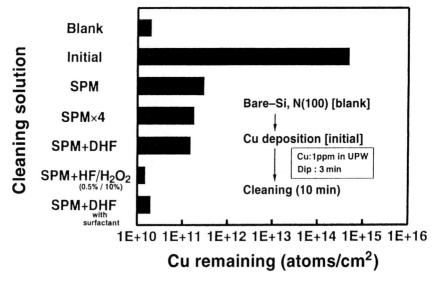

Figure 29 The effect of various cleaning methods on the removal of Cu included in the oxide. (From Ref. 33.)

Table 6 shows the amount of residual Cu and microroughness (Rms) on the Si surface. O_3 + HF/surfactant or O_3 + HF/H_2O_2 are the simplest and best cleaning method to remove Cu without increasing the surface microroughness.

There exists a single-step chemical clean that achieves a similar result. One can use a SC-1 or APM solution with a chelating agent added as a one-step

Table 6 The Amount of Residual Cu and Microroughness (Rms) When a Si Surface Contaminated with Cu Is Cleaned with Various Cleaning Methods

Chemical [conc]	Cu remaining (atom/cm^2)	Surface roughness Rms (nm)
Blank	1–5E+10	0.09–0.13
Initial contamination	1–10E+14	—
SPM	1.4E+11	0.11
HF/H_2O_2 [0.5%/10%]	1.5E+10	0.25
SPM + HF/H_2O_2 [0.5%/10%]	1.5E+10	0.12
O_3 + HF/H_2O_2 [0.5%/10%]	2.5E+10	0.12
O_3 + DHF with Surfactant	2.8E+10	0.11

Source: Ref. 33.

chemical cleaning to remove Cu. This also has all the ingredients for removing Cu without roughening the surface: high oxidation potential, oxide etching, and Cu redeposition prevention (because of the chelating agent). This was introduced by Verhaverbeke et al. [27] in 1991. Recently this single step-cleaning step has been gaining popularity. The advantage of this cleaning solution is that once the Cu is removed, the bare silicon is never further directly exposed to the solution, including the rinsing water. It remains covered with an oxide. However, even in this case, care has to taken as to the dissolved oxygen content of the rinsing water, as will be outlined in Section VI.

VI. CHEMISORPTION/DESORPTION OF METALLIC IMPURITIES ON THIN OXIDE SURFACES

Thin oxides, such as native oxides, are a special case. In this case both mechanisms as outlined in Sections IV and V can coexist. The outer surface behaves as an oxide, and therefore the mechanisms applicable to oxides take place at that surface. However, if the oxide is thin enough, metallic ions can exchange electrons over the thin oxide layer (by direct tunneling) with the silicon. Therefore, metallic ions such as Cu^{2+}, Ni^{2+}, and Fe^{2+} can still plate out on the oxide by exchanging electrons with the underlying silicon. This will happen especially with degassified solutions and thin oxides. The exchange of electrons with the underlying silicon decreases rapidly with increasing oxide thickness. Nevertheless, there are reports of Cu^{2+} and Fe^{2+} inducing pits in the silicon, even when covered with a thin oxide layer. Especially in the case of Fe^{2+}, this can happen only in the absence of dissolved oxygen. In the case of Fe^{3+}, the pitting can also happen through reduction of Fe^{3+} to Fe^{2+}. Indeed,

$$Fe^{3+} + e^- \rightleftharpoons Fe^{2+} \qquad E^\circ = 0.771 \qquad (24)$$

Therefore, Fe^{3+} is an even more powerful oxidizer than Cu^{2+}. If Fe^{3+} gets into contact with bare Si, it will immediately oxidize the Si locally. It won't plate out, as is the case with Cu^{2+}, since the reaction product, Fe^{2+}, is soluble as well. Nevertheless this local oxidation of Si could result in surface roughening. If the oxide is thin enough, electron exchange can occur by tunneling through the oxide layer and therefore, Fe^{3+} can even oxidize the silicon underneath the thin oxide layer. This can be prevented by dissolved oxygen.

As noted earlier in Section IV on oxide surfaces, and as is again clear here, it is very important to include dissolved oxygen in many solutions when dealing with oxide surfaces. This is often overlooked. Even when the surface is covered with a thin oxide, it is still important to have dissolved oxygen present to prevent electrochemical outplating and even surface roughening.

VII. CONCLUSIONS

In this chapter we reviewed the different mechanisms of metallic impurity adsorption on surfaces important in the manufacturing of semiconductors. From a technological point of view, the liquid chemical environment is the most important environment to fully understand. Wet chemistry is the most powerful and the most versatile process available today to remove metallic impurities from surfaces. The different aspects of wet chemistry were reviewed in this chapter. The state of the elements in the solution is the first important aspect of any wet cleaning chemistry. This was thoroughly reviewed together with the engineering tools available to predict any contaminant's state in any kind of solution. The next important aspect is the state of the surfaces on wafers. It was shown that almost all surfaces of interest could be divided in two main groups: the hydroxide terminated oxide and the hydride terminated silicon surface. Both surfaces react very differently with metallic impurities and, therefore, an in-depth review of the adhesion mechanisms, the adsorption forces and the cleaning solutions were given for each surface. The chapter gives the reader enough background and shows where to go in order to come up with the best process conditions for each particular contaminant of interest. Even if a particular contaminant was not reviewed as an example, the chapter gives the reader the necessary background and ground rules with which to design the best process for each situation.

ACKNOWLEDGMENTS

Many people in the industry and throughout the academic world have made tremendous contributions to the very exciting field of semiconductor cleaning. This chapter could not have been written without the contribution of many researchers, process engineers, scientists, and professors. Most of the current understanding of metallic impurity adhesion and cleaning has come together by collaboration and discussion on a global scale. I have only been able to write this chapter with the help of many of my current and past colleagues and friends in the industry. Especially, I would like to acknowledge M. Heyns who introduced me to this field; my former colleagues at IMEC, P. Mertens, M. Meuris, A. Rotondaro, H. Schmidt, and H. Bender; and Prof. G. DeClerck and Prof. R. De Keersmaecker who always supported the cleaning research at IMEC and coached me personally. Furthermore, I would like to thank Prof. T. Ohmi for guiding me with his overwhelming knowledge during my stay in Japan. The contributions of my Japanese colleagues, especially the contributions of H. Morinaga and T. Futatsuki, have been enormous in this field. Without their contributions, this chapter would not be possible. Furthermore, I would like to thank all other researchers at Tohoku University for their research efforts and their assistance. Their number is too overwhelming to list. I would like to thank T. Hurd and L. Loewenstein for their scientific contributions to this field and for the personal discussions we have had. Their work has inspired many of us. Furthermore, I would like to thank C. McConnell for giving me the opportunity to continue my scientific investigations while at CFM Technologies. I would like to thank many people in the industry: G. Higashi, Y. Chabal, F. Kern, R. Novak, K. Christenson, F. Tardif, M. Knotter, D. Levy, M. Alessandri, E. Bellandi, T. Hattori, S. O'Brien, K. Penner, F. Pipia, M. Balazs, and many more who all have helped us bring our understanding to where it is today. I cannot list all my fellow friends, researchers, scientists, and colleagues in the industry who have contributed to our knowledge as it stands today, through personal contacts, discussions, conferences, papers, and so on. But all of them may be assured that their contributions are greatly appreciated and will be welcomed in the future as our understanding continues to grow. Because of the contributions of many scientists and engineers, too many to list here, the cleaning of metallic impurities has evolved from an art into a science. The time of the magic recipes is definitely behind us. Finally, I would like to thank Bob Donovan for his persistence in publishing this book. Without his continued encouragement, this book would never have seen the press.

REFERENCES

1. PW Mertens, D Graef, S Verhaverbeke, M Meuris, M Heyns, A Schnegg, and A Philipossian. "True story about the effect of TCA on dielectric breakdown yield." First International Symposium on Ultra Clean Processing of Silicon Surfaces, Leuven, Belgium, September 17–19, 1992.
2. T Ito and R Sgino. "Dry cleaning technologies using UV-excited radicals and cryogenic aerosols." Solid State Phenomena 65–66, 1999 (Scitec Publications, Switzerland), pp. 219–224.
3. S Verhaverbeke and B Pagliaro. "Isothermal EPI-Si Deposition at 800 C." ECS Fall 1999, September 1999, Honolulu, Hawaii. In Cleaning Technology in Semiconductor Device Manufacturing VI (The Electrochemical Society, Pennington, NJ, 1999), PV99, p. 445.
4. R Moo. Materials Research Society Spring Meeting, San Francisco, CA, 1999 (Materials Research Society, Pittsburgh, PA, 1999).
5. S Verhaverbeke, PhD thesis, Katholieke Universiteít Leuven, 1993.
6. WA Kern and DA Puotinen. "Cleaning solutions based on hydrogen peroxide for use in silicon semiconductor technology." RCA Rev. 31, 1970, pp. 187–206.
7. C Werkhoven, E Granneman, M Hendriks, S Verhaverbeke, and M Heyns. "Cluster tool integrated HF vapor etching for native oxide free processing." Materials Research Society Spring Meeting, San Francisco, CA, April 12–16, 1993. In Surface

382   Verhaverbeke

Chemical Cleaning and Passivation for Semiconductor Processing. Mater. Res. Soc. 315 (Materials Research Society, Pittsburgh, PA, 1993), p. 211.

8. S Verhaverbeke, R Messoussi, and T Ohmi. "Improved rinsing efficiency after SPM (H2SO4/H2O2) by adding HF." Second International Symposium on Ultra Clean Processing of Silicon Surfaces, September 19–21, 1994, Brugge, Belgium. In Proceedings of the Second International Symposium on Ultra-Clean Processing of Silicon Surfaces (Acco, Leuven, Belgium, 1994), p. 201.

9. K Raghavachari, GS Higashi, YJ Chabal, and GW Trucks. "First-principles study of the etching reactions of HF and H2O with Si/SiO2 surfaces." Materials Research Society Spring Meeting, San Francisco, CA, April 12–16, 1993. In Surface Chemical Cleaning and Passivation for Semiconductor Processing. Mater. Res. Soc. 315 (Materials Research Society, Pittsburgh, PA, 1993), p. 437.

10. H Bender, S Verhaverbeke, and MM Heyns. "Hydrogen passivation of HF—last cleaned (100) silicon surfaces investigated by multiple internal reflection infrared spectroscopy." J. Electrochem. Soc. 141, No. 11, November 1994, p. 3128.

11. M Pourbaix. Atlas of Electrochemical Equilibria in Aqueous Solutions, Pergamon Press, London, 1966.

12. S Verhaverbeke, R Messoussi, H Morinaga, and T Ohmi. "Recent advances in wet processing technology and science." Materials Research Society Spring Meeting, San Francisco, CA, April 17–21, 1995. In Ultra Clean Semiconductor Processing Technology and Surface Chemical Cleaning and Passivation. Mater. Res. Soc. 396 (Materials Research Society, Pittsburgh, PA, 1995), p. 3.

13. GJ Norga, KA Black, HM Saad, J Michel, and LC Kimmerling. "Metal adsorption on silicon surfaces from wet wafer cleaning solutions." Second International Symposium on Ultra Clean Processing of Silicon Surfaces, September 19–21, 1994, Brugge, Belgium. In Proceedings of the Second International Symposium on Ultra-Clean Processing of Silicon Surfaces (Acco, Leuven, Belgium, 1994), p. 221.

14. CRC Handbook of Chemistry and Physics, CRC Press, Boca Raton, Fl.

15. S Verhaverbeke. PhD thesis, Katholieke Universiteit Leuven, 1993, p. 116.

16. H Hiratsuka, M Tanaka, T Tada, R Yoshimura, and Y Matsushita. Ultra Clean Technology 3, No. 3, 1991, pp. 18–27 (Ultra Clean Society, Tokyo).

17. TQ Hurd, ALP Rotondaro, J Sees, A Misra, C Appel. "Surface complexation of metals: a predictive model based on metal coordination chemistry." The Electrochemical Society Meeting, Fall 1997, Paris, France.

18. T Hiemstra, JCM Dewit and WH Van Riemsterface. J Colloid Interface Sci 133, 1989, p. 105.

19. TQ Hurd, HF Schmidt, ALP Rotondaro, PW Mertens, LH Hall and MM Heyns. "Metal interactions with silica (SiO2) surfaces: adsorption and ion exchange." The Electrochemical Society Proceedings, vol. 95-20, p. 277, 1995.

20. L Mouche, F Tardif, and J Derrien. "Mechanisms of metallic impurity deposition on silicon substrates dipped in cleaning solution." J Electrochem Soc 142(7), July 1995, pp. 2395–2401.

21. LM Loewenstein and PW Mertens. "The rinsing problem: effect of solute-surface interactions on wafer purity." Solid State Phenomena 65–66, 1999, pp. 1–6.

22. W Lee, KJ Torek, DA Palsuluch, and L Weston. "The adsorption-desorption of cations at the silica-water interface and its implications in wafer-cleaning efficacy."

Materials Research Society Spring Meeting, San Francisco, CA, April 1–3, 1997. In Science and Technology of Semiconductor Surface Preparation; Mater Res Soc 477 (Materials Research Society, Pittsburgh, PA, 1997), p. 57.

23. ALP Rotondaro, TQ Hurd, HF Schmidt, I Teerlinck, MM Heyns, and C Claeys. "Outplating of metallic contaminants on silicon wafers from diluted acid solutions." Materials Research Society Spring Meeting, San Francisco, CA, April 17–21, 1995. In Ultra Clean Semiconductor Processing Technology and Surface Chemical Cleaning and Passivation. Mater Res Soc 386 (Materials Research Society, Pittsburgh, PA, 1995), p. 183.

24. CF Baes, RE Mesmer. The Hydrolysis of Cations, John Wiley and Sons, 1976.

25. PW Schindler et al. J Colloid Sci 55, 1976, p. 469.

26. Shriver, Atkins, Langford. Inorganic Chemistry. 2nd ed., 1995, p. 192.

27. S Verhaverbeke, M Meuris, PW Mertens, MM Heyns, A Philipossian, D Graef, and A Schnegg. "The effect of metallic impurities on the dielectric breakdown of oxides and some new ways of avoiding them." International Electron Devices Meeting, Washington, DC, December 8–11, 1991. In Tech Dig IEDM (IEEE, Piscataway, NJ, 1991), p. 71.

28. S Verhaverbeke, M Meuris, P Mertens, H Schmidt, MM Heyns, A Philipossian, D Graef, and K Dillenbeck. "Advanced wet cleaning technology for highly reliable thin oxides." ECS Symposium on ULSI Science and Technology, Hawaii, May 16–21, 1993. In Proceedings of the Fourth International Symposium on Ultra Large Scale Integration Science and Technology (The Electrochemical Society, Pennington, NJ, 1993), vol. 93-12, p. 199.

29. JS Glick. SPWCC 1992, Santa Clara, CA, March 1–3, 1992. In Semiconductor Pure Water and Chemicals Conference 1992. (MK Balazs, ed.), (Balazs Analytical Laboratory, Santa Clara, CA, 1992), Volume II.

30. H Morinaga, M Aoki, T Maeda, M Fujisue, H Tanaka, and M Toyoda. "Advanced alkali cleaning solution for simplification of semiconductor cleaning process." Materials Research Society Spring Meeting, San Francisco, CA, April 1–3, 1997. In Science and Technology of Semiconductor Surface Preparation. Mater Res Soc 477 (Materials Research Society, Pittsburgh, PA, 1997), p. 35.

31. T Ohmi, T Imaoka, I Sugiyama, and T Kezuka. J Electroch Soc 139, 1992, p. 3317.

32. FW Kern, Jr., M Itano, I Kawanabe, M Miyashita, RW Rosenberg, and T Ohmi. In Proceedings of Eleventh Workshop on ULSI Ultra Clean Technology, p. 23, 1991.

33. H Morinaga, M Suyama, M Nose, S Verhaverbeke, and T Ohmi. "A model for electrochemical deposition and removal of metallic impurities on Si surfaces," IEICE Trans Electron E79-C, No. 3, March 1996.

34. T Ohmi, T Imaoka, T Kezuka, J Takano, and M Kogure. J Electroch Soc 140, 1993, p. 811.

11
Sources of Contamination and Their Control

David Jensen
Advanced Micro Devices, Inc., Austin, Texas

Contamination within microelectronics manufacturing is ubiquitous. Sources of these contaminants range from the product wafers themselves to trace levels of pollutants in the wafer environment.

The multiplicity of sources and types of contaminants in manufacturing compounds the difficulty with which the discipline of CFM must be approached. At one extreme is the idealist who assumes that every contaminant will cause a defect and thus needs to be eliminated (at least to the lower detection limit of the most sophisticated analytical apparatus). At the other extreme are those who assume no corrective action is warranted until a direct correlation between the contaminant of concern and device yield, reliability, or performance has been established.

Figure 1 attempts to illustrate the relationship between the maturity of a process and the degree of proof or experimental correlation required for investing in contamination prevention and reduction (CPR). This relationship shows that proof is no longer required to include some contamination control measures when building a new manufacturing plant (fab). When new fabs are built (i.e., very immature process), significant investments are universally made in facility design, layout, materials, installation protocol, and analytical instrumentation in order to reduce the likelihood of contaminants degrading a process. More mature processes, however, that are established and very stable require strong experimental correlation and justification before investing in CPR. These short-term vs. long-term decisions can be considered tactical and strategic, respectively. For less complex processes it's possible a greater degree of experimental correlation would be expected in order to invest in CPR. The issue is one of the costs

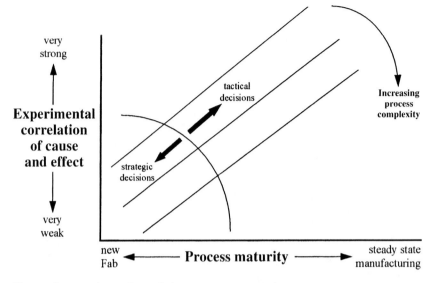

very
strong

↑

Experimental
correlation
of cause
and effect

tactical
decisions

Increasing
process
complexity

strategic
decisions

↓

very
weak

new   ←———— **Process maturity** ————→   steady state
Fab   manufacturing

Figure 1   Experimental correlation vs. process maturity.

associated with preventing or eliminating a particular contaminant vs. the costs (yield, performance, time to market, profitability) of ignoring it.

Many contamination control engineers have adopted a balanced pragmatic approach to CPR. Very simply stated: If the source of a particular contaminant is known and the contamination can be eliminated "cost effectively," then invest in prevention. This is the approach of this chapter; the objectives of which are to:

- Instill a sensitivity to the potential for virtually everything in the manufacturing environment being a source of contamination that can limit product performance, yield, and reliability
- Equip the reader with a balanced philosophy of both strategic and tactical approaches to CPR
- Provide practical recommendations for reducing sources of contamination
- Provide references for follow-up research

I.   BASIC EFFECTS OF CONTAMINATION

This section briefly summarizes product defects induced by contamination.

A. Particulate Contamination

Particulate contamination has long plagued yields within semiconductor manufacturing. Because of the isolation and couplings required between components on a chip, it's obvious that contaminants that can bridge two conduction paths, alter electrical characteristics of the conduction path, or alter the characteristics of transistors or other components will degrade device performance or reliability and ultimately impact yield and profitability. Some common effects and concerns of particulate contamination are (1) bridging of metal lines; (2) masking during ion implant, photolithographic, and etching processes, causing interlayer interactions either by shorting two layers or adding defects that grow in size from one layer to another and alter topography at the next layer; (3) reacting chemically with a film or process and altering their characteristics; (4) providing nucleation sites for unwanted deposition and growths.

Contaminant dimensions are important to consider. If two metal lines are spaced at 0.5 μm, a conductive particle $\geq$0.5 μm falling in the center of the space can introduce an unwanted conductive path; and any particle >0.5 μm, regardless of electrical properties, can mask the photo process, potentially introducing a short. However, depending on where it falls, a particle <0.5 μm may or may not cause a short between the same two metal lines. For example, it's not possible for a 0.25 μm particle to bridge a 0.5 μm space. On the other hand, a 0.5 μm particle falling in a field region, remote to the metal lines, of an even more aggressive technology (say, 0.25 μm) would likely not cause a failure unless it were to block an implant. This concept is simply pictured in Figure 2, which illustrates the concept of critical area when dealing with particulate contamination and determining critical particle size.

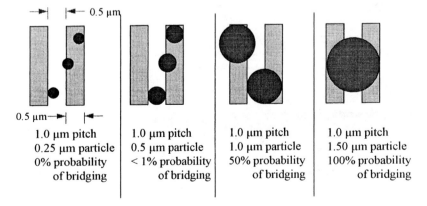

1.0 μm pitch	1.0 μm pitch	1.0 μm pitch	1.0 μm pitch
0.25 μm particle	0.5 μm particle	1.0 μm particle	1.50 μm particle
0% probability	< 1% probability	50% probability	100% probability
of bridging	of bridging	of bridging	of bridging

Figure 2 Particle size and metal bridging.

Critical area is the area of a die within which a particle must fall in order to cause a defect. For a particle which results in a short, the critical area is equal to the area of exposed resist. This area will vary dramatically from layer to layer and device to device depending on the design and stage of fabrication.

In Figure 3, only the particle at position C would result in bridging of two conductive areas. Particles at A, B, D, and E for the most part would be innocuous, not counting for potential reliability or performance effects. One can see from this figure that critical area is a function of critical dimension as well. There is only a small portion of this circuit layout where a particle of the size illustrated would actually cause a short.

Particulate contamination then must be evaluated with respect to fail probability, which is a function of particle size, location on the chip surface, and time of deposition.

Finally, with respect to particulate contamination it is important to remember size distribution. Very simply stated, ambient particle size distributions often exhibit an inverse power law distribution function such as given in Eq. (1):

$$C(x) \cong \frac{1}{x^n} \qquad (1)$$

where

C = particle concentration
x = particle size
n = an integer from 2 to 4

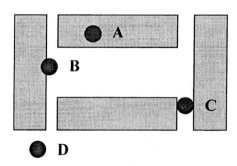

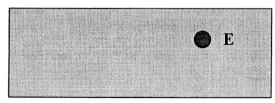

Figure 3 Particle size and critical area.

Thus, as the killer particle size continues to decrease with the decreasing critical dimensions of each new technology generation, the concentration of killer particles in the wafer ambient is likely to increase unless improved particle control measures continue to be developed and used.

B. Molecular Contamination

Molecular contamination is sometimes referred to as the invisible contamination; it can consist of no more than a single molecule. Although it's not really invisible, it is more difficult to see on the manufacturing line than particles and physical defects. The effects of molecular contamination are usually easier to detect than the molecular contamination itself. These effects are seen as voltage shifts, poor gate oxide integrity, adhesion problems, etc. The Semiconductor Equipment and Materials Institute (SEMI) has recently adopted a classification convention for molecular contaminants resident in the air within a fab. These categories are acids, bases, condensables, and dopants, defined as:

- Acid. A corrosive material reacting chemically as an electron acceptor (anion), e.g., SO_4^{2+}, Cl^-.
- Base. A corrosive material reacting chemically as an electron donor (cation), e.g., NH_4^+, Na^+.
- Condensable. A chemical substance, typically having a boiling point above room temperature at atmospheric pressure, capable of condensing on a clean surface (excluding water), e.g., IPA, methanol.
- Dopant. A chemical element that modifies the electrical properties of a semiconductive materials, e.g., As, B, P.

In this classification convention, the zero oxidation state of metallics is not considered, since metallics most often exist as cations or metal oxides (usually particles).

Molecular contamination on a surface is typically measured in terms of its areal number density—the number of contaminating atoms present per square centimeter ($atoms/cm^2$). In bulk fluids and gases molecular contamination is measured as a concentration: the number or mass of molecular contaminants per volume of host fluid or per number or mass of host fluid.

Some common effects of molecular contamination are:

- Dielectric breakdown of gate oxides due to gross defects introduced by metallic contamination
- T-topping and footing of deep ultraviolet chemically activated photoresist due to neutralization of photogenerated acids by airborne bases gases
- EPI etch pits due to oxygen contamination in the processing environment

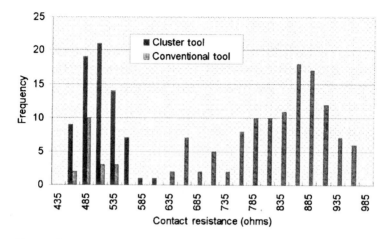

Figure 4   Polysilicon contact resistance with and without clustering. (From Ref. 4.)

- Poor oxide integrity due to moisture contamination in process gases
- Polysilicon nucleation due to metallic nucleation sites
- Uncontrolled native oxide growth upon exposure to oxygen ambient
- High contact resistance due to organic contamination
- Film adhesion problems due to organic contamination

Other processing and device impacts from airborne molecular contaminants include: shifts in the capacitance-voltage characteristics of transistors, attributable to boron contamination [1]; and changes in gate breakdown voltage, surface haze, and contact angles, attributable to hydrocarbons [2,3].

Frystak et al. [4] measured reduced contact resistance with less variability when polysilicon contacts were formed in a cluster tool rather than with ambient processing. (A cluster tool is an equipment configuration that allows multiple processing steps to be carried out in a single main low-pressure chamber with many other chambers attached and separated via a gate valve.) These results are summarized in Figure 4.

II. WHERE TO START

This chapter provides data, information, and methodology to support both tactical decisions (based on a wafer level data, strong experimental correlation, references in the open literature, and experience) and more proactive strategic decisions (based on risk management, roadmap alignment, and vision). In either case a methodical approach to problem solving is always recommended in which

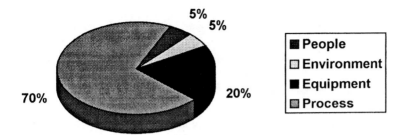

Figure 5 Sources of yield limiting contamination.

one considers information from literature reviews, basic principles, mathematical modeling, consortia, and suppliers in order to thoroughly and efficiently determine the best course of action in reducing or preventing contamination.

Figure 5 represents a consensus breakdown of the primary sources of yield-limiting (yield = functional chips/total number of chips produced per run) particulate contamination for contemporary 0.5 μm technology (minimum line spacing on a chip surface is 0.5 μm). The relative importance of the sources of contaminants changes with changing technology. For example, the 5% contribution from environmental contaminants in manufacturing a 0.5 μm technology will result in a >5% contribution in manufacturing a 0.35 μm technology, if no improvements are made in the control technology for environmental contaminants.

For effective contamination prevention and elimination, appropriate analytical technology is essential in order to ensure relevant data gathering. Some of the instruments necessary for trace impurity and surface contaminant analysis are extremely sophisticated, and contamination control engineers (and management) must realize a contamination prevention and reduction program can thrive or fail based on investment in analytical technology (Chaps. 3–5). Successful programs are those in which specifications are taken very seriously in order to meet performance, reliability, and yield metrics and are programs where management commits to the necessary analytical investment.

III. CLEANROOM CONSTRUCTION AND CLEAN BUILD

It's difficult to imagine how a manufacturing environment that has no more than a few particles >0.1 μm per ft^3 can rise from a dust bowl in the desert or a pollen-infested environment in the south or a smog-filled atmosphere in the west, and yet this is precisely what transpires when a new microelectronics manufacturing facility is constructed. This construction process results in a billionfold reduction in the concentration of aerosol particles in the manufacturing environment. The methodical approach to achieving the transition from outside ambient air contamination concentrations on the order of 10^8 particles/ft^3 to cleanroom air contamination concentrations on the order of 10^{-1} particles/ft^3 is referred to as *clean build construction*. This approach to constructing a clean-room allows for only certain types of construction activities at each successively cleaner level of protocol. Where activities such as welding, sawing, cutting, drilling, threading, tapping, concrete drilling, wire pulling, duct joining are nec-essary in clean zones of the building, clean build protocol provides methods for doing so that capture contaminants generated by these processes. Clean build construction is especially essential in the assembly and installation of the ultra-high-purity (UHP) fluid systems necessary for critical manufacturing processes. Water, chemical, and gas delivery systems require lengthy clean-up times after installation unless assembled using clean build methodology.

IV. CONTAMINATION FROM PEOPLE

People represent the greatest potential for contaminating microelectronics man-ufacturing environments and ultimately impacting yield. Figure 5 shows that contemporary technology for controlling contamination from people has suc-ceeded in making this source of contamination a small portion of yield-limiting particles in today's manufacturing environment. This section reviews contam-ination from people, how people are isolated from the product, the product environment during manufacturing, and control protocol for cleanrooms.

A. Levels of Human Contamination

In Table 1 the propensity the human body has for generating particulate con-tamination is summarized. Austin [5] provided these data for various types of activities as early as 1966 in the "Austin Contamination Control Index."

Not only do humans have a propensity for generating particulate contami-nation, but they can also shed numerous elements that have deleterious effects on device performance, reliability, and yield. Lowry et al. [6] reported the compo-sition of contamination from various human sources including numerous types of make-up (Table 2). It is now universally accepted that no make-up of any form is allowed in microelectronic manufacturing cleanrooms.

Figure 6 is a classic example of the impact of operator breath and spittle on integrated circuits [7]. The $\sim\mu$m size KCl particles shown in this 20-year-old circuit would ruin any of today's advanced circuitry, through either direct

Table 1 Levels of Human Contamination

Activity	Particles emitted/min $\geq 0.3\ \mu m$
Standing or sitting—no movement	100,000
Sitting—light head, hand, arm movement	500,000
Sitting—mod head, hand, arm movement	1,000,000
Change position, sitting to standing	2,500,000
Slow walking—2 mph	5,000,000
Slow walking—3.5 mph	7,500,000
Fast walking—5 mph	20,000,000
Calisthenics	30,000,000

Source: Ref. 5.

shorting or diffusion of mobile ionic contamination during subsequent temperature cycling.

B. Smoking

Control of contaminating emissions from smokers is always an interesting challenge for the contamination control engineer. Kozicki [8] presented data comparing the particle concentration in the exhaled breath of smokers and nonsmokers which clearly indicate the higher emissions of particulate matter from smokers. These results are shown in Figure 7. Common practice for controlling these

Table 2 Elemental Composition of Human Contamination

Contaminant	Elemental composition
Spittle	Mg, P, S, Cl, K, Ca, Na, N, O, C
Spittle (after cola drink)	Al, Mg, P, S, Cl, K, Ca, Na, N, O, C
Spittle (after eating potato chips)	Large increase in Na, Cl, and C
Sneeze	Higher Cl and Na
Perspiration	High Na, K, Cl, C, N
	Trace S, Al
Fingerprints	High C
	Also Na, K, Cl, P
Dandruff	Ca, S, Cl, C, N
Mascara	Fe, Al, C
Facial powder	Ti, Fe, Mg, Al, K
Fingernail polish	Ti, S, Mg, Al, Ca, C

Source: Ref. 6.

Figure 6 Potassium chloride spittle particles. (From Ref. 7.)

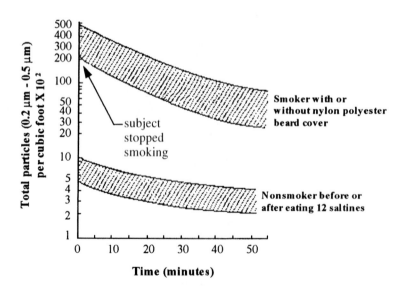

Figure 7 Particle emissions from breathing. (From Ref. 8.)

emissions includes drinking water prior to cleanroom entry, avoiding talking near the product, minimizing time between gowning and cleanroom entry, smoking cessation programs, and using bubble systems (see Sec. C below).

C. Isolation of People

Numerous means exist to isolate people and the contamination they generate from the cleanroom manufacturing environment. Even as product isolation philosophy evolves to "isolation of product from contamination" (i.e., minienvironments and the Standard Mechanical Interface [SMIF] technology; see Sec. V.D on p. 399) rather than "isolation of contamination from product," most advanced manufacturing facilities still use the established standard techniques for isolating people contamination from the manufacturing environment. These include head, beard and face covers, cleanroom garments, gloves, and shoe covers. Some facilities will even adopt an isolation strategy for the envelope immediately adjacent to the fab. In these cases, employees will often change from street clothes into dedicated building shoes and suits (much like hospital OR scrubs). These measures, although difficult to directly link to product performance, yield, or reliability, serve both as an added measure of security for the product, and, maybe more importantly, as a psychological reminder to workers that they are entering into an environment where even the smallest particles can render a circuit nonfunctional.

In addition to the additional layer of control provided by building suits and building shoes, a number of manufacturing facilities are utilizing the contamination control benefits provided by full-shield helmets (or bubble systems). These systems require a significant investment in capital as well as logistics to manage the headgear, shields, hoses, and battery packs, but they do represent a virtual impenetrable shield, isolating all operator breath from product. Dryden Engineering [9] has conducted a significant amount of this testing, as summarized in Table 3 for a nonsmoker.

D. Cleanroom Protocol

A great deal of protocol is necessary to control overall behavior in the cleanroom, as well as sources of contamination other than humans. For this purpose it is necessary to document procedures for cleaning the cleanroom, training cleanroom workers, auditing the cleanroom for compliance to protocol, and laundering cleanroom garments. Additionally, testing and specifications are necessary for virtually every support item introduced into the manufacturing environment. Section XI (p. 424) describes some of this testing and lists these materials.

The Institute of Environmental Sciences and Technology (IEST) has contributed significant standards and practices relative to contamination control and particularly cleanroom protocol. For example, IEST-RP-CC018.2 [10] provides

Table 3   Typical Oral Particle Emissions with Various Mouth Coverings*

Mouth covering	Exhaled particles (>0.5 μm)/ft^3 (measured $\frac{1}{2}$ in. from the mouth)	Coughing
No cover	~50	~1000
Foam standard mask	~600	~2000
Disposable mask	~100	~2000
Reusable mask	~100	~900
Face shield	0	0

*≥ 0.5 μm particles/ft^3, $\frac{1}{2}$ in. probe distance with probe opposite the mouth.
Source: Ref. 9.

guidance for maintaining a cleanroom and establishing housekeeping procedures and their effectiveness. Additionally, IEST-RP-CC027.1 [11] provides guidance for establishing personnel procedures and cleanroom training programs. Numerous other practices and standards have been drafted and maintained by IEST and are included in their *Handbook of Recommended Practices* [12].

V.   SOURCES OF CONTAMINATION WITHIN THE AIR ENVIRONMENT

The most fundamental concept in contamination control is likely that of establishing and maintaining a supply of clean air to the manufacturing environment. Air is often overlooked as a source of contamination, because numerous physical mechanisms act to trap particles in filter media, and yet air is the highest usage "chemical" in a cleanroom manufacturing environment.

A.   Air Flow and Cleanroom Performance

Figure 8 illustrates a typical make-up air unit. Air from the external ambient environment enters at the far left, passes through various conditioning stages and exits to the cleanroom through ULPA filters visible in a typical cleanroom ceiling.

   These filters are the last measure of defense against introduction of particulate matter in air recirculated into the manufacturing environment. This entire airflow path serves not only to filter and condition the air which enters the cleanroom, but also to provide a positive pressure differential from inside the cleanroom to the immediate envelope outside the cleanroom. This positive pressure differential is essential to maintain cleanroom cleanliness integrity. Viner [13] has published data dramatically depicting this delicate balance of positive

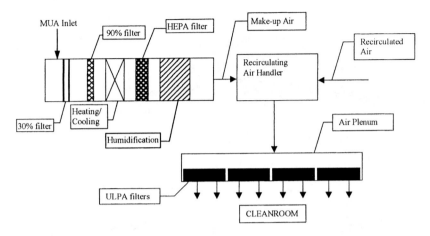

Figure 8 Typical make-up air handler configuration.

pressurization and particle control. Figure 9 shows the immediate intrusion of large particle excursions upon loss of pressurization within a cleanroom.

B. Electrostatic Discharge

Surfaces that carry a net charge can attract oppositely charged particles. Once these particles are deposited, they are very difficult to remove. Because of the number of insulating materials present in a cleanroom and the degree of human activity, the cleanroom environment offers ample opportunity for triboelectrically generated charge to accumulate. In Table 4, Tolliver [14] summarizes some activities which generate this potential.

Common methods used to reduce the buildup of electrostatic charges within a cleanroom are grounding of conductive surfaces, increasing the conductivity of materials by alternative material selections and surface treatments, humidity control, and the use of air ionization systems that can neutralize electrical charges on the surfaces of nonconductive materials.

C. Molecular Contaminants in the Outside Ambient Air

Environmental pollution and its control are established sciences that have developed a whole new language to describe the complex chemistries of the ambient air and the many species that can be transported into cleanrooms where they potentially can deposit on sensitive product surfaces. Seinfeld [15] illustrated the complexities of the atmosphere rather nicely as depicted in Figure 10. Considering just organic compounds leads to virtually an endless list of po-

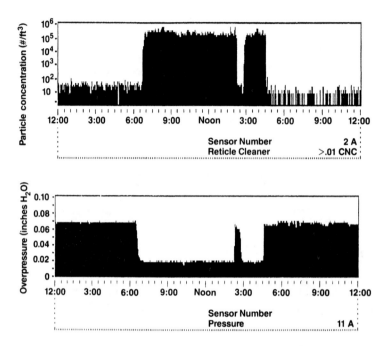

Figure 9 Cleanroom pressurization and particle performance. (From Ref. 13.)

tential contaminants that are not fully understood. Muller [16] measured the concentration of some organics found in cleanroom ambient air. Not only are outdoor organics likely to be transported into a cleanroom, but many organics are generated within the cleanroom as well. Based on these measurements, Muller [16] reported the mass of organic contaminants in typical cleanroom air to be 50,000 times greater than the mass of particulate matter. Chapter 8 of this

Table 4 Electrostatic Voltages, V, Generated by Various Activity

Means of static generation	10 to 20 percent relative humidity	65 to 90 percent relative humidity
Walking across carpet	35,000	1,500
Walking over vinyl floor	12,000	250
Worker at bench	6,000	100
Vinyl envelopes	7,000	600
Poly bag picked up	20,000	1,200
Work chair pad w/polyethylene	18,000	1,500

Source: Ref. 14.

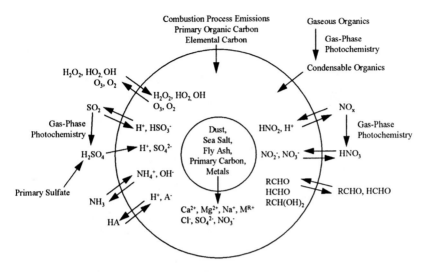

Figure 10 Typical contaminants in urban air. (From Ref. 15.)

volume reviews the science of airborne molecular contamination as it relates to cleanroom air chemistry and the deposition of contaminants onto wafer surfaces. This section focuses on product isolation technology from these contaminants.

D. Wafer Environment Control

Fosnight et al. [17] described general terminology for wafer environments (Fig. 11) and gave examples of wafer environments (Fig. 12). Very simply stated, wafer environment control considers product exposure during wafer storage, transport, and handling. This is the time that the product is exposed to the cleanroom ambient, including being loaded into process equipment. It does not include the processing environment itself.

With increased process sensitivities to numerous airborne molecular contaminants, the science of wafer environment control must extend to include time that the surface of the wafer is being chemically altered to avoid being negatively impacted by subsequent processing steps. Concepts like that of inert storage, transport, and handling as described by Kojima [18] may be necessary in future processing technologies.

There are two differing perspectives on how to maintain adequate control of surface cleanliness: global solutions and local solutions. The global solution is that in which the cleanroom is still considered as the wafer environment and product is transported throughout the cleanroom in open cassettes. In such a scenario, all materials which contact the air that contacts the product must be carefully selected so as to minimize outgassing and contamination generation. In

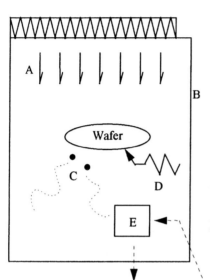

Wafer Environment
Bounded space around wafer

A - **Ambient** within space (air, N2, DI...)

B - **Boundary** of wafer environment
(cleanroom walls, storage box,
process chamber walls...)

C - **Contamination** present within wafer
environment (particles, molecular
contamination...)

D - **Devices** that physically contact
wafer (end effectors, wafer chucks...)

E - **Extraneous** objects within wafer
environment may be source of C,
may enter/leave wafer environment
(people, tables, automation...)

Figure 11 Wafer environment control definition. (From Ref. 17.)

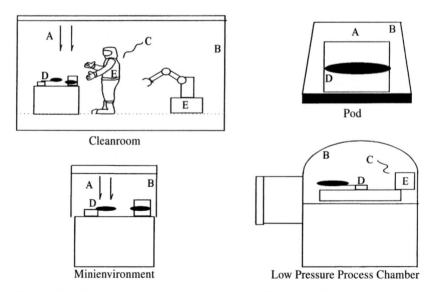

Figure 12 Wafer environment control examples. (From Ref. 17.)

the local solution scenario, the product is transported and stored in a box or pod designed to minimize outgassing and also designed to interface with handlers on each process or metrology tool. The hardware facilitating this scenario is typically referred to as a minienvironment. The minienvironment provides the clean ambient needed for the product, allowing the cleanliness classification of the surrounding cleanroom to be relaxed.

E. Airborne Particles and Yield Impact

Oftentimes particles within the cleanroom ambient are underestimated as a potential yield detractor. As a means of reducing particle excursions from normal events within cleanrooms, minienvironments have been shown to provide payback in terms of either yield benefit and/or reduced capital outlay. Jensen and Smith [19] described a process flow model which estimated die loss due to airborne particles. By estimating particle deposition velocity (Chap. 6) and measuring cleanroom particle concentrations, they calculated a particle deposition rate from Eq. (2). They estimated a ~4% yield loss due to aerosol particles in this one fab.

$$\text{Deposition velocity} = \frac{\text{deposition rate}}{\text{particle concentration}} \qquad (2)$$

Through this modeling, they were able to prioritize capital investment for wafer environmental control (i.e., minienvironments) based on the yield impact of ambient exposure at each process tool. Their pareto summary is shown in Figure 13. The abscissa, read from left to right, lists the processing stages in descending order of impact on yield.

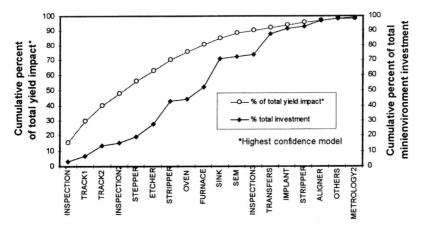

Figure 13 Cumulative yield impact and minienvironment investment. (From Ref. 19.)

VI. CONTAMINATION FROM CHEMICALS

The 1997 revision of the National Technology Roadmap for Semiconductors (NTRS) [20] relaxed impurity levels in process critical liquid chemicals previously targeted in the 1994 revision of the NTRS [21] (Table 5). The new targets are based on a better understanding of what improvements are believed to be necessary in achieving and maintaining acceptable concentrations of surface contaminants. The 1997 NTRS [20] targets for surface concentrations are expected to be tightened by ~5× over the next 10 to 15 years rather than ~10× as targeted in the 1994 NTRS [21].

Regardless of the uncertainties about future purity requirements, the importance of establishing and maintaining clean liquid chemicals and clean transport systems for those chemicals remains paramount to contamination free manufacturing. This section reviews the benefits of bulk distribution, the concept of total chemical management, filtration basics, materials, packaging, and analytical needs.

A. Benefits of Bulk Chemical Distribution

Many benefits arise from transporting liquid chemicals throughout the fab by means of bulk distribution systems. A bulk distribution system is one in which chemicals are disbursed from a central storage area through pipes to the point of use (POU) rather than transported in individual containers to each POU. Significant safety benefits arise from closed systems, which limit operator exposure to the actual chemicals. Bulk chemical distribution systems also provide containment and safety devices to monitor for leaks. Reduced cost of ownership is typically realized as well in bulk delivery of liquid chemicals. From a contamination control perspective, these systems provide for multiple-pass filtration prior to delivery to the process tool, and the systems are typically isolated from ambient atmospheres, thus reducing the introduction of chemicals into the ambient atmosphere. Some of the main benefits of a clean build construction protocol are in the area of reduced metallics and particles contributed by bulk distribution piping. Installation of bulk chemical distribution systems and piping in a controlled environment can reduce particle and metallic levels and significantly reduce the time required to achieve required purity levels when these systems are commissioned and started up. Bulk distribution systems are designed with low dead volumes.

B. Contamination Control in Bulk Chemical Distribution Systems

Bulk chemical distribution systems typically incorporate low dead volumes, the use of diaphragm valve technology that isolates all metallics from the flow-

Table 5 1994 and 1997 NTRS Chemical Purity Targets

Technology generation (half-pitch in nm)	250 nm	180 nm	130 nm	100 nm
1994 roadmap—individual metallic impurities in liquids (in ppt)	<50	<10	<10	TBD
1994 roadmap—particles >0.1 μm (/mL)	<10	<10	<1	<1
1997 roadmap—Fe and Cu impurities in liquid chemicals (in ppt)	<500	<250	<150	<100
1997 roadmap—other metallic impurities in liquid chemicals (in ppt)	<1000	<500	<300	<200
1997 roadmap—particles (#/mL > size in μm)	<0.5 @ 0.125	<0.5 @ 0.09	<0.5 @ 0.065	<0.5 @ 0.05

Source: Ref. 21.

stream, and flared connections to help maintain chemical purity. Additionally, surge suppression is necessary to eliminate pressure pulses which can dislodge particulate matter trapped in filters or on other surfaces. The goal of incorporating all these design features into a bulk chemical distribution system is to maintain both chemical purity and low-particulate concentrations during the transport of chemicals from the manufacturer to the end user, through the distribution system and finally to the actual point of use (the wafer surface). Often these are conflicting requirements, as filters (see Sec. C below) that remove particles and packaging materials (see Sec. D) that do not react with the chemical being transported can add metallic impurities. For example, Rosenfeld [22] presented typical levels of metallics contributed by bulk chemical distribution systems. These data (Table 6) emphasize the need to monitor and control the introduction of metallic contaminants by a distribution network.

C. Particle Filtration of Liquid Chemicals

Particle filtration in chemical delivery systems in essential. The primary capture mechanism in liquids is often sieving. It's important to characterize filter performance over a range of particle sizes and flow conditions due to the interaction of these mechanisms for retention and the potential also for particle shedding.

Metallic impurities are an important consideration when selecting both filter housings and filter medias. Hurd [23] measured various levels of metallic extractables from various filter types and manufacturers. These results, shown in Figure 14, indicate a vast range of impurity levels among common types of filters.

D. Chemical Packaging and Distribution Materials

The materials that contact ultraclean chemicals have the greatest potential for degrading the purity level of the chemicals. Goodman [24] collected data comparing various packaging material types and the level of extractables in nitric

Table 6 Metallic Impurities Contributed by Bulk Distribution Systems

Chemical	Elements	Total metallics (ppb)
HF	Al, Ni, Zn, Ca, Mg	$\sim$15
H_2SO_4	Al, Mg, Ni, Zn	$\sim$10

Source: Ref. 22.

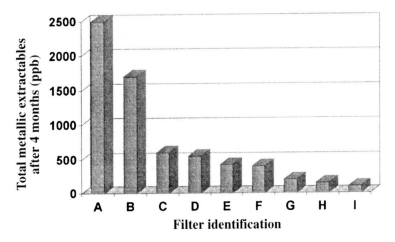

Figure 14   Metallic extractables from various filters. (From Ref. 23.)

acid. Shown in Table 7, these data indicate PFA as a preferred material for most
chemical containers.

Polymers, however, often contain leachable metallics originating with the
plasticizers, antioxidants, and stabilizers used in their manufacture. Careful selection of the polymers used in chemical transport is necessary. As such, PTFE
and PFA drum liners are now available, which can be manufactured clean and
packaged within a cleanroom environment [25]. These liners can be used in conjunction with stainless steel or HDPE drums that meet transportation regulations
for containment.

Table 7   Metallic Extractables (ppb) from Various Chemical Containers

Element	Polyethylene bottle	PFA beaker	Glass bottle	PVDF container	PTFE container
Al	5	—	—	<2	<2
Ca	7	≤0.2	<10	<10	<10
Cr	<1	≤0.1	—	—	—
Fe	5	≤6.7	—	<6	<6
K	2	≤0.2	<10	<10	<10
Mg	2	≤0.04	—	—	—
Na	3	≤0.4	<10	<10	<10
Ti	<1	≤0.006	—	—	—
Zn	<1	≤0.1	—	<2	<2

Source: Ref. 24.

Other considerations to help minimize contamination within bulk distributed chemicals include fill station design for minimizing entrainment of environmental contaminants, surface finish control, chemical compatibility, and cleaning techniques to leach out extractables [26].

E. Chemical Purity Importance

The importance of chemical purity insofar as not contributing metallic contamination into the front end of the process (i.e., the transistor) cannot be overemphasized.

Figure 15 expresses acceptable metallic impurity levels in terms of the ratio of silicon atoms to metallic atoms for successive generations of CMOS processing. Assumptions are made for equal distribution of metallics across the entire substrate, and the levels of metallic contaminants per the 1994 NTRS [21]. This presentation of NTRS data indicates that 0.18 μm technology will be able to tolerate only about 1 metallic impurity atom for every 10 million Si atoms.

F. Point of Use Generation of Process Chemicals

Some critical chemicals can be generated at their point of use by combining ultrapure DI water and cylinder gas. Ultrapure DI water is typically much cleaner as generated at semiconductor fabrication facilities than at chemical manufacturers. As such, potential exists for cleaner chemicals to be delivered more cost effectively by POU generation than by packaging at a remote plant, transporting, and

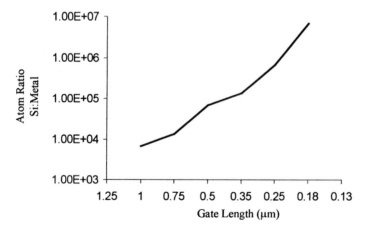

Figure 15   Silicon to metallic atom ratio per technology node. (From Ref. 21.)

then distributing within the facility. Wear [27] measured similar concentrations of metallics in bulk ammonium hydroxide and ammonium hydroxide generated at point of use. He also found that bulk hydrogen chloride had elevated levels of iron, chromium and nickel, likely from stainless steel components within the gas distribution network.

While this overview does not address waste management issues associated with chemical delivery, bulk distribution systems and POU generation are often deployed out of the necessity to reduce waste streams from semiconductor manufacturing facilities.

Finally, worthy of repeating here is the lack of thorough understanding regarding the impact of each metallic and particulate impurity on device performance, reliability, or yield. These relationships are generally understood based on phenomenological data but not to the point of having the ability to foresee exact specifications for future technology nodes. The 1997 NTRS [20] highlights the need to research and establish correlations between impurity concentrations and device electrical performance.

VII. CONTAMINATION FROM ULTRAPURE DEIONIZED WATER

Water is the cleanest liquid used in semiconductor manufacturing. By most metrics, filtration and separation technologies within semiconductor fabrication facilities produce the purest water found in any industry. Although not necessarily as sterile as water used in the pharmaceutical industry, semiconductor ultrapure deionized water (UPW) has far tighter specifications for ionic and particulate impurities. This section reviews specifications, design materials and filtration, and microbiological considerations.

A. UPW Specifications

Ultrapure water systems are part of all new manufacturing facilities. While most gases and chemicals arrive as packaged materials, UPW is manufactured on site to exacting purity specifications. These specifications cover a broad range of properties and measurements of quality including resistivity, total organic (or oxidizable) carbon (TOC), dissolved gases, anions, cations, metallics, pyrogens, bacteria (both viable and nonviable), particles, turbidity, silica, and nonvolatile residue. The highest purity water is required for final rinses prior to diffusion processes where metallic impurities are more likely to be incorporated into the bulk material and thus alter the electrical characteristics of a device. Balazs Analytical Labs [28] has developed a summary of specifications, troubleshooting guidelines, and water quality guidelines for various technology nodes. This guideline is based on water system contamination control research and field testing at numerous semiconductor manufacturing sites over the last ~20 years.

B. UPW Distribution Systems

Distribution systems are designed for continuous flow with zero dead-legs to prevent stagnation and microbiological buildup. Process equipment is often over-looked as part of the water distribution loop. However, most equipment suppliers do not design water loops within their equipment to the exacting standards of the UPW distribution systems found in a semiconductor UPW plant. In addition to the contaminants in the municipal water delivered to the semiconductor facility, contaminants may originate within the UPW itself as summarized in Table 8.

C. Microbiological Contamination

All high-purity water systems have some level of microbial contamination present. Many strains of both viable and nonviable bacteria exist in systems so that frequent sanitization is necessary to control their growth. Methods of sanitization include peroxide, ozonation, UV irradiation, and hot water rinses. It's important to keep bacteriological growth to a minimum, since microbial contamination can contribute mobile ions in addition to acting simply as physical debris. Yabe [29] provided an interesting summary of the elemental composition of bacteria (Table 9). These numbers of impurities, say for P or S, if concentrated in a ~2 μm^2 area could significantly alter surface doping levels.

Table 8 Basic Water System Contaminants and Their Sources

Contaminant	Source
Bacteria	Low velocity in pipes and tubing Stagnant water
Total organic carbon	Leaching from pipes Improper ion exchange maintenance High bacteria levels
Silica	Exhaustion of mixed bed resins
Metallics	Exhaustion of resins Degradation of metal components

Table 9 Elemental Composition of a Bacterium of Mass 1.6×10^{-13} g

Element	Content (%)	Weight (g)	Atoms
C	50	7.9×10^{-14}	3.9×10^9
O	20	3.1×10^{-14}	1.2×10^9
N	14	2.2×10^{-14}	9.5×10^8
H	8	1.3×10^{-14}	7.6×10^9
P	3	4.7×10^{-15}	9.2×10^7
S	1	1.6×10^{-15}	3.0×10^7
K	1	1.6×10^{-15}	2.4×10^7
Na	1	1.6×10^{-15}	4.1×10^7
Ca	0.5	7.9×10^{-16}	1.2×10^7
Mg	0.5	7.9×10^{-16}	2.0×10^7
Cl	0.5	7.9×10^{-16}	1.3×10^7
Fe	0.2	3.1×10^{-16}	3.4×10^7
Others	~0.3	4.7×10^{-16}	—

Source: Ref. 29.

VIII. CONTAMINATION IN GAS DISTRIBUTION SYSTEMS

Gas purity and gas system cleanliness have greatly improved in recent years. Impurities within many gases are now typically measured in parts per trillion and it is generally believed that step function improvements in gas distribution system cleanliness will not be necessary for many technology nodes to come. This section reviews the technology needed to achieve such low concentrations of contamination in gas distribution systems. Topics include system design, dead volume and purging, surface morphology, surface chemistry, electropolishing, and gas system assembly.

A. Effects of Contamination in Gases

Dillenbeck [30] reviewed the process effects of various gas contaminants (Table 10). Shapiro et al. [31] also provided a review of process problems specific to moisture. These are tabulated in Table 11.

B. Contamination Sources Overview

Virtually every aspect of gas system design, manufacturing, test, operation, and maintenance can be a source of contamination. Components that are not con-

Table 10 Process Problems Due to Gas Contamination

Gas	Contaminant	Process problem
N_2	H_2O, O_2	Deterioration of oxide integrity Nucleation with polysilicon
NH_3	H_2O, O_2	Formation of SiO_2 in nitride film
Ar	H_2O, O_2, etc.	Annealing (oxide integrity problem) Sputtering (hillocks)
Cl_2, BCl_3	H_2O	Corrosion in gas lines Metallic contaminants
N_2, AR	O_2	Silicide electrical property problems
HCl, N_2, SiH_4	Fe, Cu, etc.	Reduces minority carrier lifetime Polysilicon nucleation Formation of silicon defects

Source: Ref. 30.

structed properly can entrain particles, outgas, and leak atmospheric contaminants. The volume of gas distribution system should be as small as possible to minimize dead space that might act as virtual leaks and be difficult to purge of residual gas prior to maintenance and of atmospheric contaminants after maintenance. Assembly and installation of a gas system should be done in a clean-

Table 11 Process Problems Due to Moisture Contamination

Process	Moisture effect
PVD (incorporated in film)	Increase electrical resistance Suppress crystal orientation and grain growth
CVD	
Si epi	Interfacial oxygen
Si poly	Incorporated in film
Plasma etch and poly	Corrosion Etch rate changes Corrosion w/ residual HCl post Al etch $P + H_2O \rightarrow H_3PO_4$ Al corrosion
SF_6	Higher particle levels
Pumpdown	Particle formation

Source: Ref. 31.

room environment, using clean build assembly protocol. Otherwise the system will have high particle concentrations as well as long dry-down times due to entrained atmospheric contaminants. Similar protocol applies when a gas system must be opened for maintenance of a particular component. Oftentimes residual process gas will react with atmospheric contaminants resulting in particle formation; or residual atmospheric contamination not fully purged out after maintenance will react with process gas once it's introduced. Even the manner in which gases are sequenced can lead to contamination. If gases are not properly separated until they enter a CVD reactor, they can react with one another in the gas manifold, generating particles. Soft-starting, venting, and purging are also important so as not to reentrain particles previously deposited on surfaces within the system.

C. Gas System Design

The primary materials of construction for a high-purity gas distribution system are 316L electropolished stainless steel and low outgassing polymers for nonmetallic seals. These materials minimize particle generation and resist corrosion from acid gases. Additionally, components such as valves, regulators, mass flow controllers, etc., must use the same basic materials and be designed to minimize particle generation by the moving parts within the flowstream. One of the most important design criteria is that of low dead volume. Dead volumes can act as traps for gaseous contaminants and ultimately impact dry-down times and gas purity at point of use. Modeling is becoming more and more important in understanding the impact of dead volumes on ultimate gas system performance and dry-down times. Siefering et al. [32] described how these models can often help evaluate trade-offs in cost and ultimate performance. Coronell et al. [33] reported on correlations of this type of modeling with actual component testing. The importance of dead volume and its relationship to dry-down times and various purging conditions was investigated by Kubus and Legget [34] with dramatic differences in dry-down times, as shown in Figure 16.

Test methods for measuring contamination generation by gas-handling components have been prepared by SEMATECH, SEMI, and ASTM. Table 12 references these methods and also the recommended component specifications from the SEMI Physical Interfaces Committee [35].

D. Surface Morphology

Surface morphology is an important consideration in controlling contamination within gas distribution systems. The sawtooth surface in Figure 17 is likely to generate more particles than the sinusoidal surface even though it presents less

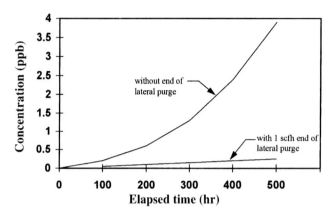

Figure 16 Predicted lateral contamination with and without purge. (From Ref. 34.)

surface area to a gas flow. However, lower surface area is important in mini-mizing attachment sites for moisture and oxygen. Electropolishing technology (Sec. E) minimizes both surface area and protrusions.

SEMATECH and SEMI [35] have developed guidelines and test methods for surface morphology, as summarized in Table 13.

Table 12 Recommended UHP Component Contamination Specifications

Parameter	Value	Ref.
Internal adsorbed moisture (hours to recover to baseline from 2 ppm spike)		
Low surf area comp	1	ASTM F1397
High surface area comp	4	
Total ionic contamination added to test water (ppm)		
Individual	≤ 0.2	ASTM D4327
Total	≤ 1	
Leak rate (He atm cm^3/s)		
Inboard	$\leq 10^{-9}$	SEMI F1
Outboard	$\leq 10^{-5}$	
Cross-seat	$\leq 4 \times 10^{-8}$	
Cycle life		
Manual valves	$\geq 25K$	ASTM F1373
Hi press pneu valves	$\geq 25K$	
Lo press pneu valves	$\geq 500K$	

Source: Ref. 35.

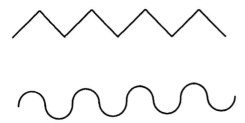

Figure 17 Surface finishes.

E. Surface Chemistry

There are three fundamental characteristics for rating a surface passivated for use in transporting high purity gases: 1) the degree of chromium enrichment at the surface, 2) the chrome to iron ratio, and 3) the absence of foreign elements. Chromium enrichment is vital for corrosion resistance in acid gas service. In fact the acid gas purity levels targeted in the 1997 NTRS are driven as much by the need to extend the service life of their distribution systems as they are for improvements in wafer processing. For example Wang et al. [36] reported an increase in the time to the onset of detectable corrosion-generated particles from 170 days to 23.3 years (extrapolated) as the moisture content of HCl decreased from 5 to 0.1 ppm. Once again, guidelines [35] for these surface characteristics and their measure have been established and are tabulated in Table 14.

The process by which a chrome-rich surface is established in stainless steel components is electropolishing. In electropolishing the part to be polished is made the anode—it is the reverse of electroplating. Material removal is greatest at surface projections where current density is the greatest, thus

Table 13 Recommended UHP Component Surface Roughness Specifications

Parameter	Value	Technique	Ref.
Surface defects (max count per photo—5 photos)	≤40	SEM	ASTM F1372
Average surface roughness ($R_a - \mu$in.)	≤7	Contact Profilometer	SEMASPEC 90120400B
Maximum surface roughness ($R_a - \mu$in.)	≤10	Contact Profilometer	SEMASPEC 90120400B
Particle contribution/ft^3			
≥0.1 μm	≤10	OPC	ASTM F1394
≥0.02 μm	≤50		

Source: Ref. 35.

Table 14 Recommended UHP Component Surface Chemistry Specifications

Parameter	Value	Technique	Ref.
Chromium oxide thickness (Å)	≥15	AUGER	SEMASPEC 90120573B
Chrome-to-iron ratio	≥1.25:1	ESCA	SEMASPEC 90120403B
Surface impurities	0	EDX	ASTM F1375

Source: Ref. 35.

providing a surface-smoothing action while increasing the chromium/iron ratio. Oxygen is liberated at the anode and reacts with the chromium in the stainless steel to form chromium oxide. This process forms a uniform passive layer. Cleanup after polishing is important to remove residual electrolyte surface contaminants.

F. Gas System Assembly

The assembly of high-purity gas distribution systems requires a total systems approach including manufacturing of components, packaging, transport, and start-up procedures. This total systems approach must include a well-designed cleanroom facility for assembly, testing, and packaging; high-purity distribution systems for the fluids (gases and liquids) that will be used to test the assembly; procedures for passing materials into the facility and into the cleanroom; tested and validated methodologies for tube cutting, cleaning, and welding; procedures for testing, packaging, and labeling procedures; etc. The bibliography at the end of the chapter describes these requirements in detail. SEMI and SEMATECH have developed guidelines [35] for the ultimate measure of gas system performance once suitable procedures and methodologies have been established and followed. Table 15 lists these performance expectations.

While these specs are essential for maintaining long-term system integrity in corrosive, toxic, and pyrophoric gas service, all distribution networks and individual assemblies do not necessarily have to exhibit these performance characteristics. For example, a utility nitrogen distribution system would certainly not need to meet such aggressive requirements. Nor would an inert cylinder gas system used to transport ~ppm quality etch gas need to achieve these levels. However, at commissioning, these performance characteristics are a very good indication of the quality of workmanship used to build the distribution network. Even in the example of the inert specialty gas system, one would not want to assemble a system that continuously outgassed ~ppm levels of moisture.

Table 15 Recommended UHP System
Performance Specifications

Parameter	Value
Moisture level (ppb)	≤ 20
Oxygen level (ppb)	≤ 10
Total hydrocarbons (ppb)	≤ 20
Inboard leak rate (He atm cm^3/s)	$\leq 10^{-10}$
Particle contribution/ft^3 ≥ 0.02 μm	
Avg	≤ 5
Max	≤ 50

Source: Ref. 35.

G. Design of Gas Distribution Systems

Maintaining low levels of moisture and oxygen in gas systems and providing a means of purging and recovering from upset events are important properties of a gas distribution system. Figure 18 shows a design by Cheung et al. [37] that compares these features with that of a more conventional design consisting of a main and laterals. In the conventional design, dry-down times are long and ultimate baseline moisture levels are significantly higher due to the dead space created by the use of two-way valves on the risers and the inherent outgassing characteristic of a dead volume lateral. Cheung employed flow-through gas-valve designs on both the main and laterals, as well as a continuous purge flow design

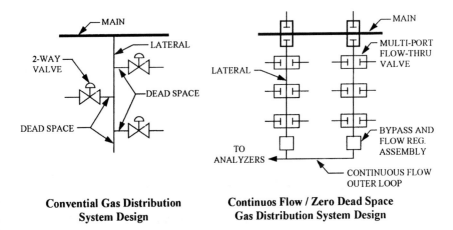

Convential Gas Distribution
System Design

Continuos Flow / Zero Dead Space
Gas Distribution System Design

Figure 18 Conventional vs. continuous flow gas distribution systems. (From Ref. 37.)

to achieve short dry-down times and low levels of gaseous impurities. Table 16 documents this systems' performance at a microprocessor fab following start-up and commissioning.

H. Gas Purification

Current practice would not be complete without mention of purification technology. Many active gas purifier technologies are now available, and can be deployed to getter or trap gaseous impurities. Bhadha and Cowan's review [38] of these technologies is summarized below:

- Organometallic resins
  Active alkali compounds on polymeric substrates
  Room temperature
  Undergo color change
- Getters
  Hot titanium sponge
  Zirconium and vanadium oxide
  Trace nitrogen/hydrogen removal at low flow rates
  Regenerable
- Palladium diffusers
  Palladium/silver alloy at 450°C
  Hydrogen purification by adsorption and diffusion
  High pressure drop
- Desiccants
  Silica gel and calcium phosphate treated with cobalt chloride (drierite)
  Silica, no indication
  Drierite, blue to pink, but at high levels of moisture
- Molecular Sieves
  Zeolite class of compounds
  Si–O and Al–O with large cavities
  Selectivity by addition of Na, K, or Ca
  Dried prior to use
  Moisture out increases as sieve is depleted
  No indication of exhaustion

IX. CONTAMINATION IN PROCESS EQUIPMENT

This section reviews practices for controlling contamination with process equipment, emphasizing certain specific tool categories (e.g., CVD, PVD, etch).

Table 16 Performance of Continuous Flow/Zero Dead Space Gas
Distribution System

Status	Date	H_2O (ppt$_v$)	O_2 (ppt$_v$)	CH_4 (ppt$_v$)	CO (ppt$_v$)	CO_2 (ppt$_v$)
Specification	07-Apr-94	<1000	<1000	<1000	<1000	<1000
Commissioning	13-Apr-94	1248	468	241	<1000	106
Sustaining	22-Aug-94	710	74	6	—	37

Source: Ref. 37.

SEMATECH has provided a document [39] outlining guidelines, methodologies, and test procedures for building and maintaining clean processing equipment. SEMATECH has also sponsored the development of courses of instruction based on these guidelines, now publicly available through various consultants. This section loosely follows this SEMATECH development, supplemented by references to the open literature.

A. Equipment Surfaces

The primary requirement of surfaces within processing equipment, beyond being nonshedding, nonparticulating, and nonoutgassing, is the need to be cleanable. Virtually all fabs have some type of standard cleanroom housekeeping protocol that dictates a wipe down of surfaces on a periodic basis. As such, equipment surfaces exposed to the cleanroom should be cleanable; these surfaces should be smooth so as not to entrap particulate matter and nonabsorbing so as not to soak up (and continuously outgas) moisture/isopropanol mixtures used for cleaning. Preferred and nonideal surfaces are listed in Table 17.

These preferred surfaces are basically resistant to chipping or peeling, do not outgas, and do not degrade significantly under normal wear.

Table 17 Preferred and Nonideal Equipment Surfaces

Preferred	Nonideal
Polished stainless steel	Decorative
Cr–Ni plated steel	Composites
Polished/treated aluminum	Galvanized
Urethanes/epoxies	Corrosion susceptible
Low outgassing	Highly volatile
Static dissipative	"Uncleanable"

B. Equipment Subsystems

Process tools consist of many subsystems assembled together. Contamination control within those subsystems is important in maintaining the cleanliness integrity of the cleanroom in which the tool is placed, especially when these subsystems come into direct contact with the wafers (e.g., wafer-handling systems) or the environments (loadlocks, minienvironments, etc.) in which the wafers will be processed or pass through.

Maintaining laminar flow within process tools can help sweep particles through the equipment and prevent stagnation points that can both generate and trap particles. This is difficult in many tool designs, but can be aided by open, "flow-through" designs that take advantage of the vertical laminar downflow characteristic of cleanrooms. Tools may often be designed without external skins to assist in this. For critical applications (e.g., minienvironments) computational airflow modeling helps to optimize the configuration of surfaces for minimizing turbulence and dead zones within the environment where the wafers will be exposed.

Heat sources (e.g., power supplies) within some of these environments can cause convective currents ($\sim$10 ft/min) that act to hold particles in the airstream longer and potentially increase product contamination. These heated surfaces will also outgas more readily and can potentially increase both particulate and molecular contamination of product surfaces. On the other hand, thermophoresis (Chap. 6) can reduce particle deposition on heated product surfaces.

Any object that moves in contact with another will generate particles. Tribology (the science of frictional abrasion and wear) is complex. Particle generation by contacting surfaces depends on the type of surrounding medium, the contact area, the bonding between the contracting materials, and the shearing forces of the surfaces in motion [40]. These mechanisms must be characterized and understood with respect to the wafer environment when designing systems that interact with the wafers; examples here include wafer-handling systems and also valve actuations within gas and chemical delivery systems.

Flexing and vibration are also important considerations in designing process tools and their subsystems. Any coated material that flexes will in time generate particulate matter and flexing often originates from vibrations within the equipment (e.g., pumps, motors). Often flexible lines (bellows material or Teflon hoses) are used to achieve vibration tolerance in the plumbing of gas or chemical distribution systems. However, build-up of material on the inner walls of these flex lines can spall as a result of the line flexing. This is problematic in vacuum plumbing where bellows are frequently used to dampen vibration from vacuum pumps.

In all these cases noted above, a rigorous materials and subsystem characterization should be accomplished to optimize designs for minimum outgassing and particulate generation. In one such example, Verma et al. [41] identified seven sources of gaseous contamination in vertical thermal reactors: 1) permeation through Teflon, 2) convection, 3) entrainment of impurity in dead volume between wafer spacing, 4) adsorption and desorption of impurities on wafer surfaces, 5) diffusion of room air impurities through quartz, 6) back diffusion, and 7) impurity leakage through elastomeric seals.

C. Vacuum Systems

Significant work has been done on particle formation, transport, and deposition in vacuum systems (see the Bibliography at the end of this chapter). Vacuum system contamination control is extremely important as vacuum is one of the most common wafer processing environments in today's semiconductor manufacturing. There are numerous forces that act on particulate matter within vacuum processing equipment, all of which can affect particle transport and deposition on wafer surfaces. Electrophoretic, fluid drag, ion drag, thermophoretic and gravitational forces all impact the manner in which particles move about in vacuum. O'Hanlon et al. [42] concluded that particles can be trapped electrostatically in plasma-processing equipment, the magnitude of the trapping varying with fluid drag, thermophoretic and gravitational forces. Unfortunately, designs that minimize particulate contamination in a vacuum process chamber are not always compatible with configurations and techniques for optimum process conditions. Researchers at Sandia (see Chap. 6) have collaborated extensively with process equipment suppliers and SEMATECH to develop guidelines that aid in minimizing particulate contamination on wafers during vacuum processing [43].

- Raise the operating pressure to increase the cutoff size of particles that deposit by inertial impaction.
- Increase the spacing between a showerhead entry port and the wafer so that fluid drag becomes more effective in transporting smaller particles out of the reactor.
- Maximize the temperature gradient adjacent to the wafer surface to thermophoretically repel particles from the wafer surface.
- Optimize the design of the gas entry ports by:
  1. Maximizing their cross-sectional area
  2. Minimizing the length of the gas accelerating path
  3. Ramping up gas flows
  4. Directing gas flows away from the front of the wafer

Rader et al. [44] provides an excellent review and summary of these effects in plasma-processing environments (see Chap. 6).

Optimal reactor design for contamination control must also consider contaminating contributions that originate with the choice of materials used in the construction of the vacuum systems. Of particular importance is the propensity of elastomeric seals to allow permeation of gaseous contaminants. Ma et al. [45] showed order of magnitude differences in the permeation coefficients of commonly used materials (PCTFE, PE, PFA, etc.) from various suppliers. They concluded, although dependent on application, that perfluoroelastomers were generally better in vacuum service than fluoroelastomers.

Additional techniques for minimum contamination in vacuum processing include controlling turbulence (slow pumping and venting), leak tight integrity and leak testing, gate valve control, and dry pumps.

D. Equipment Assembly, Cleaning, and Packaging

Process equipment intended for operation in a cleanroom environment must be manufactured and assembled under conditions that assure that the equipment will not degrade the cleanliness of that cleanroom environment. As such, all processing equipment and tools should be assembled under cleanroom conditions. When equipment is ready to be shipped, it must then be cleaned and typically triple-wrapped to provide maximum protection of the surfaces that will eventually be exposed to the manufacturing cleanroom environment. Upon arriving at the manufacturing site, the triple wrap is removed in successive stages as the equipment moves through successively cleaner loadlocks from the ambient environment to the final cleanroom environment.

E. Equipment Monitoring

As shown in Figure 5, the actual processes themselves are believed to contribute the majority of yield-limiting particulate contamination in semiconductor manufacturing. This yield-limiting contamination can often be attributed to poor maintenance procedures as films and etch by-products build up on various surfaces within the processing chamber and subsequently spall off onto the wafer surfaces. Consequently, regular preventive maintenance and chamber cleaning are critical to maintaining clean processes. Control charts must be established based on a measurable parameter indicative of this build-up. Often monitor wafer particle checks are used for this purpose. Conducted on a routine basis of every shift, or once a day, or every other lot, etc., particles per wafer per pass (PWP) checks can often isolate these "out-of-control" events from background "in-control" particle levels. In a PWP measurement, a "prechecked" particle monitor wafer with particle count N_i is run through a tool and subsequently remeasured for particle count, N_f. The total adders, or PWP, count is then calculated as simply $N_f - N_i$ and recorded on a control chart. Figure 19 graphically depicts this process.

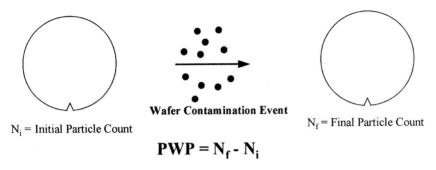

$$PWP = N_f - N_i$$

Figure 19 Particles per wafer per pass monitoring.

PWP checks can be made more valuable if conducted in deposition tools which have flowing gases and RF on to initiate an actual gas-phase reaction and film deposition. In these cases, particle generation associated with the actual process chemistry is also captured by the measurement. This is an important distinction because simple mechanical PWP checks typically result in very low particle counts and do not have statistical significance without extensive testing and large numbers (100's) of passes.

PWP testing can be utilized to isolate, or partition, a particle event to a specific chamber or condition within the process tool. In order for this testing to be effective, process tools must be baselined under known conditions, like new, or recently cleaned, or after chamber conditioning, or after/before preventive maintenance; and experiments need to account for typical sources of variation, like instrument noise and particle adders from wafer handling. These tests can include not only particles but also baseline levels of volatile species through the use of gas chromatography, mass spectrometry, or optical emission/infrared spectroscopy.

Sources of experimental error can be reduced by acquiring measurements as close to the wafer surface as possible. In situ particle monitors (ISPM) have been utilized to provide this capability (Chap. 3). Bhat et al. [46] used such a device in a nitride etch chamber. Their results are shown in Figure 20 where on-wafer counts correlated well with ISPM counts.

End users have also used ISPMs to schedule preventive maintenance by detecting the onset of chamber wall flaking before on-wafer particle counts rise out of control. In some applications ISPM counts are difficult to correlate to on-wafer counts because ISPMs count particles in just a small volume of a processing chamber. This volume may not provide a representative count of process particles. Knowing where to locate an ISPM in order to accurately measure process particle densities remains a challenge, as discussed in Chapter 3.

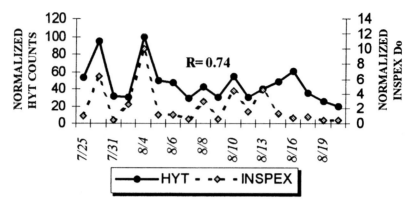

Figure 20   In situ particles (HYT) vs. on wafer particles (Inspex). (From Ref. 46.)

F.   Equipment Contamination Prevention and Reduction

All these techniques and practices (baselining, PWP, ISPM, experimental technique, etc.) can be combined to develop effective programs in contamination prevention and reduction (CPR). Uritsky et al. [47] provided process improvement through methodical characterization of particles in CVD process chambers. Their results are summarized in Table 18, where they were able to characterize both the shape and chemistry of typical particles.

Table 18   Chemistry, Shape, and Possible Root Cause of Particles

Chemistry	Shape	Possible root cause
W	Donut-shaped residue	Prior film defect
C	Minute, low contrast spheres	Oil (pump), dirty gas, bad cassette, wafer handling
O, W	Clusters or individual spheres	Contaminated gas line
Ca, Si, C, Mg	Large, irregularly shaped	Viton O-ring
Cr, Fe, Ni	Minute needles	Exposed stainless steel attacked during plasma clean
Ni, P	Tiny, oval forms	Degraded slit valve
C, O, Na, Si, S, Cl, K, Ca, Mg, Al, Fe	Miscellaneous	Variable

Source: Ref. 47.

Bilotta and Proctor [48] demonstrated good CPR technique when they reduced particle levels and defects in a series of experiments in particle sourcing and equipment modifications by doing the following (in part):

- O-ring preventive maintenance
- Slow pump/vent (flow surge suppressor)
- Pressure changes during processing
- Vacuum conductance changes

Their efforts reduced mean counts by 5× and tightened distribution by approximately 5×.

Another good example of CPR techniques is by Anderson et al. [49], who isolated a showerhead-cooling problem, introduced gas timing changes to eliminate gas-phase nucleation, eliminated pumping capacity constraints due to build-up of by products in the effluent burn box, and used H_2O_2/H_2O solutions for chamber cleans, an improvement over conventional DI/IPA mixtures.

X. WAFER CONTAMINATION

Wafer handling is common to all processing. As such, control of wafer-handling contamination is important. Wafer handling must minimize contact points; nonmetallic contact with the wafer is preferred to minimize metallic contamination—tweezers must not be utilized except in the case of removing stuck or dropped wafers from process tools, and even the use of vacuum wands should be minimized. Metallic wafer handling can produce not only significant amounts of particulate contamination through microscratching on the wafer surfaces, but also metallic contamination as measured by De Busk [50] and summarized in Figure 21, where greater concentrations of bulk iron were measured at the flats of tweezer-handled wafers than elsewhere on the wafer.

Wafer backside contamination is a critical part of wafer handling. High levels of contamination (often associated with metallics) and reproducible patterns are frequently observed on the backside of wafers sent through process equipment Although there continues to be a high emphasis on wafer front side (i.e., product side) inspection and contamination reduction, there has been comparatively little effort directed at the reduction of wafer backside contamination. SEMATECH sponsored a workshop in February of 1998 to assist in establishing priority actions with respect to backside contamination prevention. High-level themes included the importance of nonmetallic contact with the backside, minimum contact area in wafer chucks and handler end effectors, and backside cleaning and detection techniques.

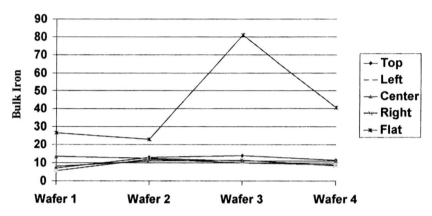

Figure 21   Iron contamination by wafer handling. (From Ref. 50.)

Even wafers can be a source of contamination as they adsorb residual reactive gases and volatile contaminants (especially moisture) that can outgas in subsequent processing. Residual films (metallic and oxide) are another potential source of backside contaminants.

XI.   MANUFACTURING SUPPORT ITEMS

A plethora of materials exist within the manufacturing environment that potentially can contaminate the wafer-processing environment and/or the wafers themselves. Methods to characterize these materials have been developed by SEMATECH [51] and include the characterization of materials such as gloves, garments/headgear, face masks, shoes/boots/covers, undergarments, garment practices, laundry, wipers/swabs, solutions, supplies, tacky mats/rollers, CR paper/pens, printers/typewriters, tape/adhesive, signs/labels, envelopes/laminators, carriers/boxes, handles, tweezers/wands, quartz, transfer systems/selectors, tables/shelves/racks, chairs, and stools/ladders.

Some of these manufacturing support items can be significant sources of contamination, both particulate and nonparticulate. Mattina and Oathout [52,53] tested 10 wiping materials and found vast differences in all criteria: sorptive capacity (roughly 10× range in both area and mass), particles (released and generated due to friction—roughly 100× difference from cleanest to dirtiest), extractible matter in IPA (roughly 10–50× range based on weight and area), inorganic matter (10× based on are and 3× based on weight).

XII. SEMATECH TEST METHODS

SEMATECH has facilitated development of numerous test methods for assessing the performance of various components and processes in the semiconductor wafer fabrication facility. These are, in summary:

- Gas components
  R. Periasamy et al., "Developing the SEMATECH Test Methods of Evaluating Cleanroom Gas-Handling Components," *Micro*, June 1994.
- Liquid components
  92010933B-STD Guide to Test Methods for UPW Distribution System Components
- Equipment
  SEMI E14-93 Measurement of Particle Contamination Contributed to the Product from the Process or Support Tool
- Manufacturing support items
  92051106A-STD SEMATECH Guide to Test Methods for Manufacturing Support Items
- Equipment clean build
  92051107A-STD SEMATECH Guide for Contamination Control in the Design, Build, Package and Shipping of Equipment
- Mass flow controllers
  92071220B-STD SEMATECH Guide to Provisional Test Methods for MFCs

SEMATECH has invested significant resources in the characterization and reduction of defects as well as providing their member companies with tools to assist in the contamination prevention and reduction process. Some of these are

- Particles per wafer per pass (PWP) test methods
- Test methods for high-purity gas-handling components
- Water component test methods
- Manufacturing support items (consumables) test methods
- Clean build equipment guidelines
- Tool accommodations standards
- In situ sensor purchasing requirements

Many other groups have prepared documents addressing various aspects of contamination control and the reader is encouraged to seek additional references from these sources: IEST (www.iest.org), FED-STD-209E, SEMI, Micro, Semiconductor International, Solid State Technology, Cleanrooms, SPWCC, Balazs Analytical Laboratory, Pitt Conference, ASMC, UCS ECS Journal, AVS Journal, AAAR Journal, UltraPure Water Journal, University of Arizona, Clarkson University, IMEC, University of Minnesota, and the Research Triangle Institute.

REFERENCES

1. Stevie, F. A., et al. Boron contamination of surfaces in silicon microelectronics processing: characterization and causes. Journal of Vacuum Science and Technology. 9(5):2813, 1991.
2. Vepa, K., et al. Role of Organics and Moisture on Silicon Wafer Surfaces. Spring ECS Meeting, 1993.
3. Tamaoki, M., et al. The Effect of Airborne Contaminants in the Cleanroom for ULSI Manufacturing Process. 1995 IEEE/SEMI Advanced Semiconductor Manufacturing Conference Proceedings, pp. 322–326.
4. Frystak, D. C., et al. Control of interfacial oxide using a novel cluster tool technology. IEEE/SEMI ASMC, November 1994.
5. Austin, P. R. Austin contamination control index. Contamination Control, vol. V, no. 1, June 1966.
6. Lowry, R. K., et al. Analysis of human contaminants pinpoint sources of IC defects. SI, July 1987.
7. Thomas, R. W. The Identification and Elimination of Human Contamination in the Manufacturing of IC's. Rome Air Development Center.
8. Kozicki, M. N. ed. Cleanrooms. p. 227, 1991, Van Nostrand Reinhold, New York.
9. Dryden Engineering testing—typical data.
10. Institute of Environmental Sciences and Technology, Recommended Practice IEST-RP-CC018.2. Cleanroom Housekeeping—Operating and Monitoring Procedures. 1992, IEST, Chicago.
11. Institute of Environmental Sciences and Technology, Recommended Practice IEST-RP-CC027.1. Personnel Practices and Procedures in Cleanroom and Controlled Environments. 1998, IEST, Chicago.
12. Institute of Environmental Sciences and Technology. Handbook of Recommended Practices, IEST, Chicago.
13. Viner, A. S. Predicted and Measured Clean-Room Contamination. In: Particle Control for Semiconductor Manufacturing. R. P. Donovan, ed., Marcel Dekker, Inc., New York, 1990, p. 139.
14. Tolliver, D. L. ed. Handbook of Contamination Control in Microelectronics. Noyce Publications, New Jersey, 1988.
15. Seinfeld, J. H. Urban air pollution: state of the science. Science, vol. 243, 10 Feb 1989, pp. 745–752.
16. Muller, A. J., et al. Measurement of airborne concentrations and surface arrival rates of organic contaminants in clean rooms. 1993 IEST ATM.
17. Fosnight, W., et al. Assessing Requirements for Wafer Environment Control. 1995 IEST ATM Proceedings.
18. Kojima, T., et al. Transportation Technology of Si Wafer Without Accompanying Carbon Contamination. Proceedings of the IEST 1995 Annual Technical Meeting, pp. 178–184.

19. Jensen, D., and Smith, M. Estimating Die Loss Due to Airborne Particle Challenge. Future Fab International, Issue 2, vol. I, pp. 185–189.
20. The National Technology Roadmap for Semiconductors. San Jose, CA, SIA, 1997.
21. The National Technology Roadmap for Semiconductors. San Jose, CA, SIA, 1994.
22. Rosenfeld, E., et al. Purity characterization of a bulk chem dist sys. Microcontamination Tutorial 107, October 1990.
23. Hurd, T. An Evaluation and Comparison of 0.05 micron Chemical Filters, Microcontamination 1993 Conf Proceedings.
24. Goodman, J. Materials for Bulk Chemical Distribution Systems. Microcontamination Tutorial 101, October 1989.
25. Lord, B., and Waldman, J. Cleanroom manufactured drum liners for advanced purity and waste minimization. IEST ATM, May 1994.
26. Dillenbeck, K., and Drab, G. Tutorial 106—Mgmt of Chemicals. 1989 Microcontamination Conference.
27. Wear, T., et al. Purity Performance of POUCG in Semiconductor Wafer Cleaning Applications. Microcontamination 1994 Conference Proceedings.
28. Pure Water Specifications and Guidelines—For Facility and Fabrication Engineers. 1993, Balazs Analytical Laboratory.
29. Yabe, K., et al. Responding to the future quality demands of ultrapure water. Microcontamination, February 1989.
30. Dillenbeck, K. Contamination Engineering Principles for Semiconductor Wafer Fabrication. Microcontamination Tutorial 3, November 1988.
31. Shapiro, A., et al. Residual Moisture: Its Role and Measurements in Semiconductor Processing Equipment. Microcontamination 1994 Conference Proceedings.
32. Siefering, K., et al. Contamination Modeling as a Tool to Improve the Cost/Performance of UHP Gas Piping Systems. IEST ATM Conference Proceedings, May 1994.
33. Coronell, D. G., et al. An Integrated Approach to Understanding Moisture Behavior in UHP Gas Delivery Systems: Component Testing and Computer Simulations. IEST ATM Conference Proceedings, May 1994.
34. Kubus, J., and Legget, G. Piping System Drydown Predictions with Field Verification. Microcontamination 1993 Proceedings.
35. SEMI Physical Interfaces Committee. Tool Accommodation Standards 4.0, chap. 2.
36. Wang, H. C., et al. Lifetime of Electropolished Stainless Steel Tubing in HCl Service. IEST ATM Conference Proceedings, May 1994.
37. Cheung, S., et al. Ultra-high purity gas distribution systems for sub 0.5 μm ULSI manufacturing. ASMC 1994.
38. Bhadha, P. M., and Cowan, C. L. Purification to ppb and ppt Levels. IEST ATM Conference Proceedings, May 1994.
39. Sylvestre, N. SEMATECH Guide for Contamination Control in Design, Assembly, and Delivery of Semiconductor Manufacturing. SEMATECH DOC ID #: 92051107A-STD.
40. Tabor, D. Friction—The present state of our understanding. Journal of Lubrication Technology, no. 103, p. 169 (1981).
41. Verma, N., et al. Modeling Outgassing in Vertical Thermal Reactors. Microcontamination Conference Proceedings, October 1994.

42. O'Hanlon, J. F., et al. Particle Traps Near Wafers Pointing Upward, Downward and Sidewise. Microcontamination Conference Proceedings, October 1994.
43. Geller, A. S., et al. Particle Reduction in LPCVD Chambers. MICRO Conference Proceedings, October 1994.
44. Rader, D. J., et al. Particle Transport in Plasma Reactors. MICRO Conference Proceedings, October 1994.
45. Ma, C., et al. An APIMS Study of Permeation Through Polymers. IEST ATM Conference Proceedings, May 1994.
46. Bhat, S., et al. Integration of an In-situ Particle Monitoring System in a Plasma Etch Tool. Microcontamination Conference Proceedings, October 1994.
47. Uritsky, Y., et al. Using an Advanced Particle Analysis System for Process Improvement. Microcontamination Conference Proceedings, October 1994.
48. Bilotta, S., and Proctor, D. Development of a Manufacturable ROXNOX Oxidation Process. IEEE/SEMI Advanced Semiconductor Manufacturing Conference Proceedings, November 1994.
49. Anderson, B., et al. Defect Density Reduction in a Tungsten Deposition Etchback. IEEE/SEMI Advanced Semiconductor Manufacturing Conference Proceedings, November 1994.
50. DeBusk, D. SPV testing—a real-time contamination monitoring technique. Semiconductor International, April 1993.
51. Sylvestre, N. SEMATECH Guide to Test Methods for Manufacturing Support Items. 92051I06A-STD.
52. Mattina, C. F., and Oathout, J. M. A Comparison of Commercial Cleanroom Wiping Materials for Properties Related to Functionality and to Cleanliness. IEST ATM Conference Proceedings, May 1994.
53. Mattina, C. F., and Oathout, J. M. Assessing Wiping Materials for Their Propensity to Generate Particles: Biaxial Shaking Versus the Construction of Characteristic Curves. IEST ATM Conference Proceedings, May 1994.

BIBLIOGRAPHY

General

1. D. L. Tolliver, ed., *Handbook of Contamination Control in Microelectronics*, Noyes Publications, New Jersey, 1988.
2. R. P. Donovan, ed., *Particle Control for Semiconductor Manufacturing*, Marcel Dekker, Inc., New York, 1990.
3. S. Kozicki, A. Hoenig, and P. J. Robinson, eds., *Cleanrooms: Facilities and Practices*, Van Nostrand Reinhold, New York, 1991.
4. W. Keeley and T. H. Cheyney, eds., "The Ohmi Papers: Challenges to Ultimate Cleanliness for Semiconductor Processing," Canon Communications, Inc., 1990.
5. W. C. Hinds, *Aerosol Technology*, Wiley-Interscience, 1982.
6. J. K. Beddow, "Particulate Science and Technology," Chemical Publishing, 1980.
7. P. Borden, "In-Situ Particle Monitoring in the 90's: The Series," Canon Communications, 1991, reprinted from *Microcontamination*, January–December 1991.

8. "Anniversary Reference Issue: Cumulative Index of the Articles and Features from Our First Decade," *Microcontamination*, December 1992.

9. W. Kern, ed., *Handbook of Semiconductor Wafer Cleaning Technology: Science, Technology and Applications*, Noyes Publications, New Jersey, 1993.

10. D. W. Cooper, "Useful Statistics for On-Line Particle Counting," IES ATM 1995 Conference Proceedings.

Construction and Clean Build

1. D. Wadkins, "Wearing Cleanroom Garments During Construction: A Use Test," *Solid State Technology*, May 1991.

2. C. Lenk, "Protocols Facilitate Upgrade Without Production Interruption," *Cleanrooms*, March 1994.

3. J. Weaver, "Installing Process Equipment Without Destroying Cleanroom Integrity," *Cleanrooms*, March 1994.

4. L. Mainers and R. Rotatori, "Clean Build Protocol—Foundation for Cleanliness at World Class Fab," *Cleanrooms*, January 1995.

5. L. C. Leone, "Planning High-Purity Manufacturing Facilities," *Cleanrooms*, December 1997.

Wafer

1. S. Tripathi, M. Moinpour, J. Li, B. Fardi, and F. Moghadam, "Criticality of Backside Contamination in 200 mm Wafer Processing," Microcontamination 1994 Conference Proceedings, pp. 319–326.

People, Protocol, and Manufacturing Support Items

1. D. A. Toy, ed., "Cleanroom Garments: Performance and Comfort," *Semiconductor International*, November 1989.

2. D. A. Toy, eds., "Cleanroom Accessories Can Impact Your Process," *Semiconductor International*, May 1990.

3. A. M. Dixon, "Cleanroom Garments: The Right Protocol Begins with You," *Semiconductor International*, September 1991.

4. R. Iscoff, eds., "Cleanroom Apparel: A Question of Tradeoffs," *Semiconductor International*, March 1994.

5. P. R. Austin, "What You Don't Know About Cleanroom Garmenting—But Should!," *Semiconductor International*, January 1995.

6. B. Y. H. Liu et al., "Development of Advanced Test Methods for Cleanroom Garments," IEST ATM 1993 Conference Proceedings.

7. H. R. Bhattacharjee et al., "The Use of Scanning Electron Microscopy to Quantify the Burden of Particles Released From Cleanroom Wiping Materials," IEST ATM 1993 Conference Proceedings.

8. J. M. Oathout and C. F. Mattina, "A Comparison of Commercial Cleanroom Wiping Materials for Properties Related to Functionality and to Cleanliness," IEST ATM 1994 Conference Proceedings.

9. C. F. Mattina and J. M. Oathout, "Assessing Wiping Materials for Their Propensity to Generate Particles: Biaxial Shaking Versus the Construction of Characteristic Curves," IEST ATM 1994 Conference Proceedings.
10. S. A. Hoenig, "Bacterial Contamination in Air Conditioning Systems: How Does It Get Into the Cleanroom?," *Cleanrooms*, November 1994.
11. R. Linke, "Cleaner, Smaller Swabs Evolve for Sensitive Applications," *Cleanrooms*, January 1994.
12. C. L. Thompson, "Procedure Measures Contamination Control Levels of Cleanroom Chairs," *Cleanrooms*, January 1994.
13. D. L. Wadkins et al., "An Evaluation of the Effectiveness of Air Showers on Removing Particles from Cleanroom Garments," IEST ATM 1998 Conference Proceedings.
14. M. Dax "Trace Contamination from Plastic Wafer Boxes," *Semiconductor International*, January 1997.
15. B. Hauff, "Controlling Colony Forming Units in Cleanrooms," *Cleanrooms*, July 1998.

Wafer Environment Control

1. W. Fosnight et al., "Assessing Requirements for Wafer Environment Control," IEST ATM 1995 Conference Proceedings.
2. P. Muller et al., "The Norcross Project: Comparing Manual and Robotic Wafer Handling Using Minienvironments in a Degraded Cleanroom," *Microcontamination*, September 1994.
3. A. Kaiser, "Achieving Chemical and Particulate Isolation Through the Use of Minienvironments," *Microcontamination*, April 1994.
4. W. Grande, "Upgrading a Class 100 Fab Through Use of Manual-Access Microenvironments," *Microcontamination*, January 1993.
5. J. J. Lee et al., "Direct Surface Analysis of Organic Contamination on Si Wafers," Micro 1994 Conference Proceedings.
6. J. Milberg et al., "Fluidic Integration of Equipment in Cleanrooms," *Solid State Technology*, August 1991.
7. J. Goodman, "Proper Care and Use of Wafer Handling Products," *Solid State Technology*, December 1991.
8. D. W. Cooper et al., "Initiating Six Sigma Statistical Quality Control Techniques for Cleanroom Settling Monitors," *Solid State Technology*, February 1993.
9. D. J. Elliott, "Contamination Control Using a Nitrogen Purged Microenvironment," *Solid State Technology*, November 1993.
10. A. Bowling and C. J. Davis, "Managing Contamination During Advanced Wafer Processing," *Solid State Technology*, February 1994.
11. P. M. Van Sickle, "Microenvironments: Particle and Outgassing Test Methods," *Solid State Technology*, May 1994.
12. A. J. Muller et al., "Volatile Cleanroom Contaminants: Sources and Detection," *Solid State Technology*, September 1994.
13. J. R. Weaver, "The Cleanroom Reverse Design Principle," *Semiconductor International*, September 1991.

14. K. C. Wiemer and J. R. Burnett, "The Fab of the Future: Concept and Reality," *Semiconductor International*, July 1992.

15. B. Newboe, ed., "Minienvironments: Better 'Cleanrooms' for Less," *Semiconductor International*, March 1993.

16. D. Jillie, "Electromagnetic Compatibility in the Semiconductor Factory," *Semiconductor International*, July 1993.

17. X. Pucel, "SEMI, JESSI, SEMATECH Drive Minienvironment Standards," *Semiconductor International*, April 1994.

18. A. Steinman, "Static Charge: It's a Killer!," *Semiconductor International*, September 1994.

19. L. A. Clark et al., "Sources of Particle Contamination in an IC Manufacturing Environment," *Aerosol Science and Technology*, vol. 16, pp. 43–50.

20. J. H. Seinfeld, "Urban Air Pollution: State of the Science," *Science*, February 10, 1989.

21. M. A. Rappa, "Future Directions in Controlling Particle Contamination in Semiconductor Integrated Circuit Manufacturing: An Industry Survey," 1994 IEEE/SEMI Advanced Semiconductor Manufacturing Conference.

22. B. Ypsilanti and B. Sanden, "Particle Reduction Through the Control of Triboelectric Charges," 1994 IEEE/SEMI Advanced Semiconductor Manufacturing Conference.

23. P. E. Carr et al., "Measured Effects of Reduced Flow Velocity in a Laminar Flow Cleanroom," IEST ATM 1993 Conference Proceedings.

24. A. J. Muller et al., "Measurement of Airborne Concentrations and Surface Arrival Rates of Organic Contaminants in Cleanrooms," IEST ATM 1993 Conference Proceedings.

25. A. C. Bonora, "SMIF Containers: Their Evolution as a Unifying Manufacturing Element," IEST ATM 1993 Conference Proceedings.

26. D. Hope and D. Milholland, "The Use of a Three Dimensional Ultrasonic Anemometer to Measure the Performance of Clean Zone Air Delivery Systems," IEST ATM 1993 Conference Proceedings.

27. J. Goodman and K. Mikkelsen, "Materials for Microenvironment Construction," IEST ATM 1993 Conference Proceedings.

28. T. Baechle and P. Mitchell, "Certification and Acceptance of Mini Environment Air Systems," IEST ATM 1993 Conference Proceedings.

29. L. B. Rothman et al., "SEMATECH Minienvironment Benchmarking Project," IEST ATM 1994 Conference Proceedings.

30. A. Y. Liang et al., "Real-time SAW Measurements of NVR in Cleanroom and in Microenvironment," IEST ATM 1994 Conference Proceedings.

31. L. A. Psota-Kelty et al., "Measuring Amine Concentrations in Cleanrooms," IEST ATM 1994 Conference Proceedings.

32. D. Kinkead, "Controlling a Killer: How to Win the War Over Gaseous Contaminants," *Cleanrooms*, June 1993.

33. K. Sakamoto et al., "Decomposition of Trace-Amount of VOC Using UV-Light and Photocatalyst," IEST ATM 1998 Conference Proceedings.

34. L. Gail and F. Ripplinger, "Correlation of Alternate Aersols and Test Methods for HEPA Filter Leak Testing," IEST ATM 1998 Conference Proceedings.

35. S. Zhu, "Modeling of Airborne Molecular Contamination," IEST ATM 1998 Conference Proceedings.
36. H. Schneider et al., "Air Flow Modeling and Testing of 300mm Minienvironment Load Port Systems," IEST ATM 1998 Conference Proceedings.
37. D. Hou et al., "Comparative Outgassing Studies on Existing 300mm Wafer Shipping Boxes and Pods," IEST ATM 1998 Conference Proceedings.
38. M. Watanabe et al., "Characteristic of Boron Free Filter," IEST ATM 1998 Conference Proceedings.
39. D. W. VanOsdell et al., "Acid Gas Tests of Gaseous Contaminant Adsorbers," IEST ATM 1998 Conference Proceedings.
40. N. Namiki et al., "Removal of Organic Gas Species by UV-Irradiated Titania Honeycombs," IEST ATM 1998 Conference Proceedings.
41. K. Takeda et al., "Evaluation of Organic Contamination in Cleanroom and Deposition onto Wafer Surface," IEST ATM 1998 Conference Proceedings.
42. K. R. Dean and R. A. Carpio, "Real-Time Detection of Airborne Contaminants in DUV Lithographic Processing Environments," IES ATM 1995 Conference Proceedings.
43. N. Streckfuss et al., "Qualification of Pods and Storage Boxes Concerning Inorganic Contamination According to SEMI Document #2389," IES ATM 1995 Conference Proceedings.
44. K. J. Budde et al., "Determination of Organic Contamination From Minienvironments According to SEMI Document #2238 First Results," IES ATM 1995 Conference Proceedings.
45. W. Fosnight et al., "Assessing Requirements for Wafer Environment Control," IES ATM 1995 Conference Proceedings.
46. C. W. Draper et al., "A Portable Nitrogen Purged Microenvironment: Design Specification and Preliminary Field Test Data," IES ATM 1995 Conference Proceedings.
47. M. Balazs et al., "Cleanroom Air Monitoring Using Scrubbing and Adsorption Methodologies," IES ATM 1995 Conference Proceedings.
48. W. S. Poon et al., "Measurement of Air Pollution Acidic Gases in a Class 10 Cleanroom," IES ATM 1995 Conference Proceedings.
49. D. Elliott, "A Nitrogen Purged Microenvironment for Wafer Isolation," *Semiconductor International*, October 1995.
50. R. M. Genco et al., "Control Microcontamination with Wafer Cassette Purging," *Semiconductor International*, April 1997.
51. D. Kinkead, "Airborne Molecular Contamination: A Roadmap for the 0.25 μm Generation," *Semiconductor International*, June 1996.
52. L. B. Levit and J. Menear, "Measuring and Quantifying Static Charge in Cleanrooms and Process Tools," *Solid State Technology*, February 1998.
53. L. Zazzera and W. Reagen, "Application of FTIR for Monitoring Cleanroom Air and Process Emissions," *Cleanroom Magazine Supplement*, May 1998.
54. C. G. Noll, "Deposits on Semiconductor Corona Emitters in Cleanroom and Simulated Air," *Solid State Technology*, July 1998,
55. F. Dwiggins, "Gas Phase Filtration for Cleanrooms," *Cleanroom*, August 1997.
56. A. Steinman, "Air Ionization: Theory, Use and Best Practices," *Cleanrooms*, July 1998.

Chemicals

1. T. Hackett and K. Dillenbeck, "Examination of Polyethylene as a Packaging Material for Electronic Chemicals," Micro 1993 Conference Proceedings.

2. D. Vernikovsky, "Sealing Technology for the Semiconductor Industry," Microcontamination 1993 Conference Proceedings.

3. D. Grant and M. Heisler, "Evaluation of a Novel PTFE Dual Asymmetric Membrane Cartridge for Filtration of Sulfuric Acid," Microcontamination 1993 Conference Proceedings.

4. T. Hurd, "An Evaluation and Comparison of 0.05 micron Chemical Filters," Microcontamination 1993 Conference Proceedings.

5. T. Wear et al., "Purity Performance of a Point-of-Use Gas-to-Chemical Generator in Semiconductor Wafer Cleaning Applications," Microcontamination 1994 Conference Proceedings.

6. T. Talasek et al., "Determination of Leachable Metallic Impurities from Semiconductor Chemical Packaging Materials," Microcontamination 1994 Conference Proceedings.

7. T. Talasek et al., "Determination of Leachable Organics from Semiconductor Chemical Packaging Material by Supercritical Fluid Extraction," Microcontamination 1994 Conference Proceedings.

8. N. Harder, "High Purity Chemical Production: Meeting the Demands of ULSI," *Solid State Technology*, October 1990.

9. S. Batchelder and M. A. Taubenblatt, "Real-time Single Particle Composition Detection in Liquids," *Solid State Technology*, October 1992.

10. G. Liebetreu III, "Gas to Chemical 'Solutions,'" *Solid State Technology*, August 1994.

11. I. Ali et al., "Charged Particles in Process Liquids," *Semiconductor International*, April 1990.

12. R. Iscoff, ed., "The Search for Ultrapure Chemicals," *Semiconductor International*, July 1990.

13. K. L. Roberts et al., "Challenging the Membrane Filter Ratings," *Semiconductor International*, November 1990.

14. G. Carr, "Continuous Chemical Repurification," *Semiconductor International*, June 1993.

15. A. Ditali and Z. Hasnain, "Monitoring Alpha Particle Sources During Wafer Processing," *Semiconductor International*, June 1993.

16. L. Peters, ed., "Point-of-Use Generation: The Ultimate Solution for Chemical Purity," *Semiconductor International*, January 1994.

17. K. Dillenbeck and S. DeVore, "Characterization of Particle Levels in Incoming Chemicals," IEST ATM 1985 Conference Proceedings, pp. 127–134.

18. D. Grant et al., "Particle Capture Mechanisms in Liquids and Gases: An Analysis of Operative Mechanisms," *Journal of Environmental Science*, vol. 42, no. 4, July/Aug 1989, pp. 43–51.

19. T. B. Hackett and K. Dillenbeck, "Characterization of Polyethylene as a Packaging Material for High Purity Process Chemical" IEST ATM 1993 Conference Proceedings.

20. D. C. Grant et al., "Issues Involved in Qualifying Chemical Delivery Systems for Metallic Extractables," IEST ATM 1993 Conference Proceedings.
21. K. Christenson and J. Zahka, "Characterization of Chemical Filter Retention Performance," IEST ATM 1993 Conference Proceedings.
22. K. T. Pate, "Evaluation of Competitive PTFE Filters," IEST ATM 1994 Conference Proceedings.
23. F. C. Wang et al., "Design, Certification and Verification of Technology for Delivering sub-ppb, Low-Particle Chemicals to Semiconductor Cleaning Baths in Wafer Fabs," IEST ATM 1994 Conference Proceedings.
24. J. Zahka et al., "Chemical Filtration—Designing for OEE," IEST ATM 1998 Conference Proceedings.
25. M. Xu et al., "Particle Sheddgin Characteristics of Membrane Filters Under Pulsed Flow Conditions in Liquid Filtration," IEST ATM 1998 Conference Proceedings.
26. B. Y. H. Liu et al., "Characterization of Slurry Particles Used in Chemical Mechanical Polishing (CMP) of Wafers," IEST ATM 1998 Conference Proceedings.
27. J. Zahka et al., "Design Considerations for Point-of-Use Dilute HF Purification System," IES ATM 1996 Conference Proceedings.
28. T. Hackett and Z. Hatcher, "Chemical Compatibility and Materials of Construction for High Purity Chemical Packaging," IES ATM 1996 Conference Proceedings.
29. L. Mouche et al., "Photoresist Filtration Performance of UPE and PTFE Filters," IES ATM 1996 Conference Proceedings.
30. B. Parekh et al., "Performance of a POU Purifier in Ionic Contamination Removal," *Solid State Technology*, April 1996.
31. R. DeJule, "Bulk Chemical Distribution Addresses Tightening Specs," *Semiconductor International*, August 1996.
32. S. A. Sturm and P. M. Van Sickle, "Contamination Control in Liquid Chemical Distribution Systems," *Cleanrooms*, October 1998.
33. R. T. Talasek et al., "Evaluating Reusable HDPE Containers for Delivery of High-Purity Hydroflouric Acid," *Microcontamination*, July/August 1997, pp. 95–96, 98, 100.

DI Water

1. I. L. Pepper et al., "Measuring Bacterial Contaminants in Ultrapure Water: A Rapid Analytical Method," *Microcontamination*, October 1994.
2. M. Homnick et al., "Assessing the Design and Performance of a Critical Hot Ultrapure Water System," *Microcontamination*, February 1993.
3. C. Raghunath et al., "Interaction of Soluble Silicate with Wafers," Microcontamination 1994 Conference Proceedings.
4. G. Husted et al., "Evaluation of Current Limits of Detection for Particles, Macromolecules, and Bacteria in Ultra Pure Water," Microcontamination 1994 Conference Proceedings.
5. D. Sinha, "Total Organic Carbon Reduction in Ultrapure Water Processing," *Solid State Technology*, March 1992.
6. M. K. Balazs, "Ultrapure Water: Friend or Foe?," *Solid State Technology*, October 1993.

7. C. F. Frith, "Water Quality in the Semiconductor Rinsing Process," *Semiconductor International*, September 1991.
8. P. Singer, ed., "DI Water Filters: The Last Defense Against Microcontamination," *Semiconductor International*, December 1994.
9. R. Hango, "DI Water Polishing Experience and System Performance," Pure Water 1985 Conference Proceedings.
10. M. Logan et al., "Control of Water Borne Contaminants in VLSI Device Fabrication," IEST ATM 1984 Conference Proceedings, pp. 152–159.
11. A. Gough et al., "Microbial Contamination in Ultra-Pure Water," *Solid State Technology*, February 1986, pp. 139–142.
12. W. Harned, "Bacteria as a Particle Source in Wafer Processing Equipment," *Journal of the IEST*, June 1986.
13. D. C. Grant, "A Comparison of the Particle Shedding Characteristics of High Purity Water Filtration Cartridges," Proceedings of the 5th Annual Semiconductor Pure Water Conference, pp. 252–265.
14. C. W. Smith, "Control of TOC in High Purity Water Systems," *Ultrapure Water Journal*, December 1993.
15. M. Patterson et al., "Isolation, Identification, and Microscopic Properties of Biofilms in High Purity Water Distribution Systems," *Ultrapure Water Journal*, May/June 1991.
16. R. A. Governal and F. Shadman, "Design of High-Purity Water Plants: Fundamental Interactions in Removal of Organic Contamination," *Ultrapure Water Journal*, September 1992.
17. M. Henley, "An Overview of Colloidal and Soluble Silica," *Ultrapure Water Journal*, December 1992.
18. V. Anantharaman et al., "Detection and Characterization of Organics in Semiconductor DI Water Processes," *Ultrapure Water Journal*, April 1994.
19. D. Blackford et al., "The Measurements of Nonvolatile Residue in High Purity Water and Clean Liquids," *Ultrapure Water Journal*, July/August 1994.
20. G. Chen and F. Shadman, "Photocatalytic Oxidation of Chlorinated Hydrocarbons in Ultrapure Water," IES ATM 1996 Conference Proceedings.
21. R. Mohindra, "New Process for Producing Particle-Free Deionized Water," *Semiconductor International*, July, 1997.

Gas

1. S. D. Cheung et al., "Ultra-High Purity Gas Distribution Systems for sub 0.5 micron ULSI Manufacturing," 1994 IEEE/SEMI Advanced Semiconductor Manufacturing Conference.
2. R. Periasamy et al., "Developing the SEMATECH Test Methods for Evaluating Cleanroom Gas-Handling Components," *Microcontamination*, June 1994.
3. S. Chesters and H. Wang, "Using Atomic Force Microscopy to Evaluate Alloys for Corrosive Gas Systems," *Microcontamination*, June 1994.
4. S. N. Ketkar et al., "Using APIMS to Certify Ultra-High-Purity Gas Distribution Systems in an Operational Fab," *Microcontamination*, January 1994.

5. W. Plante et al., "Using All-Stainless-Steel Filters in Hydrogen Chloride Gas Lines: An Investigation of Contamination Effects," *Microcontamination*, May 1993.

6. Y. Shirai et al., "Specialty Gas Distribution System Free from Corrosion, Gas Decomposition Reaction—Perfect Cr_2O_3 Treated Tubing System," Microcontamination 1994 Conference Proceedings.

7. A. Ohki et al., "Quick Inspection Technology of External Leakage for Total Gas Delivery System by APIMS," Microcontamination 1994 Conference Proceedings, pp. 298–307.

8. C. Ma et al., "An APIMS Study of Back Diffusion of Moisture and Oxygen," Microcontamination 1994 Conference Proceedings.

9. S. Ketkar et al., "Evaluation of High Purity Gas Cabinets—A Study of Moisture Dry Down and Particle Generation Characteristics," Microcontamination 1994 Conference Proceedings.

10. W. Plante, "A Contamination-Based Method for Corrosion Testing of Gas System Components and Its Application to All-Metal Filters," Microcontamination 1994 Conference Proceedings.

11. S. Chesters et al., "Component Failure and Materials Selection for ESG Gas Systems," Microcontamination 1994 Conference Proceedings.

12. R. Periasamy et al., "Evaluation of Contamination Performance of Mass Flow Controllers," Microcontamination 1994 Conference Proceedings.

13. Athalye et al., "APIMS Characterization and Performance Modeling of Ultra-High-Purity Special Gas Cabinets," Microcontamination 1994 Conference Proceedings.

14. D. S. Zuck, "Particle Control in the Construction of a 1 Mbit DRAM Gas Distribution System," *Solid State Technology*, November 1989.

15. S. Chesters et al., "A Fractal Based Method for Describing Surface Texture," *Solid State Technology*, January 1991.

16. H. Tomari et al., "Metal Surface Treatment for Semiconductor Equipment: Oxygen Passivation," *Solid State Technology*, February 1991, pp. 51–55.

17. M. Amari et al., "Understanding Gas System Contamination Through the Evaluation of Used Filters," *Solid State Technology*, June 1991.

18. S. Chesters et al., "Atomic Force Microscopy of Gas-Surface Corrosion in Stainless Steel," *Solid State Technology*, June 1991.

19. E. Flaherty et al., "Particle and Metal Contamination in Gas Cylinders," *Solid State Technology*, January 1992.

20. P. M. Bhadha and E. R. Greene, "Joule-Thomson Expansion and Corrosion in HCl Systems," *Solid State Technology*, July 1992.

21. R. Duguid et al., "HCl Gas Distribution Systems: The Effect of Surface Finish and Point-of-Use Purification," *Solid State Technology*, July 1993.

22. D. A. Toy, ed., "Purifying at the Point of Use Keeps Your Gases at Their Cleanest," *Semiconductor International*, June 1990.

23. J. C. Oswalt and B. Todd, "How to Deliver High Purity Gases to Process Tools," *Semiconductor International*, June 1990.

24. P. Burggraaf, ed., "Nitrogen is Just Nitrogen, Right?," *Semiconductor International*, June 1990.

25. D. A. Bohling and M. A. George, "Controlling Contamination in WF_6 Applications," *Semiconductor International*, September 1991.

26. R. Iscoff, ed., "Specialty Gas Cylinders: Help Wanted," *Semiconductor International*, April 1992.
27. M. George et al., "Minimizing System Contamination Potential from Gas Handling," *Semiconductor International*, July 1993.
28. H. Kobayashi, "How Gas Panels Affect Contamination," *Semiconductor International*, September 1994.
29. K. Siefering et al., "Modeling of Moisture Transport in Ultra-Clean Gas Distribution Systems," IEST ATM 1993 Conference Proceedings.
30. A. M. Haider, C. Ma, and F. Shadman, "Interactions of Ceramic, Metallic, and Polymeric Filters with Gaseous Contaminants," IEST ATM 1993 Conference Proceedings, pp. 158–164.
31. H. C. Wang, "Lifetime of Electropolished Stainless Steel Tubing in HCl Service," IEST ATM 1994 Conference Proceedings.
32. R. Richardson, "Maintaining Gas Purity—A Systems Approach to Contamination Control," *Cleanrooms*, January 1995.
33. S. Nijhawan et al., "Particle Measurements and Transport in Silane LPCVD," IEST ATM 1998 Conference Proceedings.
34. C. Ma et al., "Moisture Dry-Down in High Purity Hydrogen Chloride," IEST ATM 1998 Conference Proceedings.
35. S. Azuma et al., "New Tubing Material for Specialty Gases—Ferritic Stainless Steel Tubing with Inert CR_2O_3 Surface," IEST ATM 1998 Conference Proceedings.
36. A. Miyake et al., "High Purity Hydrogen and Oxygen Generator by Water Electrolysis Using Solid Polymer Electrolyte," IEST ATM 1998 Conference Proceedings.
37. S. N. Ketkar et al., "Analysis of Trace Metals in Chlorine Triflouride (CIF3)," IEST ATM 1998 Conference Proceedings.
38. K. Siefering and W. Whitlock, "Analysis of PPB-Level Impurities in Silane Gas Using APIMS: Comparison of Purified and Unpurified Silane Products," IES ATM 1995 Conference Proceedings.
39. H. Wang et al., "Comparative Corrosion Studies for HCl and HBr Gas Distribution Systems," IES ATM 1995 Conference Proceedings.
40. S. Krishnan et al., "Case Study: Ultraclean Gas Delivery," *Semiconductor International*, April, 1995.
41. R. W. Rosenberg, "The Advantages of Continuous On-line RGS Monitoring," *Semiconductor International*, October 1995.
42. S. M. Fine, "Design and Operation of UHP Low Vapor Pressure and Reactive Gas Delivery Systems," *Semiconductor International*, October 1995.
43. B. Jurcik and A. Norvilas, "Modeling Moisture Transport in UHP Distribution Systems," *Solid State Technology*, November 1995.
44. S. Krishnan et al., "Site-specific Corrosion in Gas Delivery Tubing Exposed to Semiconductor Grade HCl," *Cleanrooms Magazine Supplement*, October 1995.
45. M. Conroy et al., "Analysis of Condensable Corrosive Gases for Metals Contamination," *Cleanrooms Magazine Supplement*, October 1995.
46. J. Cestari et al., "The Next Step in Process Gas Delivery: A Fully Integrated System," *Semiconductor International*, January 1997.
47. D. Laureta, R. Sallot, and E. Robinson, "New Torque Suppressor Cures Problems of Metal Face Seal Fittings," *Solid State Technology*, April 1997, pp. 89–90, 92–93.

48. B. Gotlinsky et al., "PPT Purification Eliminates Process Variables," *Solid State Technology*, April 1998.

Equipment

1. S. Y. Lynn, B. Huling, and M. Su, "Characterization of Gaseous Contamination in a Multi-Chamber Etch Tool," Microcontamination 1993 Conference Proceedings, pp. 113–117.
2. M. Moinpour, M. Sherasa, W. O'Toole, B. Stueve, R. Wilkes, and J. Reece, "Contamination Control and Defect Reduction Through Process Characterization and Equipment Modification: A Case Study of a LPCVD Nitride Vertical Thermal Reactor," Microcontamination 1993 Conference Proceedings, pp. 170–180.
3. Y. Uritsky et al., "Using an Advanced Particle Analysis System for Process Improvement," *Microcontamination*, May 1994, pp. 25–29.
4. F. Lee et al., "Detecting and Reducing Particles for LPCVD Silicon Nitride Deposition," *Microcontamination*, March 1994, pp. 33–37, 76.
5. G. J. Nestle, "Reducing Photolithography Defects to Meet Submicron Device Requirements," *Microcontamination*, February 1993, pp. 35–38.
6. A. Geller et al., "Particle Reduction in LPCVD Chambers," Microcontamination 1994 Conference Proceedings.
7. D. Rader et al., "Particle Transport in Plasma Reactors," Microcontamination 1994 Conference Proceedings.
8. L. J. Arias, Jr., "A Quadrupole Mass Spectrometer Analysis of a Plasma Enhanced Chemical Vapor Deposition of Undoped Silicate Glass," Microcontamination 1994 Conference Proceedings.
9. S. A. Bhat et al., "Integration of an In Situ Particle Monitoring System in a Plasma Etch Tool," Microcontamination 1994 Conference Proceedings.
10. J. Sasserath and R. Yenchik, "Superior Particle Control for PVD TiW Processes Through Improved Chamber Shield Design," Microcontamination 1994 Conference Proceedings.
11. F. Lee et al., "Defect Reduction and Monitoring in Reactive Ion Etch Equipment," Microcontamination 1994 Conference Proceedings.
12. A. Fuerst et al., "Vibration Analyses to Reduce Particles in Sputtering Systems," *Solid State Technology*, March 1993.
13. M. Hill et al., "Quartzglass Components and Heavy Metal Contamination," *Solid State Technology*, March 1994.
14. D. A. Toy, ed., "Designing Particles Out of the CVD System," *Semiconductor International*, March 1990.
15. B. C. Smoak, Jr., et al., "Gas Control Improves Epi Yield," *Semiconductor International*, June 1990.
16. P. Burggraaf, ed., "Equipment Generated Particles: Ion Implantation," *Semiconductor International*, September 1991.
17. A. Busnaina, "Solving Process Tool Contamination Problems," *Semiconductor International*, September 1993.
18. J. Hunter and H. K. Nguyen, "In-Situ Particle Monitoring Reduces Wafer Defects," *Semiconductor International*, November 1993.

19. R. Iscoff, ed., "Are Ion Implanters the Newest Clean Machines?," *Semiconductor International*, October 1994.
20. T. Hattori and S. Koyota, "Detecting and Identifying Equipment-Generated Particles for Yield Improvement," *Microcontamination*, September 1989.
21. J. Horvath and P. Borden, "Particle Control Methods for High Current Ion Implantation," IEST ATM 1991 Conference Proceedings.
22. L. Peters, ed., "20 Good Reasons to Use In-Situ Particle Monitors," *Semiconductor International*, November 1992.
23. J. F. O'Hanlon and H. G. Parks, "Impact of Vacuum Equipment Contamination on Semiconductor Yield," *Journal of Vacuum Science and Technology*, vol. 10(4), July/August 1992.
24. G. S. Selwyn et al., "In-Situ Laser Diagnostic Studies of Plasma Generated Particulate Contamination," *Journal of Vacuum Science and Technology*, vol. 7(4), July/August 1989.
25. G. S. Selwyn et al., "In-Situ Plasma Contamination Measurements by HeNe Laser Light Scattering: A Case Study," *Journal of Vacuum Science and Technology*, vol. 8(3), May/June 1990.
26. D. Hemmes, "Microcontamination Reduction in PECVD Systems—Parts I/II," *Semiconductor International*, May/July 1987.
27. B. Anderson et al., "Defect Density Reduction in Tungsten Deposition and Etchback," 1994 IEEE/SEMI Advanced Semiconductor Manufacturing Conference.
28. D. J. McCarron and M. A. Jones, "Achieving Full Capability from In-Situ Particle Monitors," IEST ATM 1993 Conference Proceedings.
29. M. Reath et al., "Use of Residual Gas Analysis in Low Pressure Semiconductor Process Reactors," IEST ATM 1993 Conference Proceedings.
30. B. Huling et al., "Characterization and Improvement of Gaseous Contamination Levels in a Multi-Chamber Etch Tool," IEST ATM 1993 Conference Proceedings.
31. H. Fitch and K. L. Mittal, "Surface Contamination Control Primer," *Cleanrooms*, April 1993.
32. G. DePinto, "Optimizing Polysilicon Deposition for Film Uniformity and Particles," *Journal of the IEST*, July/August 1993.
33. T. Yamamoto et al., "Model Study of Contaminant Flow in the Vicinity of Semiconductor Processing Equipment," *Journal of the IEST*, July/August 1990.
34. J. J. Wu et al., "Process Improvement Using In-Situ Particle Detection," *Journal of the IEST*, May/June 1993.
35. R. Markle et al., "Employing Yield Models, In-Situ Monitoring to Predict and Achieve Defect Density Goals," *Microcontamination*, February 1992.
36. H. Chein et al., "Application of an ISPM on an Lam Alliance 9400 Etching Process," IEST ATM 1998 Conference Proceedings.
37. G. Vereecke et al., "In Situ Gas Analysis on an RTP Tool with APIMS," IEST ATM 1998 Conference Proceedings.
38. R. Alchalabi, "In Situ Gas Analysis with Contamination Modeling for Diffusion Furnaces," IEST ATM 1998 Conference Proceedings.
39. P. J. Ziemann et al., "Particle Beam Mass Spectrometry of Submicron Particles Formed During LPCVD of Polysilicon Films," IES ATM 1995 Conference Proceedings.

40. C. E. Utter et al., "An Investigation into Particle Transport Within Plasma Etchers," IES ATM 1995 Conference Proceedings.

41. S. Fang et al., "In-Situ Particle Monitor at LPCVD Poly Furnace," IES ATM 1995 Conference Proceedings.

42. P. Singer, "Electrostatic Chucks in Wafer Processing," *Semiconductor International*, April 1995.

43. P. Singer, "LPCVD Tube Cleaning: A Dirty Job Gets Easier," *Semiconductor International*, April 1995.

44. H. D. Pham et al., "LSI Logic Proves Yield Gain Using In-Situ Monitor," *Semiconductor International*, April 1995.

45. R. W. Campbell and W. Tan, "A Guide to Static-Safe Polymers for Automated Handling Equipment," *Solid State Technology*, February 1997.

46. R Jarvis and L. L. Armentrout, "Full-Fab Surface Particle Detection Improves Yields," *Semiconductor International*, June 1997.

47. J. McAndrew, "Progress in In Situ Contamination Control," *Semiconductor International*, May 1998.

48. Y. Uritsky et al., "Investigating the Effect of Plasma Processes on Ceramic Materials," *Microcontamination*, March 1995.

49. J. Asbell et al., "Improving Tungsten CVD Performance with In situ Particle Monitoring," *Microcontamination*, July/August 1997.

Vacuum

1. J. F. O'Hanlon et al., "Particle Traps Near Wafers Pointing Upward, Downward and Sidewise," Microcontamination 1994 Conference Proceedings.

2. N. Verma et al., "Kinetics and Mechanisms of Gas Phase Purging in Process Tools and Components," Microcontamination 1994 Conference Proceedings.

3. A. Shapiro et al., "Residual Moisture: Its Role and Measurement in Semiconductor Processing Equipment," Microcontamination 1994 Conference Proceedings.

4. A. Haider et al., "In Situ Tungsten Etch Back Process Analysis Using a Residual Gas Analyzer," Microcontamination 1994 Conference Proceedings.

5. C. Johnston and F. Tapp, "In Situ Analysis Used to Optimize a Titanium Silicide Reactor," Microcontamination 1994 Conference Proceedings.

6. J. Zhao et al., "The Formation of Water Aerosols During Pump Down of Vacuum Process Tools," *Solid State Technology*, September 1990.

7. J. H. Singleton, "10 Ways to Ensure Vacuum Integrity," *Semiconductor International*, March 1991.

8. J. Harvell and P. Lessard, "Managing Water Vapor in a Vacuum Process," *Semiconductor International*, June 1991.

9. G. S. Selwyn, "The Unconventional Nature of Particles," *Semiconductor International*, March 1993.

10. R. K. Waits, "Residual Gases: The Invisible Process Variables," *Semiconductor International*, September 1993.

11. P. Carr et al., "RTP Characterization Using In-Situ Gas Analysis," *Semiconductor International*, November 1993.

12. B. Y. H. Liu, "How Particles Form During Vacuum Pump Down," *Semiconductor International*, March 1994.
13. R. K. Waits, "Controlling Your Vacuum Process: Effective Use of a QMA," *Semiconductor International*, May 1994.
14. C. E. Bryson, III, "The Problem of the 'East' Vacuum Specification," *Semiconductor International*, August 1994.
15. D. Friede and P. Lessard, "Improve Your Sputter Process by Better Water Vapor Pumping," *Semiconductor International*, August 1994.
16. A. G. Mathewson et al., "Comparison of Chemical Cleaning Methods of Aluminum Alloy Vacuum Chambers for Electron Storage Rings," *Journal of Vacuum Science and Technology*, vol. 7(1), January/February 1989.
17. N. Yoshimura, "Water Vapor Permeation Through Viton O-ring Seals," *Journal of Vacuum Science and Technology*, vol. 7(1), January/February 1989.
18. H. C. Hseuh and X. Cui, "Outgassing and Desorption of the Stainless-Steel Beam Tubes After Different Degassing Treatments," *Journal of Vacuum Science and Technology*, vol. 7(3), May/June 1989.
19. D. Chen and S. Hackwood, "Vacuum Particle Generation and the Nucleation Phenomena During Pumpdown," *Journal of Vacuum Science and Technology*, vol. 8(3), March/April 1990.
20. K. Sugiyama et al., "Low Outgassing and Anticorrosive Metal Surface Treatment for Ultrahigh Vacuum Equipment," *Journal of Vacuum Science and Technology*, vol. 8(4), July/August 1990.
21. D. Cheng et al., "Dynamic Particulate Characterization of a Vacuum Load-Lock System," *Journal of Vacuum Science and Technology*, vol. 7(5), September/October 1989.
22. R. A. Bowling and G. B. Larrabee, "Behavior and Detection of Particles in Vacuum Processes," *Journal of the Electrochemical Society*, vol. 136, no. 2, February 1989.
23. J. F. O'Hanlon, "Contamination Reduction in Vacuum Processing Systems," *Journal of Vacuum Science and Technology*, vol. 7(3), May/June 1989.
24. J. J. Wu et al., "An Aerosol Model of Particle Generation During Pressure Reduction," *Journal of Vacuum Science and Technology*, vol. 8(3), May/June 1990.
25. D. Rader et al., "Application of Numerical Models to Reduce Particle Contamination in Semiconductor Process Environments," IEST ATM 1994 Conference Proceedings.
26. K. Roubik et al., "Monitoring a Vacuum Tool Using In-situ Particle Monitoring (ISPM)," IEST ATM 1998 Conference Proceedings.
27. S. Boumsellek and R. J. Ferran, "Detection of 1 PPM Contamination Levels Using the Micropole^{TM} RGA," IES ATM 1996 Conference Proceedings.
28. R. Heyder et al., "Nonevaporable Gettering Technology for In-situ Vacuum Processes," *Solid State Technology*, August 1996.

Index